Avtandil Okrostsvaridze · Daniel Tormey
Editors

Geoheritage of Georgia

The Caucasus

Springer

Editors
Avtandil Okrostsvaridze
Institute of Earth Sciences
Ilia State University
Tbilisi, Georgia

Daniel Tormey
Catalyst Environmental Solutions
Santa Monica, CA, USA

ISSN 2363-765X ISSN 2363-7668 (electronic)
Geoheritage, Geoparks and Geotourism
ISBN 978-3-032-27638-4 ISBN 978-3-032-27639-1 (eBook)
https://doi.org/10.1007/978-3-032-27639-1

The book was prepared with the support of the Shota Rustaveli National Science Foundation of Georgia (Project FR-21-10905).

This Springer imprint is published by the registered company Springer Nature Switzerland AG
The registered company address is: Gewerbestrasse 11, 6330 Cham, Switzerland

Katitsvera Mountain (Fold), Greater Caucasus, Georgia. Photograph by Ani Baidashvili

„ქვა იცვითება წყალზედა, მაღლითა მთა მდინარესგან“
“Stone yields at last to water—to rivers spilling down from the high mountain.”

Shota Rustaveli - *The Knight in the Panther Skin*

Preface

The Caucasus mountain region lies between the Black and Caspian Seas, extending approximately 1200 km in a southeast–northwest direction and covering an area of about 440,000 km^2. The name of the region is closely associated with its mountains and was first mentioned in the tragedy *Prometheus Bound* by the ancient Greek dramatist Aeschylus (525–456 BC). According to the Roman naturalist Pliny the Elder (24–79 AD), the origin of the name may be traced to the Scythian language, in which "Caucasus" means "white as snow." Despite its long history, the term "Caucasus" became widely established as the general name for the region only in the twentieth century.

The Caucasus comprises the mountain systems of the Greater and Lesser Caucasus, as well as the intermontane depressions between them. Georgia occupies the central part of the region and contains a remarkable diversity of rock types and geological formations representing different ages and origins within this orogenic system. As a result, the country is rich in geoheritage of global significance, which inspired the preparation of this book. It is also noteworthy that in Georgian, the Earth is referred to as "Mother Earth," reflecting a deep spiritual and cultural connection between the Georgian people and their geoheritage.

Many of Georgia's Earth scientists contributed to the preparation of this book, and we express our sincere gratitude to all of them. We have made a deliberate effort to present each chapter in clear and accessible language, and we hope that this book will appeal not only to Earth scientists but also to nature enthusiasts and a broader audience.

Tbilisi, Georgia — Avtandil Okrostsvaridze
Santa Monica, CA, USA — Daniel Tormey

Acknowledgements

We would like to express our sincere gratitude to all organizations and individuals who generously granted permission to use their photographs. Our special thanks go to the talented photographers Guram Kharshiladze, Ani Baidashvili, Giorgi Tsiklauri, Mykola Sakovych, Rudu Gapel, Andrew Staniland, Mikhail Golubev, Elene Goldman, and Eleonora Zaiceva for allowing us to include their outstanding work. We are also grateful to the travel agencies Georgian Travel Guide, Georgian Travel, Geotrip.ge, Travel Man, Viva Georgia, CarTravelPaths, Turpomru, and TutTravelRu for kindly providing access to a number of their photographs.

We would also like to extend our special thanks to Bagrat Razmadze, Project Manager at the Shota Rustaveli National Science Foundation of Georgia, whose encouragement, unwavering support, and professional assistance accompanied us throughout the project and the preparation of this book.

Finally, we would like to express our sincere gratitude to the entire team of the Springer Conservation and Management Series for their outstanding professionalism, guidance, and continuous support throughout the preparation and publication of this book. Their dedication and collaborative spirit have been invaluable in bringing this volume to publication.

Contents

About the Editors

Avtandil Okrostsvaridze is a Professor at Ilia State University (Tbilisi, Georgia) and a leading Georgian geoscientist whose work integrates magmatism, geochronology, metallogeny, and geoheritage. He graduated from Tbilisi State University and obtained his Doctor of Geological Sciences degree from the Geological Institute of the Georgian Academy of Sciences. His research focuses on collision-zone magmatism and volcanism, regional geological evolution, and the identification, assessment, and conservation of key geoheritage sites that underpin sustainable geotourism and emerging geopark initiatives in Georgia. Dr. Okrostsvaridze has led and contributed to numerous national and international research projects. He has authored more than sixty peer-reviewed publications and several books, and has presented widely at international scientific congresses. His work has significantly advanced understanding of the geology of Georgia and the Caucasus, particularly in relation to geoheritage and geoconservation.

Daniel Tormey is Technical Director at Catalyst Environmental Solutions Corporation, a US-based environmental consulting firm specializing in protected areas, water resources, energy, and infrastructure. He is a member of the IUCN World Commission on Protected Areas and has extensive experience in environmental assessment and management of protected and conserved areas. He has undertaken numerous World Heritage-related evaluations and contributions, including field missions and co-authoring IUCN guidance on World Heritage volcanoes and geoconservation in protected and conserved areas. Since 2019, he has served as a lead reviewer for multiple UNESCO Global Geopark applications, supporting both the development and evaluation of geopark proposals. Dr. Tormey has conducted fieldwork in Georgia since 2010, focusing on volcanology, tectonics, and related geoheritage themes. He holds a Ph.D. in Geology and Geochemistry from MIT, and Bachelor's degrees in Civil Engineering and Geology from Stanford University.

Geoheritage of Georgia: Where Continents Meet and Culture Endures (Introduction)

Avtandil Okrostsvaridze and Daniel Tormey

Abstract

The Caucasus is one of the world's classic collision zones, and Georgia occupies its central segment, where the southern slopes of the Greater Caucasus, the northern slopes of the Lesser Caucasus and the intervening intermontane basins meet in a compact, highly diverse territory. This introduction sets the stage for the volume *Geoheritage of Georgia* by explaining why Georgia's geoheritage matters now and how it can be framed as a strategic national asset. It first presents Georgia as a "collision-zone country" whose geology, from Paleozoic basement and ophiolitic fragments to Neogene volcanic plateaus and coastal plains, underpins an exceptional concentration of landforms, resources and hazards. It then shows how this geological variety supports world-class geotourism: in a small area, visitors can move from glaciers to semi-deserts, from volcanic gorges to Black Sea coasts, and from karst caves to tower villages and cave monasteries, making Georgia a natural destination for "living classroom" experiences. The chapter also highlights the tight coupling between geodiversity and biodiversity in the Caucasus Biodiversity Hotspot, where rock types, relief and soils generate sharp environmental gradients and a mosaic of habitats, and where conserving geoheritage is inseparable from conserving biological diversity. Finally, it introduces the "Georgian soul" as a geocultural phenomenon, emphasising how landscapes, myths, epics, music, architecture and everyday practices are rooted in the physical fabric of the land. Throughout, the introduction outlines the organising frameworks used in the book; Key Geoheritage Areas, UNESCO Global Geoparks and related international tools; which are developed further in subsequent chapters to provide a coherent pathway from scientific understanding to conservation, education and sustainable development in Georgia.

Keywords

Georgia · Caucasus · Geoheritage · Geotourism · Key Geoheritage Areas (KGAs) · UNESCO Global Geoparks · Biodiversity hotspot · Cultural landscape

1 Why Geoheritage Matters Now for Georgia

Georgia's geoheritage matters now because it offers the country a way to link its outstanding geology with biodiversity, culture and sustainable development. Geological sites and landscapes in Georgia provide essential evidence for the history of the Earth, life and climate, and they underpin many of the country's most important ecosystems, agricultural systems and cultural traditions. At the same time, geoheritage is one of Georgia's clearest pathways to international recognition and visibility: by framing Key Geoheritage Areas and aspiring UNESCO Global Geoparks around world-class mountains, volcanic plateaus, karst systems and mining landscapes, Georgia can present itself as a collision-zone country with a distinct story and a strong future role in global conservation. In this volume we express that strategic value through the language of geoheritage, using Key Geoheritage Areas and candidate geopark territories as the backbone for the chapters that follow.

A. Okrostsvaridze (✉)
Institute of Earth Sciences, Ilia State University, Tbilisi, Georgia
e-mail: avtandil.okrostsvaridze@iliauni.edu.ge

D. Tormey
Catalyst Environmental Solutions, Santa Monica, USA

IUCN World Commission on Protected Areas, Gland, Switzerland

A. Okrostsvaridze and D. Tormey (eds.), *Geoheritage of Georgia*, Geoheritage, Geoparks and Geotourism,
https://doi.org/10.1007/978-3-032-27639-1_1

2 Collision Zone: Geological Foundation of Georgia's Landscapes

Georgia's landscapes are born of collision, both in the literal tectonic sense and in the way different climates, cultures and ecosystems meet. The country sits within the Arabia–Eurasia collision zone, where the Greater Caucasus, the Inner Caucasian (Transcaucasian) Microplate and the Lesser Caucasus volcanic–metallogenic belt have been built and reworked over hundreds of millions of years. This long history has produced a compact mosaic of rock types and landforms: high crystalline massifs, volcanic plateaus, folded sedimentary belts, karst platforms, basins and coastal plains.

This collision-born complexity is the bedrock of Georgia's geoheritage. It means that in a small territory, visitors can encounter the signatures of ancient oceans, island arcs, continental assembly, post-collisional volcanism and ongoing uplift and erosion. Later chapters in this volume explore these belts in more detail, showing how the Greater Caucasus high peaks, the Inner Caucasian Microplate, the Lesser Caucasus volcanic highlands and the Black Sea–Colchis margin each support distinctive Key Geoheritage Areas and candidate geopark territories.

3 Geotourism Destination: Georgia's Landscapes as Living Classrooms

Georgia is emerging as a premier geotourism destination, where visitors can travel from alpine glaciers to semi-deserts, from volcanic plateaus to Black Sea coasts, and from deep caves to high ridges within a few days. A defining feature of Georgian geotourism is the short distance between very different landscapes, and the way these landscapes are embedded in living culture: highland villages with stone towers and transhumance routes, vineyards on volcanic and alluvial soils, spa towns built around mineral springs, and cave monasteries carved into cliffs and ignimbrites. Increasingly, these places are being organised into geotourism routes and aspiring geopark territories that link geosites, trails, viewpoints and communities into coherent narratives of land and people, offering visitors not only scenery but also stories of how geology, biodiversity and culture interact.

4 Biodiversity: Life Woven into the Geology of Georgia

Georgia sits at the core of the Caucasus Biodiversity Hotspot, one of only a few dozen such regions recognised globally for their high levels of endemism and threat. Its forests, grasslands, wetlands and semi-deserts owe much of their distinctiveness to the underlying geology: the distribution of rock types, landforms and soils controls where Colchic rainforests thrive, where alpine meadows develop, where steppe and shrublands appear, and where wetlands and peatlands accumulate. Because of these tight linkages, conserving geoheritage is also part of conserving biodiversity. Many of the Key Geoheritage Areas identified in this volume coincide with, or overlap, Key Biodiversity Areas and important protected areas, and future management will need to integrate geodiversity and biodiversity so that the full fabric of Georgia's natural heritage is maintained.

5 The Georgian Soul: Where Land, Culture, and Spirit Converge

Perhaps the most compelling reason to protect and interpret Georgia's landscapes is not scientific or economic, but cultural. The Georgian soul is a composite of geology, ecology, and human history. For millennia, mountains have been both fortress and muse, shaping not just livelihoods but also moral and aesthetic codes. Legends of the Argonauts and the Golden Fleece, the poetry of Shota Rustaveli, and the polyphonic harmonies of mountain choirs all carry echoes of place. In highland villages, defensive towers rise from bedrock as symbols of both resilience and belonging. From the vine terraces of Kakheti laid out on alluvial fans to the tower villages of Svaneti built from local schist, everyday life is grounded, literally, in stone and soil. To separate culture from landscape here would diminish both; the Georgian soul is expressed as much through rock and earth as through song, dance, and shared wine. The supra, with its ritual toasts, binds people to the land that yields the wine and bread shared at the table.

In this worldview, the peaks of Georgia are not inert, they are ancestors; rivers are not mere channels, they are lifelines. This intimacy with the land gives Georgia's geotourism a depth that goes beyond scenic appreciation. Visitors are invited into a relationship, not just an itinerary. A walk through a volcanic gorge becomes an encounter with the deep time of both Earth and culture; tasting wine from a basaltic slope is a communion with the elements that formed it. In this sense, the Georgian soul offers something rare in the modern world: a lived example of how humans can be both of the land and its stewards.

This fusion of land and identity is mirrored in myth and epic. The Argonauts' quest for the Golden Fleece, set in ancient Colchis, has roots in placer gold mining along western Georgian rivers, where fleeces were used to trap flakes of alluvial gold. The medieval poem *The Knight in the Panther's Skin* by Shota Rustaveli celebrates ideals of loyalty, courage, and love that remain core to Georgian self-understanding. Polyphonic song, recognized by UNESCO as intangible cultural heritage, layers voices like ridgelines, its harmonies echoing the country's complex topography and long memory of place.

The Georgian soul also resides in resilience. In this active collision zone, earthquakes, landslides, and shifting rivers are part of lived experience. Communities adapt and rebuild, maintaining rituals and languages while engaging with modern life. Soviet industrial centers now sit alongside revived craft workshops; ancient monasteries still host pilgrims and curious visitors. This capacity to endure change without losing essence reflects the geological dynamism beneath Georgian feet: just as plates converge and mountains rise, society absorbs shocks and continues.

For visitors, geotourism in Georgia is therefore more than a catalogue of sites; it is an invitation into a worldview in which mountains are ancestors, rivers lifelines, and sharing bread and wine a daily reaffirmation of belonging to the land. A walk through a volcanic gorge becomes an encounter with both deep time and living tradition; tasting wine from basaltic or limestone slopes is a quiet communion with the materials that shaped the nation. In this sense, the Georgian soul is a geocultural phenomenon, born of tectonic, ecological, and historical convergence, and Georgia's geoheritage is its most enduring, tangible expression.

The various chapters of this book provide detailed descriptions of nearly all major geosites of Georgia, each shaped by the long-lasting geological processes that have taken place across its territory. A significant number of Georgia's Earth scientists contributed to the preparation of this book, and we extend our sincere gratitude to all of them. At the same time, we strived to ensure that each chapter was written in a clear and accessible language, with the aim of engaging not only specialists but also a wide audience of geology and nature enthusiasts.

Geoheritage Pathways for Georgia: Key Geoheritage Areas and The UNESCO Global Geopark Framework

Daniel Tormey and Avtandil Okrostsvaridze

Abstract

Georgia has several outstanding areas of globally significant geoheritage with strong cultural and biological connections, yet to date it has no UNESCO Global Geopark. This chapter sets out a framework for organising Georgia's geoheritage into Key Geoheritage Areas (KGAs) and for using the UNESCO Global Geopark (UGGp) and World Heritage (WH) instruments in a complementary way. At the scientific core are KGAs, areas whose geological features and processes are of international significance and provide essential evidence for the history of the Earth, life and climate. On top of this backbone sit UGGps as territory-based designations focused on sustainable development, and World Heritage Sites as apex recognitions focused on Outstanding Universal Value and long-term protection. The chapter explains why UGGps align particularly well with Georgia's geological and social setting: many of the country's most distinctive landscapes are territorial and community-based; mountain corridors, volcanic plateaus, sandstone canyon systems, karst belts and desert–delta margins; where geology, culture and livelihoods are tightly interwoven. In a European context where mountain tourism is shifting away from crowded, highly regulated destinations, Georgia offers authenticity, wildness and cultural richness, yet remains largely unknown to global audiences. A network of geoparks built on KGAs is therefore presented as a strategic mechanism to connect Georgia's world-class geology with geotourism, local economies, education and international visibility. The chapter then summarises the UNESCO Global Geopark concept and evaluation criteria, structured around four essentials: geological heritage of international scientific value; management and governance; visibility, education and communication; and networking and community engagement.

Keywords

Georgia · Geoheritage · Key Geoheritage Areas (KGAs) · UNESCO Global Geoparks

1 Introduction

Georgia has several outstanding areas of globally significant geoheritage with strong cultural and biological connections. Landscapes and geological formations here are key witnesses to the evolution of the Caucasus collision zone and determinants of future sustainable development. For Georgia, UNESCO Global Geoparks offer a tested international framework to connect world-class geoheritage with national identity, rural livelihoods, and international visibility. UNESCO Global Geoparks are living, working landscapes where geological heritage is conserved and interpreted in close partnership with local communities, and used as a basis for sustainable economic activity, especially geotourism.

This chapter sets out how that framework can be applied. Section 1.1 introduces the three main pathways relevant to Georgia's geoheritage, Key Geoheritage Areas, UNESCO Global Geoparks and World Heritage. Section 1.2 explains why the geopark model in particular is strategically important for Georgia at this moment. Sections 2, 2.1, 2.2 and 2.3 and then summarise the UNESCO Global Geopark process. Chapter Georgia's Candidate Geopark Territories: Portraits and Readiness applies this framework to a set of candi-

D. Tormey
Catalyst Environmental Solutions, Santa Monica, USA

D. Tormey (✉)
IUCN World Commission on Protected Areas, Gland, Switzerland
e-mail: dtormey@ce.solutions

A. Okrostsvaridze
Institute of Earth Sciences, Ilia State University, Tbilisi, Georgia

A. Okrostsvaridze and D. Tormey (eds.), *Geoheritage of Georgia*, Geoheritage, Geoparks and Geotourism,
https://doi.org/10.1007/978-3-032-27639-1_2

date territories, providing portraits and a comparative readiness assessment that together outline a national blueprint for future geoparks.

1.1 Geoheritage Pathways: Key Geoheritage Areas, UNESCO Global Geoparks and World Heritage

Georgia's geoheritage can be understood as a three-part pathway. At the scientific core are Key Geoheritage Areas (KGAs), areas whose geological features are of international significance and provide essential evidence for the history of the Earth, life and climate. KGAs are the backbone used throughout this volume: they are the units of geoheritage that matter globally, whether or not they yet carry an international designation.

On top of this KGA backbone sit two complementary UNESCO tools. UNESCO Global Geoparks (UGGps) are primarily a territory model focused on sustainable development. They are living designations for sizable areas where notable geology is woven together with education, geohazard literacy, local enterprise, geotourism, and community partnerships. In Georgia, geoparks are especially relevant for mountainous regions and volcanic plateaus where multiple geosites and communities share one story.

World Heritage Sites are primarily a protection model focused on "Outstanding Universal Value". They inscribe discrete properties (or serial series) that meet stringent criteria and long-term management tests. World Heritage emphasizes conservation at the highest global bar and is used sparingly, to recognize the most exceptional places on Earth.

For Georgia, these instruments are complementary rather than competing. UGGps act as an on-ramp: they build capacity, partnerships, inventories and management systems across territories that contain one or more KGAs. Over time, the same scientific evidence and management practice can support World Heritage nominations or extensions for those KGA cores that meet the OUV threshold. In this sense, a simple triad captures Georgia's strategic opportunity:

- **KGA**—the scientific backbone: where Georgia's geology is globally significant.
- **UGGp**—the territory frame: where communities and institutions learn to manage that geoheritage for development.
- **World Heritage**—the apex recognition: where the most outstanding KGAs also demonstrate OUV and long-term protection.

Building on this KGA–UGGp–World Heritage triad, UNESCO Global Geoparks emerge as the most immediately useful tool for Georgia's geoheritage. Georgia's most exceptional geological assets are territorial in character: they extend across mountain corridors, volcanic plateaus, karst belts, desert and delta systems, and they are inhabited by communities whose futures are closely tied to those landscapes. Unlike many cultural monuments or small natural sites, Georgia's geology is directly embedded in regional development, mountain economies, and tourism. This makes the UNESCO Global Geopark framework not only appropriate, but strategically essential.

1.2 Why UGGps Align Well with Georgia's Geological Setting

The UGGp programme carries the UNESCO name, which is important for international recognition, but unlike World Heritage—where the focus lies on inscription and Outstanding Universal Value—the geopark model focuses on using geological heritage as a driver for sustainable regional development. In practice this means emphasising geotourism and branding, education and community engagement, local entrepreneurship, inter-municipal cooperation and capacity-building in governance, rather than on a single inscription event.

Georgia's geological story is inherently territorial and continuous. Many of its strongest landscapes are already organised into coherent corridors and mountain blocks: the Greater Caucasus high peaks, the Inner Caucasian intermontane basins, the Lesser Caucasus volcanic highlands and the Colchis and semi-desert margins, all of which blend geology, culture and living communities.

Across Europe, a generational shift in mountain tourism is underway. Increasingly, visitors with both resources and curiosity seek destinations that offer genuine exploration, local culture and affordability. The Alps, while still iconic, are often crowded, expensive, highly regulated and diminished in their sense of discovery. Georgia offers the exact qualities this demographic now pursues: peaks and ridges as storied as the Alpine giants, with deep folklore, religious history and mountaineering roots; authentic mountain villages still shaped by traditional lifeways; dramatic gorges, glaciers, caves, canyons and volcanic terrains; and affordable, accessible adventures in landscapes that remain largely unknown to global audiences. Yet the world does not yet appreciate the scale, drama and uniqueness of Georgia's mountains.

A network of UNESCO Global Geoparks is therefore one of the most effective mechanisms available to

Georgia to signal international legitimacy and global-quality geology, build branding around its mountain regions, attract sustainable tourism and investment, create long-term rural economic opportunities and raise Georgia's visibility as a major emerging geotourism destination. In practical terms, UGGps give Georgia a UNESCO-backed platform to present its mountains to the world while also building management capacity at regional and municipal levels, strengthening geoscience monitoring and data, training local guides and educators, improving interpretive materials and visitor infrastructure, and deepening governance experience across multiple jurisdictions. These are precisely the foundations required for a successful World Heritage nomination under natural criteria. Rather than delaying World Heritage ambitions, UGGps prepare the ground for them: they are the territory-scale learning spaces in which Georgia and its partners can practise integrated heritage management while bringing tangible benefits to the people who live within them.

1.3 A Network Model: From Individual Landscapes to National Geostory

Georgia is uniquely positioned to develop a coherent national network of Geoparks, each representing a different facet of its geological diversity. Strong candidates include:

- Vardzia and the volcanic plateau systems
- The Enguri–Svaneti glacier, canyon, and high-mountain corridor
- Javakheti volcanic uplands and lake basins
- Dariali and the central Greater Caucasus tectonic corridor
- Eastern sandstone canyon systems (e.g., Vashlovani, Gareja landscapes)
- Imereti karst and cave systems.

Together, these areas illustrate how Georgia's geology naturally groups into corridors and clusters that can form the backbone of a future geopark network, further developed in Chap. Georgia's Candidate Geopark Territories: Portraits and Readiness.

2 Overview of UNESCO Global Geopark Decision Process

2.1 What Is a UNESCO Global Geopark?

UNESCO Global Geoparks are single, unified geographical areas where sites and landscapes of internationally significant geology are managed with a holistic concept of protection, education, and sustainable development (UNESCO 2025). Rather than being "geology museums," they are living territories in which rocks, landforms, ecosystems, and communities are managed together. A UNESCO Global Geopark uses its geological heritage, in connection with all other aspects of the area's natural and cultural heritage, to help people understand key issues facing society today—from using natural resources wisely to adapting to climate change and reducing natural hazard risk (UNESCO 2025).

A defining feature of these areas is their bottom-up character. Geoparks are not imposed from outside; they grow from local initiative. UNESCO Global Geoparks are established through partnerships between municipalities, protected area administrations, universities, schools, tourism operators, community groups and, where present, Indigenous or minority communities (UNESCO 2025). This collaborative governance is formalized in a management body recognized under national legislation, with a shared strategy that covers conservation, geotourism, education, infrastructure, and finance. For Georgia, success depends not only on outstanding geology, but also on building strong, long-term alliances among scientific institutions, local authorities, and communities in candidate areas.

UNESCO Global Geoparks are also explicitly about local pride and livelihoods. By raising awareness of the importance of an area's geological heritage in history and daily life, they strengthen people's sense of belonging to their region (UNESCO 2025). Geotourism, tourism that focuses on experiencing the landscape and its stories, becomes a tool to stimulate small enterprises, new jobs, and training opportunities while encouraging protection of the geological resource base. In many existing geoparks, local food, crafts, and festivals are branded together with the geological story, so visitors experience a whole territory rather than isolated sites. For Georgia, with its emerging wine routes, mountain guesthouses, and cave and canyon attractions, the geopark approach offers a tested international model for linking geoheritage to sustainable local economies.

Finally, UNESCO Global Geoparks are part of a global knowledge network. All geoparks belong to the Global Geoparks Network (GGN), which connects sites on every inhabited continent that are working on similar challenges: natural hazard education, climate change interpretation, sustainable resource use, youth engagement, and more (UNESCO 2025). Membership in this network would give Georgia's candidate areas access to peer learning, joint projects, and international visibility. For Georgian institutions, a geopark is therefore not only a local initiative, but also an entry point into a worldwide community of practice that celebrates "Earth heritage, sustaining local communities."

2.2 Four Essentials—And What They Mean for Georgia

When UNESCO evaluates an application for a Global Geopark, it looks for four fundamental features. In practice, these are the questions that will be asked of any Georgian candidate area.

(1) **Geological heritage of international scientific value**

UNESCO Global Geoparks are built on geology that matters globally, not only locally. Evaluators rely on peer-reviewed scientific publications and comparative analysis to decide whether the rocks, landforms, or processes in an area truly stand out at the international scale. For Georgia, this means organizing and presenting existing scientific work on key themes such as Caucasus mountain building, Quaternary volcanism, karst and caves, or early human evolution, in a way that clearly shows how Georgian sites compare with classic areas elsewhere in the world.

(2) **Management: a functioning, legally recognized body and plan**

Every UNESCO Global Geopark must be managed by a body with legal standing under national law, responsible for the entire territory and backed by a comprehensive management plan. That plan must cover governance, conservation, education, tourism, communications, infrastructure, finances, and partnerships. In Georgia, this implies bringing together protected area agencies, municipalities, scientific institutions, tourism boards, and community representatives into a single structure for each aspiring geopark, and giving that structure a clear mandate and resources through national or regional instruments.

(3) **Visibility: a clear identity and communication system**

Geoparks must be easy to find and understand. UNESCO expects a dedicated website, maps, leaflets, and signage that connect the area's geosites and explain how geology relates to other values. A recognizable logo and coherent visual identity help visitors and residents alike see the geopark as a unified territory. For Georgia, this means investing not only in trails and visitor centers but also in branding, interpretation, and digital communication so that a teacher in Tbilisi, a tourist in Kutaisi, or a local family in Javakheti can quickly see what the geopark is, where it is, and why it matters.

(4) **Networking and community engagement**

UNESCO Global Geoparks are expected to work in two directions: inward, with their own communities, and outward, with other geoparks worldwide. Locally, they must actively involve communities as key stakeholders, drawing on local and Indigenous knowledge alongside science in planning and management. Internationally, they are required to participate in the Global Geoparks Network, sharing experience and collaborating on joint projects. For Georgia, this means that candidate areas will need demonstrable support from municipalities, community groups, and businesses, as well as a plan to engage with existing geoparks in Europe and beyond from the earliest stages of preparation.

Taken together, these four essentials provide the framework we use throughout this volume to assess Georgia's candidate areas. In Chap. Georgia's Candidate Geopark Territories: Portraits and Readiness, we apply them, along with Georgia-specific considerations such as national identity and geotourism potential, to build an overview and prioritisation of sites most ready to move toward UNESCO Global Geopark status. In Georgia we also express these essentials through the language of Key Geoheritage Areas (KGAs): for each candidate territory we identify one or more KGAs of international significance, and then ask whether a potential geopark can provide a coherent, well-managed wrapper around those KGA cores.

We use these four UNESCO essentials as the backbone for analysing Georgia's potential UNESCO Global Geoparks. For each candidate territory in Chap. Georgia's Candidate Geopark Territories: Portraits and Readiness we consider four core questions: (1) geological heritage of international scientific value; (2) management and governance; (3) visibility, education and communication; and (4) networking and community engagement.

2.3 From Framework to Territories

The sections above outline Georgia's geoheritage pathways and the UNESCO Global Geopark process in principle. They define how Key Geoheritage Areas, UNESCO Global Geoparks and World Heritage relate to one another, set out the four essentials used by UNESCO to evaluate applications, and describe how Georgia can translate these into a practical evaluation framework for its own candidate territories.

Georgia's Candidate Geopark Territories: Portraits and Readiness

Daniel Tormey and Avtandil Okrostsvaridze

Abstract

Building on the geoheritage and UNESCO Global Geopark framework introduced in Chap. 2, this chapter applies the evaluation approach to a representative set of Georgian landscapes that could form the backbone of a future geopark network. Twelve candidate territories are considered, spanning the full range of Georgia's Key Geoheritage Areas (KGAs): volcanic plateaus and cave cities of the south (Vardzia and the Javakheti Plateau), karst and cave systems of Imereti (Sataplia and Prometheus), high-mountain regions in the Greater Caucasus (Kazbegi, Upper Svaneti, Tusheti and Racha), forested parks and volcanic uplands in the Lesser Caucasus (Borjomi–Kharagauli, Mtirala, Javakheti Highlands), and lowland and semi-desert systems at the margins (the Colchis Lowlands and Vashlovani). For each territory, the chapter provides a concise portrait that highlights its international geological value, links to biodiversity and cultural landscapes, and current status in terms of protection, access and community engagement. To move beyond descriptive comparison, the chapter synthesises these portraits into three readiness tiers for UNESCO Global Geopark nomination, based on five dimensions: geological significance, integrity and management, education and interpretation, geotourism and local economy potential, and community engagement and identity. Eight territories are used to illustrate this analysis in detail. Tier 1 ("first wave") sites, Sataplia–Prometheus, Dmanisi and Vardzia, already combine outstanding geoheritage with relatively strong management and interpretation and are identified as realistic near-term candidates. Tier 2 areas, Kazbegi, Upper Svaneti and Borjomi-Bakuriani, are medium-term candidates that would benefit from further work on governance, branding and geotourism products. Kolkheti Wetlands and the Kakheti Volcanic Highlands are treated as Tier 3, long-term strategic opportunities where early investment in inventories, management and local partnerships can keep future geopark options open.

Keywords

Georgia · Candidate territories · Readiness assessment · Geotourism · The Caucasus

1 Introduction

Chapter Geoheritage Pathways for Georgia: Key Geoheritage Areas and The UNESCO Global Geopark Framework set out Georgia's geoheritage pathways and the UNESCO Global Geopark framework. It introduced Key Geoheritage Areas (KGAs) as the scientific framework, explained how UNESCO evaluates geopark applications, and described the four essentials and rating approach we use to assess candidate territories.

In this chapter we apply that framework to a representative set of Georgian landscapes that could form the backbone of a future network of UNESCO Global Geoparks. These candidate territories are distributed across the whole country. Together they show that Georgia's internationally significant geoheritage is not confined to a single flagship site, but forms a national pattern of outstanding territories with strong geological, ecological and cultural stories.

In this Chapter we consider their strengths and gaps based on the four UNESCO essentials and Georgia-specific criteria and outline a national blueprint for a phased network

D. Tormey
Catalyst Environmental Solutions, Santa Monica, USA

D. Tormey (✉)
IUCN World Commission on Protected Areas, Gland, Switzerland
e-mail: dtormey@ce.solutions

A. Okrostsvaridze
Institute of Earth Sciences, Ilia State University, Tbilisi, Georgia

A. Okrostsvaridze and D. Tormey (eds.), *Geoheritage of Georgia*, Geoheritage, Geoparks and Geotourism,
https://doi.org/10.1007/978-3-032-27639-1_3

of UNESCO Global Geoparks. The chapter concludes with implications for institutions, partnerships and heritage diplomacy.

2 Candidate Areas within the Republic of Georgia

Georgia already has a rich portfolio of landscapes that could form the backbone of a future network of UNESCO Global Geoparks. In this chapter we review a representative set of candidate areas distributed across the whole country, building on the overview provided in the Introduction. Together they show a national pattern of outstanding territories with strong geological, ecological and cultural stories.

For each of these areas, the following sections provide a short, accessible portrait: where it is, what makes its geology internationally important, how geoheritage connects with biodiversity and cultural traditions, and what opportunities and challenges exist for a UNESCO Global Geopark designation. This overview offers a first comparative picture of where Georgia stands now, which territories are closest to meeting UNESCO's expectations, and where targeted investment and capacity-building could bring additional candidates to the same level over time.

2.1 Vardzia

Carved into the sun-warmed cliffs of southern Georgia, Vardzia sits in thick welded ignimbrite that enabled the rock-cut city. Its rock-cut halls, chapels, and chambers are etched into thick volcanic tuff, the legacy of massive eruptions from the Javakheti volcanic plateau. Layers of welded ignimbrite tell the story of pyroclastic flows that once roared through the valleys, cooling into stone soft enough for medieval tools yet strong enough to withstand centuries of wind and rain. The landscape around it is a living classroom in ignimbrite stratigraphy, erosion, and rock-stability conditions that require thoughtful conservation.

As a geopark, Vardzia is a rare fusion of geological heritage and living culture. Once a fortified city-monastery built by Queen Tamar in the twelfth century, it embodies the deep intertwining of faith, defense, and landscape. The surrounding valleys shelter endemic plants and traditional orchards, while the Mtkvari River below sustains both wildlife and agriculture. Pilgrims, historians, and hikers converge here not just for its beauty but for its layered narratives: ancient trade routes, Mongol invasions, monastic persistence. Its potential as a geopark lies in this holistic story—a place where volcanic fire became a sanctuary, and where geological interpretation can be seamlessly woven with local community engagement, artisanal craft revival, and sustainable tourism.

2.2 Sataplia and Prometheus Caves

In the Imereti region, Sataplia and Prometheus Caves open a window into the underground world of karst. The limestones here, deposited in warm shallow seas during the Cretaceous, have been sculpted over millions of years into vast chambers, stalactites, stalagmites, and underground rivers. At Sataplia, fossilized dinosaur footprints are preserved on bedding planes, linking the cave's subterranean beauty to the surface's ancient life.

As a geopark, the Sataplia–Prometheus complex could weave paleontology, speleology, and karst hydrology into a compelling narrative. The sites are already major tourist attractions with interpretive centers and guided tours, yet they have untapped potential to connect visitors more deeply to the geological processes that created them. Educational programs could draw on the dinosaur story to inspire young visitors, while advanced cave science could appeal to more specialist audiences. By linking these sites to the broader karst belt of western Georgia, the geopark could promote conservation of fragile cave ecosystems alongside sustainable tourism.

2.3 Dmanisi

In the rolling hills of southern Georgia, where the Mashavera and Pinezaouri rivers meet, lies Dmanisi, recording early human dispersal from Africa in its fossil-bearing sediments capped and protected by basalt flows. Beneath the medieval fortress walls, paleontologists have uncovered skulls and tools over 1.8 million years old, belonging to some of the earliest known humans in Eurasia. The geology of basalt capping and valley incision help explain why the fossil horizons are so well preserved and accessible for interpretation. The volcanic rocks speak of fire and upheaval, while the river valleys record quieter cycles of deposition, erosion, and life.

As a geopark, Dmanisi's draw is profound: it unites geology, archaeology, and anthropology into a single narrative about movement, adaptation, and belonging. Visitors can walk from a medieval church into an active dig site where our ancestors' faces emerge from the earth, then gaze out across landscapes little changed in form since those ancient days. The potential for community engagement is rich, from heritage guides drawn from local villages to partnerships with schools in nearby Tsalka and beyond. In Dmanisi, the rocks do not just tell a geological tale; they tell the human story, reminding us that every landscape is also a migration map.

2.4 Kazbegi (Stepantsminda/Mount Kazbek Region)

Rising like a sentinel on Georgia's northern frontier, Mount Kazbek is a volcano crowned with ice. At 5047 meters, it is one of the great peaks of the Caucasus, its flanks sculpted by glaciers that flow into deep gorges. The mountain's core is built from andesitic and dacitic lavas erupted thousands of years ago, standing atop an older foundation of folded sedimentary rocks. Kazbek's slopes are a living textbook in alpine geomorphology, cirques, moraines, and periglacial landforms speaking to the dance of fire and ice over deep time.

In geopark terms, Kazbegi is both a geological icon and a cultural crossroads. The Georgian Military Highway passes through, carrying centuries of trade, conflict, and pilgrimage. The Holy Trinity Church of Gergeti, perched on a grassy knoll against the glaciered massif, is as much a part of the landscape's identity as its rock and ice. The surrounding Kazbegi National Park shelters rare alpine flora and endangered wildlife like the Caucasian tur, while local villages sustain traditions of mountain herding and hospitality. A geopark here would link high-altitude ecology, volcanic history, and living culture into a single arc, attracting mountaineers, trekkers, and cultural travelers alike.

2.5 Upper Svaneti

In the remote northwest, Upper Svaneti is a fortress of mountains and memory, where geology and cultural continuity are inseparable. The Greater Caucasus rises here in a wall of metamorphic schists and gneisses, intruded by granite and sculpted into jagged ridges and glacier-hung valleys. Seasonal snowmelt feeds torrents that carve through ancient moraines and debris flows, revealing the restless power of mountain landscapes. Glacial retreat in recent decades offers a visible, urgent lesson in climate change, one that is already reshaping both ecosystems and livelihoods.

Upper Svaneti's human story is equally compelling. Its medieval stone towers, UNESCO World Heritage listed, stand in clusters above the villages, once serving as family fortresses in times of conflict. The Svans' language, songs, and craftsmanship remain vibrantly alive, creating a living cultural layer over the deep-time geology. As a geopark, Upper Svaneti could be a beacon for integrating high-mountain conservation, glaciology research, and cultural preservation. Tourism is already strong, but a geopark framework could guide it toward sustainability, ensuring that the region's geological and cultural treasures endure for generations.

2.6 Javakheti Volcanic Plateau

High in southern Georgia, the Javakheti Plateau spreads like a basalt crown between the Caucasus and the Armenian Highlands. This is a land born of fire: extensive lava flows, volcanic cones, and maars shaped by both explosive and effusive eruptions during the Quaternary. Lakes such as Paravani and Sagamo fill volcanic depressions, their waters reflecting rolling grasslands. Beneath the surface, the region's geothermal potential bubbles quietly, a reminder that the volcanic story is not yet fully written.

As a geopark candidate, Javakheti offers a vast, coherent geological theme, volcanism in a high-plateau setting, tied to distinctive biodiversity. The area's wetlands are vital for migratory birds, and its pastoral communities maintain a resilient, self-sufficient way of life. Linking geotourism with birdwatching, agritourism, and volcanic interpretation could create a year-round destination that strengthens both conservation and livelihoods. Here, visitors could trace lava's path from crater to plain, then share cheese and bread with shepherds who have walked the same slopes for centuries.

2.7 Tusheti

Reached only by a high, winding pass open for part of the year, Tusheti is a remote highland realm where geology dictates both access and culture. Schists and sandstones of the Greater Caucasus are folded and faulted into ridges that rise above deep river valleys, while periglacial and fluvial processes actively sculpt the landscape. Avalanches, rockfalls, and seasonal floods remind residents and visitors alike of the mountains' power.

Tusheti's isolation has preserved both its ecosystems and its traditions. Stone-built villages cling to slopes above the Alazani and Gometsari valleys, their defensive towers echoing those of Svaneti. The pastures above are rich in wildflowers, grazing sheep, and endemic plant species. As a geopark, Tusheti could integrate geology, pastoral heritage, and high-mountain biodiversity into an experience of true remoteness.

2.8 Racha

On the northern edge of central Georgia, Racha is a land of limestone cliffs, karst caves, and glacial valleys framed by the Greater Caucasus. The region's Mesozoic limestones, uplifted and fractured by mountain-building, are riddled with sinkholes and underground rivers. The Shaori Reservoir mirrors the forested hills, while the Tskhrajvari and Shovi areas reveal volcanic centers and mineral springs.

Racha's charm lies in its blend of geology and rural authenticity. Its wine, particularly the semi-sweet Khvanchkara, carries the taste of the land's soils and climate. Villages are scattered among orchards and chestnut groves, and the caves shelter bat colonies alongside speleothems. A geopark here could highlight karst science, glacial geomorphology, and volcanic heritage while promoting agro-tourism and local crafts. Racha's relatively low visitor numbers today mean there is an opportunity to plan geotourism sustainably from the outset, keeping the region's landscapes pristine.

2.9 Colchis Lowlands

Stretching along Georgia's Black Sea coast, the Colchis Lowlands are a vast alluvial plain where rivers descending from the Caucasus deposit layer upon layer of sediments. Beneath the fertile soils lie deep sequences of Quaternary sands, silts, and clays interbedded with peat deposits, a testament to repeated cycles of river migration, sea-level change, and wetland formation. The coastal margin bears the geomorphic traces of ancient shorelines, barrier spits, and lagoons, revealing a dynamic interplay between fluvial and marine forces.

From a geopark perspective, the Colchis Lowlands offer the chance to tell a story of life in a landscape constantly remade by water. As part of the Colchic Rainforests and Wetlands World Heritage Site, they already hold exceptional biodiversity; the geopark framework could layer on the geological dimension, sediment transport from the high Caucasus, delta dynamics, and the role of wetlands in carbon storage. Visitors could trace the journey from mountain to sea, exploring river deltas, floodplain forests, and coastal dunes, with interpretation tying together geology, ecology, and cultural traditions of fishing and reed harvesting.

2.10 Vashlovani

In Georgia's far southeast, Vashlovani National Park is a landscape of bare badlands, sheer cliffs, and semi-desert steppes unlike anywhere else in the country. Here Miocene and Pliocene sediments—soft clays, marls, and sandstones, have been carved by wind and water into labyrinths of canyons, hoodoos, and mesas. The geology records the closing of the Paratethys Sea and the uplift of the Caucasus, with marine fossils and ancient shorelines preserved in the rock.

For geopark interpretation, Vashlovani's interpretive power lies in exposed Neogene sediments carved into badlands that record Paratethys closure and uplift. Visitors can literally see the layers of time in the exposed strata and walk through landscapes that resemble those of the American Southwest. This is also a place where geology shapes biodiversity: arid-climate plants, vultures nesting on cliffs, and the elusive striped hyena all depend on the terrain's peculiarities.

3 Comparative Assessment: Readiness Tiers

To compare the candidate areas in a way that is transparent and useful for decision-makers, we applied the evaluation framework described in Chap. Geoheritage Pathways for Georgia: Key Geoheritage Areas and The UNESCO Global Geopark Framework to a representative subset of eight territories. These were: Sataplia–Prometheus, Dmanisi, Kazbegi, Vardzia, Upper Svaneti, Kolkheti Wetlands, the Kakheti Volcanic Highlands and the Borjomi–Bakuriani region. For each area we considered five dimensions: international geological value, integrity and management, education and interpretation, geotourism and local economy potential, and community engagement and identity. We then synthesise these into an overall view of "readiness" for UNESCO Global Geopark nomination. Rather than reproducing a detailed rating table here, we present the results as three readiness tiers that can guide national planning.

Tier 1—Near-term nomination candidates ("first wave")

Territories in this group already combine outstanding international geoheritage with relatively strong management, interpretation and geotourism infrastructure. In Georgia, three areas fall into this category: Sataplia–Prometheus, Dmanisi and Vardzia. All three have well-defined stories of global significance (karst caves and dinosaur footprints, early hominin occupation, and a "living ignimbrite" rock-cut city), existing protected-area frameworks and substantial visitor facilities. Their main needs are targeted investments in governance, interpretation, community engagement and branding, rather than basic readiness. These are the most realistic near-term candidates for UNESCO Global Geopark nomination.

Tier 2—Medium-term candidates (5–10 years)

A second group shows medium–high overall readiness and strong potential to move toward nomination over the next five to ten years: Kazbegi, Upper Svaneti and Borjomi–Bakuriani. These territories all possess high geological value and strong cultural landscapes, but require more focused work on management structures, interpretation and geotourism products. For example, Kazbegi and Upper Svaneti are already major mountain destinations, but would benefit from clearer territorial definition, integrated management planning and strengthened community partnerships. Borjomi–Bakuriani has internationally recognisable mineral waters and a long tourism history, but needs stronger integration of geoscience, hazard interpretation and local business networks. With sustained effort, these areas could form a second wave of nominations that extend Georgia's presence along both the Greater and Lesser Caucasus.

Tier 3—Long-term strategic opportunities

Finally, Kolkheti Wetlands and the Kakheti Volcanic Highlands currently sit at a lower overall readiness level. Their geological value is clear, coastal and deltaic systems in the west and extensive volcanic plateaus in the east, but management structures, interpretation, access and community engagement are less developed. For the purposes of this volume, these territories are best regarded as long-term strategic opportunities: places where early work on geosite inventories, protection, monitoring and local partnerships can lay the foundations for future geoparks, even if formal UNESCO nomination is some distance away.

Taken together, these three tiers outline a phased approach: a first wave of near-term nominations where Georgia can realistically meet UNESCO expectations within the next decade; a second wave of medium-term candidates that can advance as governance, funding and community partnerships mature; and a set of longer-term strategic territories where early investment in geoheritage and geotourism can keep future options open. The following sections use the KGA framework (Sect. 3.1) and the national blueprint (Sect. 3.2) to interpret these tiers at the scale of the whole country.

3.1 Key Geoheritage Areas (KGAs): A New IUCN Recognition Layer

As introduced in the chapter Geoheritage Pathways for Georgia: Key Geoheritage Areas and The UNESCO Global Geopark Framework, Key Geoheritage Areas (KGAs) are an emerging IUCN framework recognising areas whose geological features and processes are of global significance. The KGA initiative fills a gap between site-based inventories and high-level international designations. Many geosites of genuine global significance are not, and may never be, recognised as World Heritage Sites or UNESCO Global Geoparks. KGAs provide a science-based recognition layer that sits underneath those more demanding labels. They make it possible to see, on a single map, where the world's most important geoheritage is found, regardless of whether formal designations have yet been pursued.

For Georgia, this is particularly valuable. The country already hosts hundreds of documented geosites and natural monuments, along with candidate territories for geoparks and possible World Heritage extensions. The KGA framework offers a way to:

- **Cluster related geosites** into coherent areas that tell one geological story (for example, a volcanic plateau, a high-peak belt, a karst system or a desert basin).
- **Prioritise among many strong candidates**, by highlighting where Georgian geoheritage is of clear international significance.
- **Align geoheritage planning with biodiversity**, by developing a KGA map in parallel with the existing Key Biodiversity Area (KBA) network.
- **Support future geopark and World Heritage initiatives**, by making the scientific case for which areas are most suitable for more demanding international designations.

Within this book's logic, Key Geoheritage Areas are the broad base layer: almost all sites discussed in thematic and regional chapters can be grouped into a finite set of KGAs that together express Georgia's role in the Caucasus collision zone. Above that KGA layer sit the proposed UNESCO Global Geoparks and, in some cases, potential World Heritage properties or extensions. The comparative assessment in this chapter can therefore be read in two ways: as a readiness analysis for UNESCO Global Geoparks, and as an early sketch of Georgia's future national KGA map.

3.2 Blueprint for a National Geopark System

The assessment shows that Georgia's strengths are widely distributed across the country. KGAs underpinning Vardzia, Sataplia–Prometheus, Dmanisi, Kazbegi, Upper Svaneti, Javakheti, Tusheti, Racha, Borjomi–Kharagauli, Mtirala, the Colchis Lowlands and Vashlovani express different facets of the Caucasus collision zone and its active processes. This diversity argues in favour of a network of geoparks over time, rather than a single site.

Across many candidate territories, the geological story is already of international calibre, but there are recurring gaps in: (1) formalized governance structures and clear territorial boundaries; (2) integrated management plans that link geoheritage with biodiversity, cultural heritage and local development; (3) consistent branding, on-site interpretation and digital communication; and (4) mechanisms for structured community and small-enterprise engagement.

The assessment highlights the value of a unified KGA inventory and framework. By explicitly identifying the KGAs hosted within each candidate territory, for example, the Vardzia–Javakheti volcanic plateau, the Greater Caucasus high-peak belt, the Imereti karst and cave system, or the desert and badlands of Vashlovani, Georgia can ensure that geopark proposals are grounded in internationally recognized geoheritage, and that future World Heritage nominations or extensions build logically from the same scientific base.

4 Conclusions

Georgia is exceptionally well placed to benefit from the UNESCO Global Geoparks programme. Few countries of comparable size combine such a concentration of internationally significant geoheritage, from high peaks and active glaciers to volcanic plateaus, karst systems, desert badlands and coastal wetlands, with such a rich tapestry of cultural landscapes and living traditions. The candidate territories analysed in this chapter show that Georgia already has a portfolio of places where geology, biodiversity and culture can be woven together into a powerful story that supports sustainable development, education and national identity.

The comparative assessment demonstrates that Georgia's opportunity is not limited to a single "showcase" geopark. Instead, the pattern of strengths suggests a phased network of UNESCO Global Geoparks built on clearly defined Key Geoheritage Areas. In the near term, a first wave of territories can move toward nomination, while medium–and longer-term candidates are supported to build capacity, partnerships and infrastructure. Over time, this approach would give Georgia a coherent, country-wide geopark system that reflects the diversity of its geology and regions.

Realising this vision will require more than strong individual dossiers. It calls for a national framework that aligns scientific work on KGAs, protected area and cultural landscape management, destination planning, and community-led development. A practical step would be the establishment of a small National Geopark Council or consortium, hosted under the Georgian National Commission for UNESCO and bringing together key ministries, municipal representatives from candidate territories, academic and heritage experts, and geotourism entrepreneurs. Such a body could steer a national roadmap, maintain a shared readiness matrix, coordinate applications and revalidations, and ensure that geopark development reinforces rather than duplicates existing World Heritage Sites and protected areas.

Finally, geoparks give Georgia a distinctive tool for heritage diplomacy. By demonstrating how geoheritage can anchor risk reduction, climate resilience, education and local livelihoods, Georgian geoparks would speak directly to the priorities now shaping decision-making across UNESCO bodies and IUCN. A well-designed geopark network, grounded in Key Geoheritage Areas and linked to both World Heritage and national georoutes, can therefore strengthen not only Georgia's rural economies but also its voice in international discussions on conservation and sustainable development.

The chapters that follow provide the necessary thematic and regional detail to underpin this trajectory. They document the individual landscapes that make up Georgia's Key Geoheritage Areas and show how, taken together, they can support a future in which geology, culture and community thrive side by side.

A Brief Overview of Georgia's History and Cultural Heritage

Otar Janelidze

Abstract

Georgia is located in the South Caucasus, at the crossroads of two continents–Europe and Asia, and is one of the oldest countries in the world. Its position between the Greater and Lesser Caucasus and along the Black Sea coast meant that mountain passes, river valleys, and coastal plains shaped the routes of trade, migration, and invasion that define much of its history. It is known as the birthplace of the "Golden Fleece", grapevine and wine, as well as one of the oldest centers of origin of wheat. On the territory of Georgia (in the city of Dmanisi) the remains of an ancient man aged 1.8 million years were discovered. The Dmanisi hominins are considered the oldest inhabitants of the European continent and are also called the First Europeans. Georgia as a state has a long history, which begins with Diadochus in the twelfth century BC and continues with Colchis-Colchida, the kingdoms of Iberia (Kartli) and Egrisi and ends with the monarchy of the united Georgia. The country has long been the subject of struggle and confrontation between the great powers–Rome, Persia, Byzantium, Arab Caliphate, Mongolia and the Ottoman Empire. In the nineteenth century, Georgia was conquered by Russia and subordinated to its rule. On May 26, 1918, Georgia reappeared on the political map of the world as an independent state, but in 1921, its free development was stopped by the force of Soviet Russia. After 70 years of being part of the totalitarian Soviet Empire, the country finally gained its statehood on April 9, 1991.The Christian faith has been spreading in Georgia since the first century, and in 326 it was declared the state religion of Iberia (Kartli). During their long history, Georgians have created a unique, rich and diverse culture. The Georgian people have enriched the treasury of human culture with numerous achievements of civilization. This chapter examines the main stages of the political history of Georgia and presents the cultural heritage of the Georgian people.

Keywords

Georgia · History · Golden Fleece · Iron metallurgy · Christianity · Cultural heritage

1 History

Georgia is one of the oldest countries in the world. Its territory covers 69,700 km^2 and is located at the junction of two continents, Europe and Asia, in the South Caucasus. Due to its location, Georgia historically served as a bridge connecting Western and Eastern civilizations. The transcontinental trade and caravan route, the Great Silk Road, passed through this region, linking East Asia with the Mediterranean and Black Seas coasts. This trade corridor contributed to the socio-political and economic development of the region, but also facilitated the movement of external conquerors.

Georgia is known as the origin of the 'Golden Fleece', a myth likely rooted in ancient placer gold mining along western Georgian rivers, as well as for endemic wheat varieties and viticulture. In the Dmanisi region of Georgia, fossilized remains of the oldest known hominins in Europe, dating to 1.8 million years ago, have been discovered (Gabunia, Vekua 1995). Today, the Dmanisi hominin (Homo erectus) is regarded as the earliest known inhabitant of the European continent, often referred to as the "first European." These remains were preserved in volcanic and sedimentary layers that also record the evolution of the Caucasus landscape, making Dmanisi a key geoarchaeological site.

O. Janelidze (✉)
Ilia State University, Institute of Political Science, Tbilisi, Georgia
e-mail: otar_janelidze@yahoo.com

A. Okrostsvaridze and D. Tormey (eds.), *Geoheritage of Georgia*, Geoheritage, Geoparks and Geotourism,
https://doi.org/10.1007/978-3-032-27639-1_4

Numerous archaeological sites from the Paleolithic and subsequent periods identified in Georgia confirm continuous human habitation and cultural inheritance in the region. By the Early Bronze Age (4th–3rd millennium BCE), the territory of Georgia was extensively developed. During this period, the local population reached a high level of development, as evidenced by the abundance and quality of artifacts from the Mtkvari-Araksi culture and its later Colchian phase (2nd millennium BCE), including agricultural tools, gold and silver vessels, and intricate jewelry.

Georgia has a long history of statehood. The earliest known political union of Georgian tribes, Diaokhi, emerged in the southeastern Black Sea region as early as the twelfth century BCE. Diaokhi, referenced in Assyrian, Greek, and Urartian inscriptions, existed until the 8th century. In the same period, the Kingdom of Colchis arose, extending along the eastern coast of the Black Sea and encompassing a large part of western Georgia, including the present-day provinces of Abkhazia, Svaneti, Samegrelo, Imereti, Guria, and Racha.

The successor to Colchis was the Kingdom of Colchis, known in ancient Greek sources as the "golden country", formed in the 6th–5th centuries BCE. It nearly unified all of present-day western Georgia. Colchis is regarded as one of the early centers of coinage civilization, with its own silver currency, the Colchian tetri, in circulation. The kingdom also featured several urban centers, including Kutaia, Vani, Tsikhe-Goji, and Sairkhe. Colchis maintained strong connections with the Greek world, as evidenced by Greek-founded settlements along the Black Sea coast, such as Phasis, Gienos, and Dioscuria etc.

The Kingdom of Iberia (Kartli), a Hellenistic state, emerged in eastern Georgia during the 4th—3rd centuries BCE. The supremacy of its first king, Parnavazi, was also recognized by Kuji, the ruler of western Georgia, leading to the political unification of Colchis and Iberia into a single Georgian state. The capital of Iberia was Mtskheta. Parnavazi, credited with the foundation of Georgian literary tradition, also introduced significant administrative and religious reforms and established the Parnavazid royal dynasty.

In 65 BCE, the Roman general Pompey invaded Georgia. Colchis first became part of the Roman Empire and later the Byzantine Empire. Meanwhile, the Kingdom of Iberia (Kartli) maintained its formal independence, at times allying with Rome and at other times confronting Rome, Parthia, and Sassanid Persia.

The first third of the 4th century was a crucial point in Georgian history. Christianity, which had spread across the region since the 1st century, was declared the state religion of Iberia (Kartli) and later of Colchis (Egrisi) in 326 CE. This event had profound significance, not only shaping the spiritual life of the country but also influencing its cultural and state development (Lomouri 2009). Saint Nino, revered as Equal to the Apostles, is credited with converting Georgia to Christianity.

Vakhtang Gorgasali made a significant contribution to the struggle for liberation from external enemies and the reconstruction of the country. He ruled Kartli in the second half of the 5th century and successfully incorporated Western Georgia and part of South-Western Georgia into the Kingdom of Kartli. During his reign, the Kartli Church gained autocephaly, and he is also regarded as the founder of Tbilisi. His successor, Dachi, later moved the capital from Mtskheta to Tbilisi. The geopolitical struggles between Byzantium and Persia had a profound impact on both Western and Eastern Georgia. In 523 CE, Persia abolished the Kingdom of Kartli, while by the end of the century, the Byzantine Empire dismantled the statehood of Egrisi, reducing it to a Byzantine province.

From the late 7th century, long-term Arab rule was established in Eastern Georgia. The Arabs made repeated attempts to gain a foothold in Western Georgia (Egrisi-Abkhazia) but were unsuccessful. Their influence in the mountainous regions of the country remained minimal.

At the turn of the 8th–9th centuries, several independent political entities emerged in Georgia, including the Abkhazian Kingdom with its center in Kutaisi, the Kingdom of Hereti, and the principalities of Kakheti and Tao-Klarjeti. Their gradual unification culminated in the formation of the Georgian Kingdom at the end of the 10th century. The first ruler of this unified state was Bagrat III Bagrationi (975–1014), whose royal residence was in Kutaisi. Bagrat III was titled "King of the Abkhazians, Georgians, Rans, and Kakhs".

In the history of Georgia, the role of King David IV (1089–1125), known as the Builder, is particularly significant. To expand and strengthen the state, David curtailed the privileges of noble feudal lords and implemented several key reforms, including those in law and coinage, as well as the creation of a standing army. The victory achieved under his leadership in the Battle of Didgori (1121), where a powerful coalition of Muslim forces was defeated—a triumph described by the Georgian chronicler as a "miraculous victory"—held immense political and military significance. The following year, David the Builder moved the capital from Kutaisi to Tbilisi, gradually extending his political influence over neighboring peoples and establishing Georgia as the dominant power in the region. This was reflected in his official title: "King of the Abkhazians, Georgians, Rans, Kakhs, and Armenians, Sharvansha, and Shahansha." He also proclaimed himself the protector and supporter of Christians. An Arabic inscription on a coin minted during his reign reads: "King of Kings David, son of George, sword of the Messiah" (Metreveli 2002).

The reigns of his successors, Demetrius I and George III, were successful, but Georgia reached even greater

Fig. 1 The Caucasus region at the beginning of the thirteenth century, according to Eastmond (1998)

strength under a female ruler, Queen Tamar (1184–1207) (Fig. 1). Notably, during the Crusades, Georgia served as an outpost and defender of Christian civilization against the Muslim world. Under Tamar's leadership, Georgia secured several significant military victories, including the battles of Shamkhori (1195) and Basiani (1203) (Javakhishvili 1983).

The upward development of Georgia was hindered by constant Mongol invasions and domination, which began in the 1220s. The country managed to recover only in the first half of the fourteenth century, largely due to the efforts of King Giorgi V the Magnificent, who successfully restored Georgia's unity once again.

However, the campaigns of Timur the Lame proved equally devastating. This formidable conqueror invaded Georgia eight times between 1383 and 1403, inflicting immense destruction upon the country, though he ultimately failed to subjugate it. By the late fifteenth century, the once-unified Georgian kingdom had fragmented into the separate kingdoms of Kartli, Kakheti, and Imereti, along with the principality of Samtskhe-Javakheti. The disintegration continued, as the principalities of Guria, Samegrelo, Abkhazia, and Svaneti emerged within Imereti kingdom, while Kartli and Kakheti were further divided into smaller feudal entities.

The following period was marked by relentless invasions from Persia and the Ottoman Empire, alongside frequent raids by North Caucasian mountaineers and internal conflicts among Georgian political entities. These challenges led to economic decline and a significant decrease in population. After the fall of the Byzantine Empire, Georgia was effectively cut off from the Western world, making it impossible to secure meaningful European assistance against Persian and Ottoman aggression. Faced with an increasingly dire situation, Georgia turned toward Russia, a fellow Orthodox Christian state, for support. The Treaty of Georgievsk signed in July 1783, establishing a formal alliance and Russian protection over the Kingdom of Kartli-Kakheti.

In the nineteenth century, Georgia's history is marked by incorporation into, and eventual emergence from, the Russian and then Soviet empires. The kingdoms and principalities of eastern and western Georgia were gradually annexed by Tsarist Russia, bringing the country into a new imperial framework but also triggering resistance and cultural revival. On 26 May 1918, in the aftermath of the collapse of the Russian Empire, Georgia declared independence and briefly reappeared on the political map as a democratic republic (Fig. 2), before Soviet forces occupied the

Fig. 2 Present-day political map of Georgia. Modified after "Freedom Quest"

country in 1921 and incorporated it into the USSR for the next seventy years (Asatiani and Janelidze, 2009; Janelidze, 2018). During the Soviet period, Georgia underwent rapid industrialisation and urbanisation, alongside episodes of repression and cultural control. Yet Georgian language, literature, music and religious traditions remained strong, and landscapes—from highland villages to wine-growing valleys—continued to anchor identity.

On April 9, 1991, Georgia restored its state independence once again. Since then, the country has faced numerous difficulties and challenges, including struggles for territorial integrity, a severe economic crisis, a fratricidal civil war, ethnopolitical conflicts, and Russian military aggression in August 2008. The most serious consequences of these events remain unresolved: 20% of Georgia's territory is still occupied. Abkhazeti and the Tskhnvali Region are temporarily outside Georgia's control.

2 Cultural Heritage

Throughout its long history, Georgia has developed a unique, rich, and diverse culture. This remarkable heritage is a defining characteristic of the country today. The ancestors of the Georgian people contributed significantly to the treasure of human culture, making early advancements in civilization. These traditions emerged in particular river valleys, foothill zones, and coastal areas where climate, soils, and geology combined to favor early agriculture, settlement, and exchange.

Historical, archaeological, and ethnographic evidence confirms that Georgia is the birthplace of viticulture and winemaking. The oldest evidence of vine cultivation and wine production dates back to the Neolithic era, around the 6th millennium BC. In Georgia, the wine was regarded as a sacred, divine plant, and wine had a ceremonial role. According to some researchers, the term "wine" entered world languages from the Georgian language as a descriptor for this alcoholic beverage.

Georgia is also recognized as one of the oldest centers of wheat cultivation. Among the 20 known wheat species globally, five are endemic to Georgia. Georgian wheat varieties, such as Zanduri, Makha, Dika, and Tavtukhi are known for their exceptional adaptability to the environment, resistance to pests, and other notable qualities.

Bread and wine are the central part of Georgian identity and are widely seen as symbols of the ancient national tradition of hospitality. Local cuisine, as well as the Georgian table, is a unique cultural phenomenon, often presided over by the "Lord of Bread"—the toastmaster. It is also important that Georgia (Colchis) is one of the regions where, according to Greco-Roman sources, iron metallurgy originated. The Georgian tribe of the Khalibs, who lived on the southeastern Black Sea coast, was famous for this (there is an opinion that the name of the Khalibs should be associated with the name for iron in the ancient Greek language—"χᾶλυψ").

Since the Hellenistic period (4th century BC. 30s–1st century) urban life was widely developed in Georgia, as evidenced by the existence of such cities as: Mtskheta, Vani, Kaspi, Nokalakevi, Uplistsikhe and others. Built according to the rules of architectural art (Strabo), these centers of trade, crafts and culture had fortification and cult buildings (Surguladze 1988), artistic (theater building) or hygiene facilities (bathrooms, the floors of which were painted with mosaics).

The high level of Georgian culture of this era is represented by the magnificent bronze, gold and iron pieces of work, including artistically decorated cult objects, military and agricultural tools, exquisitely crafted jewelry, and other artifacts, which were created here since ancient times and have been discovered in large quantities throughout the territory of Georgia. The rise of Georgian culture was given a new impetus by the spread of Christianity and its declaration as the state religion, which fundamentally changed the worldview and aesthetic taste of Georgians. Christian art and literature shifted the emphasis from the beauty of the body to the beauty of the soul. Both types of the main genre of sacred writing—hagiography—were created and developed: martyrdom and life-citizenship. They describe the lives and martyrdom of both male and female saints and martyrs. The earliest Georgian hagiographic work that has come down to us, "The Martyrdom of Shushaniki" (author Iakob Tsurtaveli), dates back to the fifth century.

The literary heritage of Georgia is deeply rooted in its original script, which is among the oldest writing systems in the world (Fig. 2). According to national historical tradition, the Georgian alphabet was created by the first king of Iberia (Kartli), Parnavaz, in the 4th–3rd centuries BCE. He is credited with "spreading the Georgian language" and "establishing Georgian literacy." However, archaeological discoveries in 2015 on Grakliani Hill near the village of Igoeti may challenge this traditional perspective. A unique inscription found at the site has been dated to the 11th–10th centuries BCE by renowned laboratories in the United States, Germany, and Italy, making it the oldest known script in Georgia (Lichelli 2018).

The Georgian writing system consists of three distinct alphabets that differ from each other by letter-contouring and graphical forms: Mrglovani, Nuskhuri, and Mkhedruli. The first two are ancient, while Mkhedruli serves as the modern script (Fig. 3). All three have been recognized as intangible cultural heritage and are included in the UNESCO World Heritage List. Churches and monasteries served as significant cultural and educational centers throughout history. The efforts of scribes—including translators, calligraphers, and book editors—who worked within these institutions have contributed to the preservation of a vast collection of ancient manuscripts and handwritten books, often adorned with unique miniatures. Today, major libraries in Georgia and abroad hold approximately 11,000 Georgian manuscripts on parchment or paper. Notably, more than 500 Georgian manuscripts, including examples of ancient palimpsests, are inscribed in the UNESCO of Memory of the World International Register as significant manuscript monuments. The first printed Georgian book—„The Georgian Alphabet with Prayers", was published in Rome in 1629.

Fig. 3 The modern Georgian alphabet. https://expathub.ge/learn-georgian-language/

At the zenith of its power, Georgia maintained several cultural and historical centers beyond its borders. Among them were the Djvari Monastery in Palestine, the Iveron Monastery on Mount Athos (Byzantium), the Black Mountain Monastery in Syria, and the Petritsoni Monastery in Bulgaria, where extensive literacy and translation activities were carried out. Georgian clergy, literators and theologians–such as Ioane and Ekvtime of Athos, George of Mtatsminda, Ephrem Mtsire, and Ioane Petritsi–translated numerous biblical texts, writings of church fathers, and other theological works.

The Golden Age of Georgia, spanning from the late 11th to the 13th century, was a period marked by remarkable achievements in architecture, painting, poetry, and both ecclesiastical and secular literature. During this time, the Gelati and Ikalto academies were established, fresco painting experienced a revival, and outstanding examples of engraving and enamel art were produced. One of the most significant historical compilations of the era, The Life of Kartli, was assembled. This period also saw the construction of monumental churches, including Alaverdi, Svetitskhoveli, and Gelati, which stand as enduring symbols of Georgia's artistic and spiritual heritage. A defining literary achievement of the era was the creation of the epic poem The Knight in the Panther's Skin, an indelible monument of poetic culture distinguished by its original structure, epic form, and masterful use of artistic expression. The poem was authored by the great humanist poet and philosopher Shota Rustaveli.

In 1918, Tbilisi University was founded and soon transformed into a state university following the emergence of the Democratic Republic of Georgia. In 1941, the Academy of Sciences was established. Several renowned scientific schools were developed, including those in mathematics (led by academicians Niko Muskhelishvili, Ilia Vekua, and Viktor Kupradze), psychology (Dimitri Uznadze), geology (Alexander Janelidze and Aleksandre Tvalchrelidze), and physiology (Ivane Beritashvili), among others.

The richness of Georgia's spiritual heritage is reflected in its oral tradition, which dates back to ancient times. Georgian choreography is widely celebrated, and the polyphony of Georgian folk songs is truly remarkable. Notably, the Georgian folk song "Chakrulo" (performed by Ilia Zakaidze and Rostom Saginashvili) was sent into space in 1977 aboard the Voyager interplanetary station, alongside several other musical masterpieces. The achievements of Georgian theater, cinema, opera, ballet, and other arts—which continue to enrich and diversify the cultural heritage of the civilized world—are universally recognized.

These historical and cultural traditions have always evolved in close relationship with Georgia's mountains, river valleys, and coasts; later chapters in this volume return to these same landscapes as geoheritage, showing how the physical foundations of the country continue to shape its identity and future.

References

Asatiani N, Janelidze O (2009) History of Georgia. Publishing house "Petite", Tbilisi:211–213

Eastmond A (1998) Royal imagery in medieval Georgia. The Pennsylvania State University Press, pp 1–178

Gabunia L, Vekua A (1995) A Plio-Pleistocene hominid from Dmanisi, East Georgia, Caucasus. Nature 373:509–512. https://doi.org/10.1038/373509a0

Janelidze O (2018) The democratic republic of Georgia 1918–1921. Publishing house "Petite", Tbilisi, pp 1–245

Javakhishvili I (1983) Works in twelve volumes, vol 2. Science Publishing House, Tbilisi, pp 1–175

Licheli V (2018) Grakliani hill–a unique archaeological monument. http://online.tsu.edu.ge/ge/science/20560/?p=6

Lomouri N (2009) The spread and establishment of Christianity in Georgia. Georgian Patriarchate Publishing House, Tbilisi, pp 5–7

Metreveli R (2002) David IV the builder. Queen Tamar. Tbilisi University Publishing House, pp 1–285

Surguladze A (1988) Essays on the history of Georgian culture (Book 1). Art Publishing House, Tbilisi, pp 68–69

Geological History of the Caucasus: How Georgia's Geoheritage Was Created

Daniel Tormey, Avtandil Okrostsvaridze,
and Giorgi Boichenko

Abstract

This chapter briefly discusses the long, almost 400 million-year history of the formation of the Caucasus Collisional Orogen. It is shown that this process was caused by the movement of the Arabian plate of Gondwanan origin towards the Scythian platform and the closure of the Paleo-, Meso-, and Neo-Tertian ocean basins. As a result of these long and complex processes, in this wide collision zone, in the interval from about ten million to five million years, a modern continental-type crust was formed, consisting of 3 large geotectonic domains: Anatolian, Iranian, and Caucasian. The Caucasus is the northernmost domain that stretches across the Scythian Platform from the Caspian Sea to the Black Sea and is built from rocks of a wide age range and diverse genesis. Georgia is located in the central part of this orogen, which is why its small territory (69,700 km^2) includes a diverse geoheritage of different types and ages. Some of them represent geological sites of global importance, which are carriers of great scientific and educational information. This chapter describes how Georgia, located in the heart of the Caucasus Collision Orogen, has developed such a diverse geological geoheritage, which represents a unique component of the world's geological heritage. In addition, this chapter shows how Georgia's geology does more than provide background context. It explains why the country's geoheritage looks the way it does, and it points directly to how a national geoheritage conservation program can grow.

Keywords

The Caucasus · Geological history · Georgia · Geoheritage · Conservation program

1 Introduction

Georgia sits within the wide zone of continent–continent collision between the still-converging Eurasian and Africa–Arabian lithosphere plates, and its geology is therefore best understood as a compact, readable collision zone through time (Gamkrelidze et al. 2021). In a small territory, Georgia brings into close contact a four-part Caucasus architecture: the Greater Caucasus to the north, the intervening Transcaucasian domain/microplate (including basement windows such as the Dzirula massif), and the Lesser Caucasus to the south, and the Black Sea–Colchis margin. This framework is widely used in modern Caucasus syntheses because it captures the country's first-order structural organisation: two collisional mountain belts separated by an intervening microplate that contains Mesozoic–Cenozoic sediments underlain by pre-Jurassic crystalline basement, exposed locally in key basement windows (Okrostsvaridze et al. 2024; Okrostsvaridze and Tormey 2013). A fourth element, the Black Sea–Colchis lowlands and coastal margin, is treated in this chapter as the principal sedimentary and geomorphic expression of the same system at the western edge of Georgia: the mountain-to-sea receiving basin where rivers, floodplains, wetlands and coastal processes integrate the erosional output of the three-part orogenic architecture.

This chapter is dedicated to the memory of the outstanding Georgian geologist,
Academician Irakli Gamkrelidze.

D. Tormey (✉)
Catalyst Environmental Solutions, Santa Monica, USA
e-mail: dtormey@ce.solutions

IUCN World Commission on Protected Areas, Gland, Switzerland

A. Okrostsvaridze · G. Boichenko
Institute of Earth Sciences, Ilia State University, Tbilisi, Georgia

A. Okrostsvaridze and D. Tormey (eds.), *Geoheritage of Georgia*, Geoheritage, Geoparks and Geotourism,
https://doi.org/10.1007/978-3-032-27639-1_5

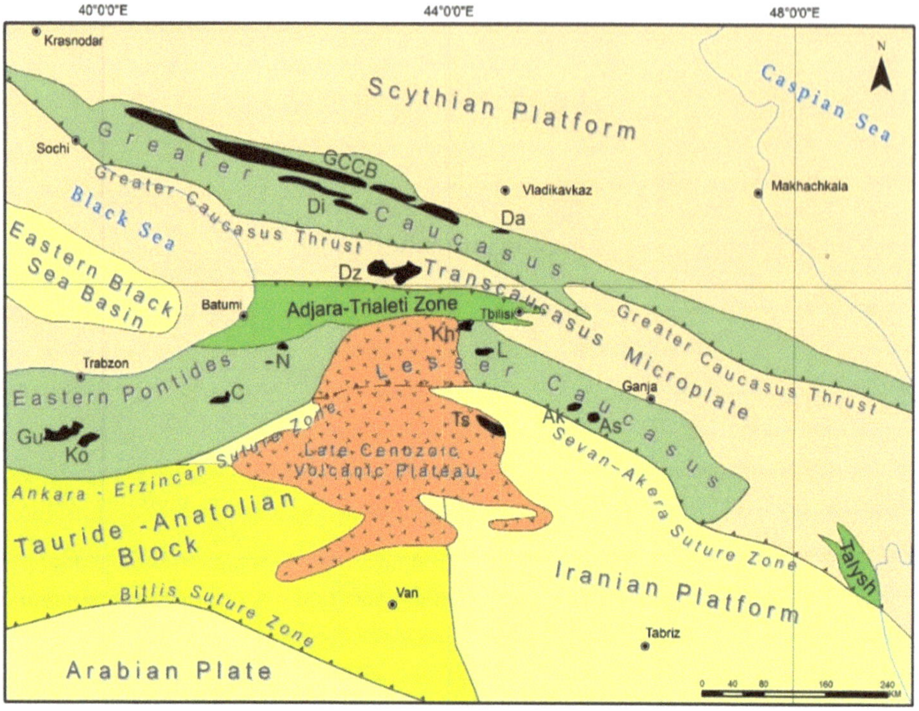

Fig. 1 Simplified schematic geological map of the Arabian plate–Scythian platform collision zone. Pre-Jurassic outcrops: GCCB–Greater Caucasus crystalline basement; Di–Dizi series. Massifs: Da-Daraiali, Dz-Dzirula, Kh-Khrami, L-Loki, Ak-Akhum, As-Asrikchai, Ts-Tsakhuniats, N-Nartic, C-Camukaua, Ko-Kose, Gu–Gumushane (Okrostsvaridze et al. 2022)

This chapter is not a conventional "Geology of Georgia" textbook overview. Its purpose is to present the geological skeleton and evolution that allow the reader to understand the geoheritage case studies throughout this volume in 3 spatial dimensions and through time: why high peaks cluster where they do; why ignimbrites host rock-cut cultural monuments; why karst and caves form major belts; why certain basins preserve key fossil and ash archives; and why metallogenic belts and mining landscapes recur along specific structural and magmatic trends. Georgia's story is not merely ancient; it is also active and ongoing, with deformation, uplift, volcanism, erosion and hazard processes that are directly relevant to geopark practice and heritage stewardship.

In the following discussion, we first describe the geology of the Caucasus collisional mountain belt at a territory wide scale. We then describe each of the elements of the framework individually.

2 Brief Geological History of the Caucasus

Within the Caucasus collisional orogen, the formation of new continental crust has taken place over more than 400 million years, driven by the northward movement of the still-converging Arabian Plate, of Gondwanan origin, toward the relatively stable Eurasian (Scythian) platform (e.g. Şengör 1984; Gamkrelidze 1986). As a result of these long and complex processes, a modern-type landmass was assembled in this collision zone over an interval of approximately 5–15 million years, which can be divided into three large segments: Anatolia, Iran and the Caucasus (e.g. Lin et al. 2020). Notably, this collision zone represents the second most powerful and active collision system in the Earth's Phanerozoic crust after the Himalayan orogen. It currently extends SE–NW from southwest of the Caspian Sea to the eastern Black Sea coast, exceeding 1200 km in length, with a width of ~300–500 km (Fig. 1; Fig. 2).

Geological studies of the Caucasus began in the nineteenth century and became particularly active from the second half of the twentieth century, when the region became a major scientific and educational focus within the former Soviet Union. Soviet geologists often referred to it as a "geological museum in the open air." Among the works of this period, the fundamental monograph *The Geological Structure of the Caucasus* (Milanovsky and Khain 1963) is noteworthy; it describes the region in detail against the backdrop of the then-dominant geosyncline concept. In the late twentieth century, pioneering articles on the structure and evolution of the Caucasus orogen appeared in the international literature (e.g. Zaridze and Shengelia, 1978; Gamkrelidze 1986; Philip et al. 1989), framing the Caucasus within plate-tectonic theory. The work of Philip et al. (1989) is particularly important for identifying the Borjomi–Kazbegi strike-slip fault of regional extent in Georgia, linked by the authors to the dynamics of the Arabian Plate during the early stages of continental collision.

In general, the Caucasus is characterised by a complex geological structure, as Gondwanian and Laurasian terranes

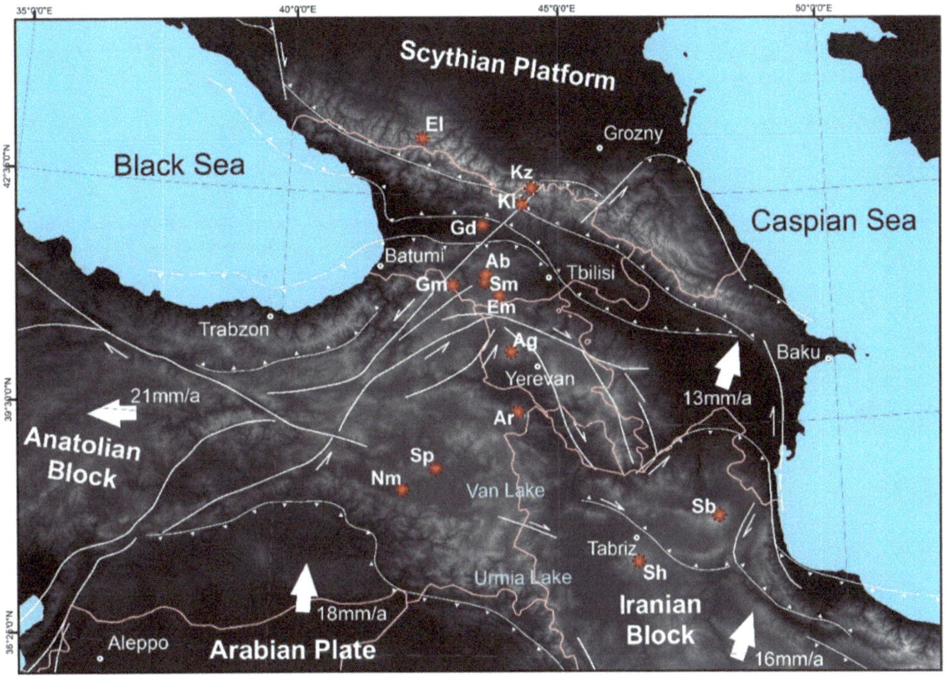

Fig. 2 Simplified structural map of the Arabian plate–Scythian platform collision zone. Some volcanoes of the region are marked with red stars: Ab-Abuli; Ar-Ararat; Ag-Aragats, Gm-Gumbati; El-Elbrus; Em-Emlikli; Kl-Keli; Md-Goridziri; Nm-Nemrut; Sb-Sabalan; Sh-Sahand; Sp-Suphan. The territory of Georgia is marked with a pale yellow line (The map modified after Philip et al. 1989; Reilinger et al. 2006; Avagian et al. 2010)

converge and participate in a collisional geodynamic regime. The origin and configuration of these terranes remain the subject of active discussion, and sharply differing views exist. For example, in a terrane-analysis framework (Gamkrelidze 1997) distinguishes a Laurasian-origin Greater Caucasus, a Black Sea–Central Transcaucasian terrane, and Gondwana-origin Bayburt–Sevan and Iran–Afghanistan terranes. In more recent work, Gondwanan-origin terranes are gradually being traced toward the north, to the Scythian platforms. For example, some authors consider the Early Paleozoic gneissic-migmatite complex of the crystalline basement of the Greater Caucasus to be of Gondwana-derived (Somin 2011; Cowgill et al. 2016).

Taken together, these studies show that a full understanding of the Caucasus requires integrated structural-geological, petrological, geochemical, isotopic and palaeontological data. Recent syntheses, such as Mumladze et al. (2015) and Cowgill et al. (2016), attempt to combine these strands into coherent models of the Phanerozoic geodynamic evolution of the collision zone (Fig. 3). In this chapter, we adopt the widely used four-part architecture introduced in Sect. 1, the Greater Caucasus, the Inner Caucasian (Transcaucasian) Microplate, the Lesser Caucasus and the Black Sea–Colchis margin, as a practical framework for interpreting how this long and complex evolution is expressed today in Georgia's geoheritage.

By this model, the Caucasus rifted from Gondwana around the Ordovician (~480). The Dizi Basin opened within the Caucasus by the Devonian (~380 Ma). South-directed subduction led to Carboniferous accretion of the Caucasus on the Laurussian margin (~330–300 Ma). Mesozoic changes in subduction dynamics within the Tethyan system caused alternating periods of rifting and basin inversion within the Caucasus Basin (~190–135 Ma), with the former separating the Lesser and Greater Caucasus basements and the latter likely resulting in formation of the Main Caucasus Thrust. The Miocene-Pliocene (~10–5 Ma) closure of the Caucasus Basin, collision between the Lesser and Greater Caucasus, and rapid exhumation of the Greater Caucasus basement occurred following the Eocene-Miocene collision of Arabia with the Anatolian-Tauride-Armenian (ATA) block and the Bitlis-Pötürge (B-P) block (Wang et al. 2015, Cowgill et al. 2016).

3 How Georgia's Geoheritage Was Created

The territory of Georgia covers all major tectonic units of the Caucasian orogeny: The Greater and the Lesser Caucasus mobile belts and Transcaucasian Microplate (intermountain depressions) lying between them (Fig. 4). This area represents a real "natural geological laboratory" exposing magmatic, sedimentary and metamorphic rocks, ranging wide on the geologic time scale (from the Neoproterozoic to the Quaternary inclusive). From the Late Miocene (10–8 Ma), intensive exhumation of the Caucasus orogen began, during which molasse, lacustrine and subaerial volcanic formations were being accumulated (Gamkrelidze et al. 2021). As a result of these long-term and complex geological processes, a number of geoheritage

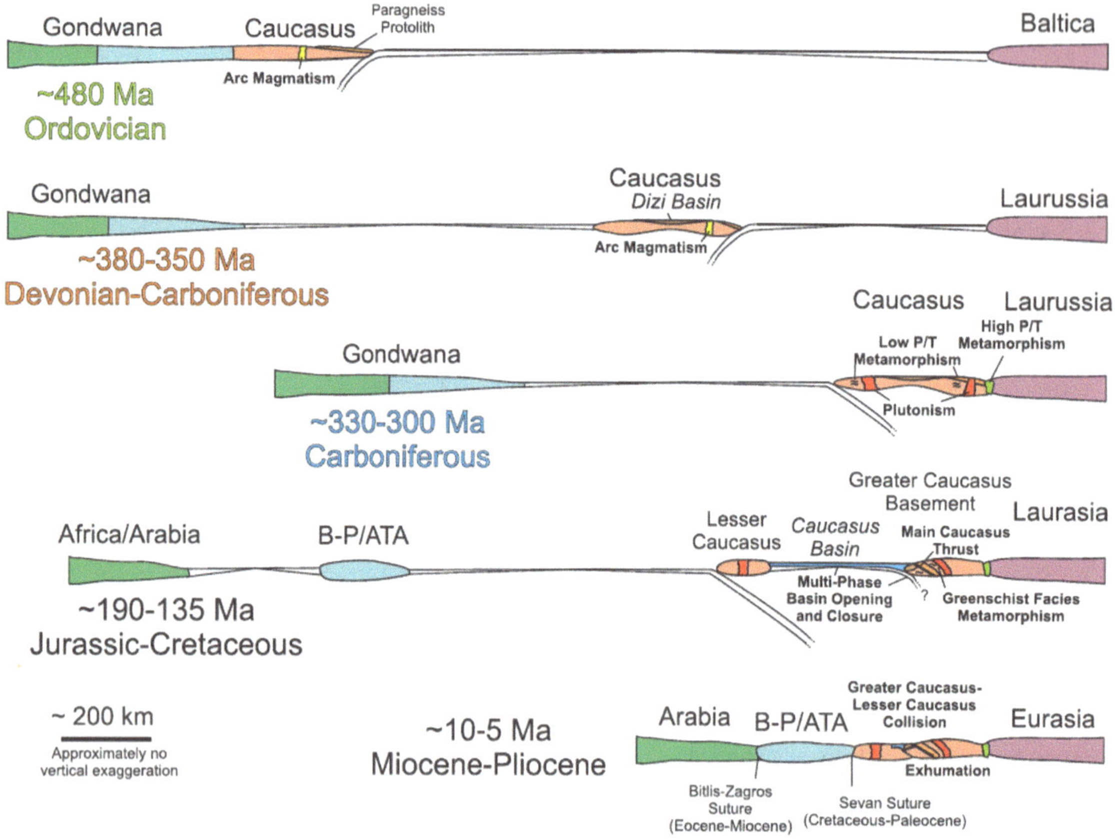

Fig. 3 Schematic model of Phanerozoic geodynamic evolution of the Caucasus collisional orogen according to Cowgill et al. (2016). ATA–Anatolian-Tauride-Armenian block; BP–Bitlis-Pötürge block

sites of global importance were formed on the territory of Georgia (Fig. 5), which we will discuss.

3.1 Greater Caucasus

The Greater Caucasus fold–thrust belt is the northernmost expression of the Caucasus orogen, extending approximately 1200 km NW–SE between the Black and Caspian seas and accreted to the southern margin of the Precambrian Scythian platform. In its simplest geological architecture, two major zones are commonly distinguished: (1) a pre-Jurassic crystalline basement and (2) an overlying Meso–Cenozoic magmatic and sedimentary cover (e.g. Okrostsvaridze and Tormey 2011).

The pre-Jurassic basement complex is structurally zoned (from south to north) into the Southern Slope, Main Range, Front Range and Bechasyn zones, with the Main Range subdivided into the Pass and Elbrus subzones along the Alibek–Urukh regional fault. Along its southern margin, the basement complex is carried southward on the Main Thrust of the Greater Caucasus (MTGC), overthrusting Lower Jurassic formations (Gamkrelidze and Shengelia 2005; Gamkrelidze et al. 2020; Okrostsvaridze et al. 2023). The story that matters for geoheritage is how this architecture was built, overprinted and then exposed.

3.1.1 Paleozoic Margin and Basement Growth

In the Paleozoic, the pre-Jurassic basement is interpreted as an active continental margin, where Paleo-Tethys oceanic crust subducted northward beneath the Eurasian plate, establishing a long-lived convergent-margin setting for crustal growth. Within this basement, zircon LA–ICP–MS U–Pb geochronology from the Svaneti segment resolves three principal pulses of plutonic magmatism tied to major orogenic events: Ordovician (Caledonian), Late Carboniferous (Variscan) and Middle Jurassic (Cimmerian) (Okrostsvaridze et al. 2023).

The first anchor in that record is Early Ordovician crust formation. Granitoid protoliths crystallised at 488–475 Ma and are now preserved as biotite orthogneisses and migmatites in the Greater Caucasus crystalline basement. A second Caledonian component occurs in the Late Ordovician (~455 Ma) as Szgimazuki plagiogneisses, interpreted as possible accretionary-wedge material and tied to Late Caledonian orogenesis. As the Paleozoic progressed,

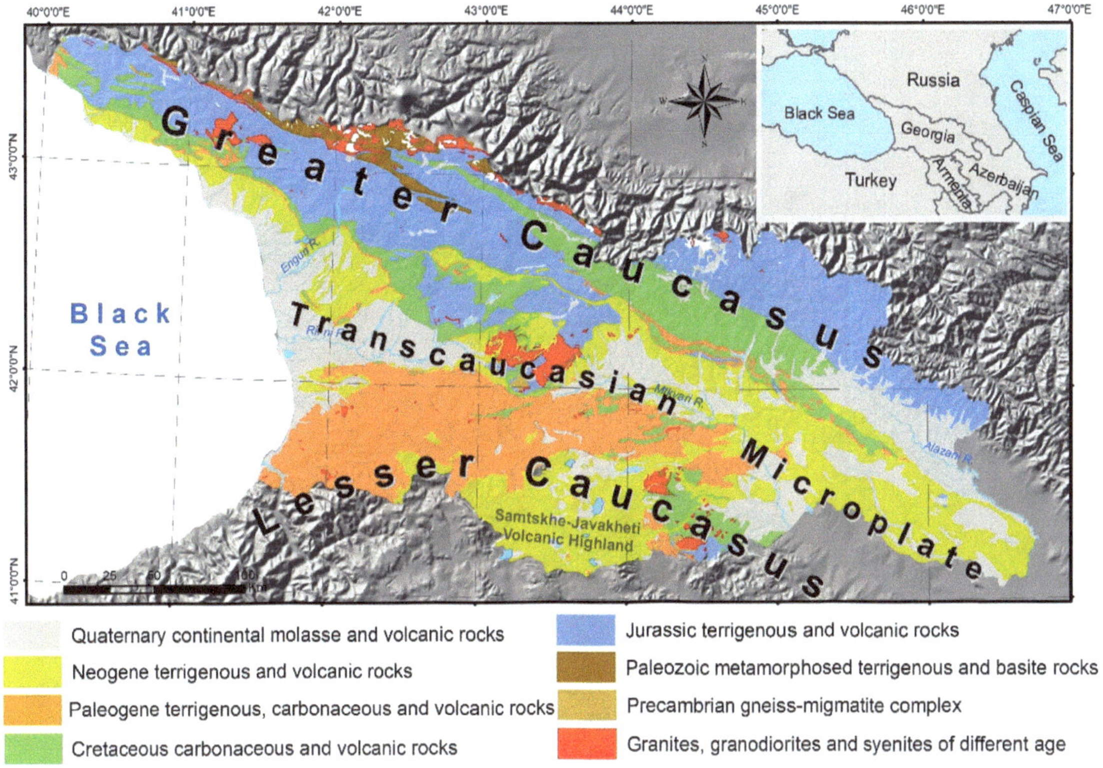

Fig. 4 Simplified geological map of Georgia with major tectonic units. (The map was modified after Gudjabidze et al. 2003)

the southern slope of the Greater Caucasus accumulated a long sedimentary record. In Georgia this Southern Slope zone is exposed only in the Svaneti segment, where it is represented by the Dizi Series: terrigenous, intensely folded sedimentary rocks of ~2000 m thickness, metamorphosed to greenschist facies and deposited continuously from the Early Devonian through the end of the Triassic (e.g., Somin 2011). This Dizi succession is a key host and recorder for later tectono-magmatic change along the southern Greater Caucasus margin.

3.1.2 Variscan and Early Mesozoic Transitions

The next major pulse is Late Carboniferous (Variscan) plutonism. Zircon U–Pb ages show widespread emplacement of granodiorite–granite plutons at ~320–312 Ma, intruding Caledonian migmatised gneissic infrastructure across both the Pass and Elbrus subzones; these plutons are later cut by leucocratic vein systems at ~310–305 Ma (Gamkrelidze et al. 2023).

The Ordovician and Variscan stages are likely supra-subduction, syn-accretional crust-building events: the Greater Caucasus basement is a crust assembled through repeated convergent-margin magmatism rather than a single inherited block. By the end of the Triassic, the system shifted again. This is interpreted as closure of the oceanic basin of the Southern Slope of the Greater Caucasus at the end of the Late Triassic (Early Cimmerian/Indosinian), preceding Jurassic extension. The sedimentary record captures the transition into the Mesozoic: the Greater Caucasus cover begins with Jurassic marine transgression and deposition of ~3000 m of black shales and sandstones. These Jurassic strata (and the broader sedimentary cover) are then intruded by mafic to intermediate suites including gabbros, diorites, syenites and quartz-diorites (Gamkrelidze et al. 2023).

3.1.3 Middle Jurassic Plutonism and Belt Architecture

The third magmatic pulse, the one that most clearly ties structure, basin history and timed intrusions into a single narrative, is Middle Jurassic plutonism, preserved in the Paleozoic–Triassic Dizi Series and in Lower Jurassic clay shales.

New U–Pb ages document a systematic pattern in Svaneti: Middle Jurassic plutons become younger to the south, with a representative sequence of Ushba (~177 Ma), Dizi (~168 Ma) and Jorkvali (~164 Ma) (Okrostsvaridze et al. 2022). This polycyclic story is especially clear in Svaneti because the belt's architecture is unusually well exposed. The segment covers >7000 km^2, exposes crystalline basement beneath Mesozoic cover and hosts multiple plutons of different ages and genetic types

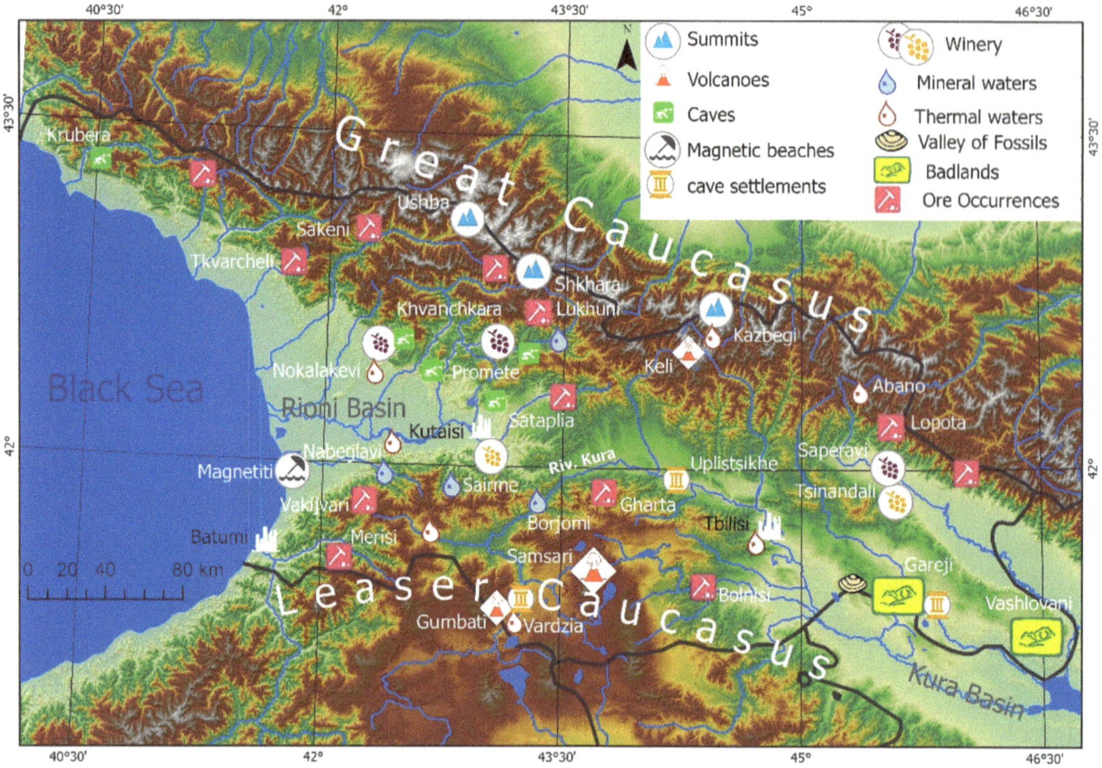

Fig. 5 Some Georgian geoheritege sites on the geographical map of the Caucasus. Territory of Georgia is outlined by a black line

(Shkhara, Szgimazuki, Memuli, Sakeni, Ushba, Dizi, Jorkvali) (Okrostsvaridze et al. 2022, Gamkrelidze et al. 2023).

In other words, Svaneti functions as a "core sample" of the Georgian Greater Caucasus: Ordovician basement foundations, Variscan plutonism intruding migmatised infrastructure, Devonian–Triassic sedimentation (Dizi), Jurassic transgression and basin fill, and Middle Jurassic intrusions tied to basin-scale tectonic change can all be read within one mountainous landscape.

3.1.4 Cenozoic Collision and Active Overprint

The Greater Caucasus story continues into Cenozoic collision and neotectonics, overprinting the older basement–cover architecture with a strongly expressed syn- and post-collisional landscape. In regional syntheses, the Caucasus is framed as the northern segment of the broader Arabia–Eurasia collision system, in which continued convergence drives ongoing uplift, basin inversion, and deformation of young sedimentary successions (Gamkrelidze, 1986; Reilinger et al., 2006).

Cenozoic basin fill in Georgia is commonly expressed as thick Neogene–Quaternary successions in intermontane depressions, and Miocene–Quaternary time is also when many of the landscapes most visible today, terraces, uplifted surfaces, and basin-margin deformation, become strongly developed. In this interval, volcanism remains an important part of the regional overprint: Miocene and Quaternary volcanic activity is treated as widespread in the South Caucasus, including the formation of volcanic plateaus in southern Georgia and young volcanic centres in the broader Greater Caucasus region (Gamkrelidze et al. 2021; Lin et al. 2020).

For the purposes of this chapter, the key point is that Cenozoic deformation is not only recorded in mountain-scale structures but also in the deformation of young basin deposits and geomorphic surfaces at the belt margins, providing direct, observable evidence that the Greater Caucasus remains an actively evolving fold–thrust system within the ongoing collision zone (Gamkrelidze, 1986; Reilinger et al., 2006).

Modern deformation of the Greater Caucasus is driven by ongoing Arabia–Eurasia convergence; regional estimates summarize shortening and convergence on the order of ~10–20 mm/yr. (Gamkrelidze et al., 2021), with the remaining relative motion accommodated chiefly by crustal shortening in the South Caucasus. Seismicity is a primary surface expression of this active shortening. Regional syntheses describe the Caucasus as capable of generating strong earthquakes with magnitudes up to approximately M 7.0–7.5 and intensities reaching ~9 (MSK scale), with destructive historical events documented within Georgia (Gamkrelidze et al., 2021).

Finally, the Greater Caucasus "roof beam" is not only a geomorphic expression but a crustal-scale one: crustal thickness reaches ~55–60 km beneath the Greater Caucasus, compared with ~40–45 km beneath the Transcaucasian intermontane lowland. This provides a physical basis for why the Greater Caucasus dominates Georgia's topography and hydrology.

3.1.5 Geoheritage Implications: Svaneti and the Greater Caucasus KGA

Within the KGA framework used in this volume, the Greater Caucasus is expressed as a Greater Caucasus High Peaks Key Geoheritage Area, with Svaneti as one of its clearest windows. Here, visitors and scientists can trace the full crustal story in a single territory: ancient basement, repeated magmatic pulses, long sedimentary records, Jurassic transgression, and the effects of Cenozoic uplift and glaciation, all framed by contemporary hazards and climate change.

For geoheritage and geopark practice, this means that landscapes such as Upper Svaneti and Kazbegi do not only offer dramatic scenery and cultural heritage; they also provide a uniquely legible record of how a collision-zone mountain belt is built and still evolving. Later chapters on peaks, hazards and cultural landscapes draw directly on this Greater Caucasus stage (Gamkrelidze et al. 2015).

3.2 Transcaucasian Microplate (Inner Caucasian Microplate)

The Caucasus orogen is commonly framed as two mobile belts, the Greater and Lesser Caucasus, separated by an intervening Transcaucasian Microplate. This microplate is not simply a gap between mountain ranges. It is a structural hinge and archive: it contains Mesozoic–Cenozoic sediments underlain by pre-Jurassic crystalline basement, and within it Georgia exposes rare basement windows that preserve the deeper story of how the country's crust was built.

In the terrane framework used by Okrostsvaridze and Tormey (2013), the key unit is the Black Sea–Central Transcaucasian terrane, interpreted as the Inner Caucasian Microplate, situated between the Greater and Lesser Caucasus belts. Its defining architecture is explicit: pre-Alpine basement exposed in the Dzirula and Khrami massifs is overlain by an Alpine volcano-sedimentary cover nearly 10 km thick. This juxtaposition, old basement plus thick young cover, is why the microplate matters for geoheritage: it allows Georgia to tell both a deep-time crustal story and a Cenozoic basin-archive story within a single national stage.

3.2.1 Neoproterozoic Foundations and Ophiolitic Collage

The Dzirula massif is the strongest single expression of that basement story. It occupies nearly 1200 km^2 at the present erosion level and is described as the widest outcropped part of the pre-Alpine crystalline basement in the Inner Caucasian Microplate. The massif is presented as a coded record of regional geodynamic and magmatic events from the Neoproterozoic through Alpine time and as a key object for understanding continental crust evolution in a collisional setting.

The story of the Dzirula record begins in the Neoproterozoic, where juvenile crustal formation and metamorphic reworking dominate. The most widespread unit is a gneiss–migmatite complex, dominated by quartz-dioritic gneisses (~70%), described as metaluminous I-/H-type granitoids with isotopic signatures interpreted as juvenile crust with a substantial mantle component. An Rb–Sr crystallisation age of 686 $\pm$ 74 Ma is linked to Pan-African tectono-thermal events (Okrostsvaridze and Tormey 2013).

Within the same Neoproterozoic framework, the massif also preserves ophiolitic fragments in the Chorchana–Utslevi ophiolitic zone. Harzburgites are interpreted as melting restites of tholeiitic basalt and carry an Nd model age of 810 $\pm$ 100 Ma (Zakariadze et al. 1998). The structural relationships, ophiolite lenses overthrusted by and locally overlain by the gneiss–migmatite complex, support the interpretation that Dzirula is a collage assembled through accretion rather than a single coherent basement block (Okrostsvaridze and Tormey 2013).

3.2.2 Cambrian and Variscan Crustal Reworking

The record then moves into the Cambrian, when crustal melting and granitoid intrusion continue. Cambrian tonalites intrude the gneiss–migmatite complex, with the largest exposures described along the Macharula and Kvirila river gorges over ~800 m. Contact zones show selective melting/migmatization and inclusions of gneiss and migmatite within tonalite bodies. These tonalites yield an Rb–Sr isochron age of 538 $\pm$ 53 Ma and are interpreted to reflect strong upper-crustal partial-melt influence (Okrostsvaridze and Tormey 2013).

In the Late Paleozoic, the Dzirula story becomes a classic Variscan crustal-thickening and granite-generation episode. Late Paleozoic microcline granites are widely distributed and described as products of Variscan tectono-thermal events, occurring as numerous intrusions across the massif. The largest body (the Rkvia intrusion) is ~65 km^2 with a NW–SE lens geometry. These granites are characterised as peraluminous S-type with isotopic signatures consistent with upper-crustal melting, and an Rb–Sr age of 351 ± 21 Ma is attributed to Variscan events (Okrostsvaridze and Tormey 2013).

3.2.3 Cimmerian Mafic Intrusion and Vertical Accretion

By the Upper Triassic, the microplate basement is intruded again, this time by mafic bodies that carry their own interpretive power. Orthoclase gabbros ("rikotites") are described from the Rikotula river gorge as two small bodies cross-cutting quartz-dioritic gneisses, with contact-zone partial melting and abundant leucocratic quartz–orthoclase inclusions interpreted as xenoliths. Intrusion ages are tightly constrained to ~210–220 Ma (Rb–Sr ~211 ± 11 Ma; ^40Ar–^39Ar hornblende ~219 ± 4 Ma), interpreted as Upper Triassic magmatism associated with Cimmerian tectono-thermal events (Okrostsvaridze and Tormey 2013).

Taken together, these elements: juvenile Neoproterozoic crust, accreted ophiolitic fragments, Cambrian tonalites, Variscan S-type granites and Upper Triassic mafic intrusions, are why Dzirula is more than "just" a basement exposure. In synthesis, the massif is presented as a vertically accreted structure composed of two large formations (an upper Gondwana-derived gneiss–migmatite complex and a lower ensialic island arc), with collision–accretion processes generating partial melts that produced granitoid intrusions and, by late Variscan time, a crust resembling typical Phanerozoic continental crust (Okrostsvaridze and Tormey 2013).

In a single Georgian massif, the microplate therefore records a long multi-cycle history of crust building and reworking, exactly the deep-time scaffolding needed for a national geoheritage narrative.

3.2.4 Cenozoic Hinge and Archive Between the Belts

The microplate story then becomes explicitly Cenozoic, not through more basement assembly, but through its role as a hinge and archive between the two active belts. In the Late Alpine cycle, regional syntheses describe inversion and collision reorganising the Caucasus into fold–thrust belts (Greater and Lesser Caucasus) separated by a Transcaucasian intermontane depression superimposed on a rigid platform/massif zone. (Okrostsvaridze et al. 2023) use chronology of volcanic ash to illustrate how this hinge role is made operational. Exhumation began at approximately 10–9 Ma, exposing both Greater and Lesser Caucasus, which were separated by a Neogene marine sedimentary basin subdivided into eastern and western parts by the pre-Jurassic Dzirula crystalline massif: eastern basins (Kartli, Kura, Alazani) and western basins (Rioni, Enguri). The Eastern Caucasus intermontane Neogene basin then preserves a detailed molasse history (Lower Miocene to Upper Pliocene) and, critically, Upper Miocene–Pliocene deposits containing multiple volcanic ash layers.

Those ash layers provide the cleanest national "connector thread" between belts. Zircon U–Pb ages show that they were deposited at approximately 7.5 Ma, and multiple lines of evidence support a common source with ignimbrites of the Goderdzi Volcanic Complex (Samtskhe–Javakheti volcanic highlands), including the Mtkvari (Kura) ignimbrite. In other words, the microplate and its basins preserve the archive, while the Lesser Caucasus provides the volcanic source.

3.2.5 Geoheritage Implications: Dzirula and the Inner Caucasian Microplate KGA

Within the KGA framework used in this volume, the Transcaucasian (Inner Caucasian) Microplate is expressed as an Inner Caucasian Microplate Key Geoheritage Area, with Dzirula as its core exposure. Here, Georgia's deep crustal assembly is most directly observable (a 1200 km^2 pre-Alpine basement window recording Neoproterozoic–Triassic tectono-magmatic events), and Georgia's Cenozoic basin archives are most systematically organised (intermontane basins subdivided by the Dzirula massif, preserving 10–9 Ma exhumation context and 7.5 Ma ash-marker horizons sourced from Lesser Caucasus volcanism).

For geoheritage, this hinge zone is where many of the country's most integrative narratives come together: corridors and intermontane plains, deserts and badlands, fossil-bearing molasse, ash markers, mineral waters and cultural landscapes. In this chapter, the Transcaucasian Microplate is therefore the belt where deep basement, basin evolution and volcanic signals from the Lesser Caucasus can be read as parts of one coherent system, a structural stage that later chapters on basins, corridors, deserts and volcanic geoheritage build upon.

3.3 Lesser Caucasus

The Lesser Caucasus orogenic belt lies south of the Transcaucasian Microplate and records the transition from Mesozoic subduction of northern Neo-Tethyan oceanic lithosphere to Cenozoic collision and reactivation. Much of the discussion in this section is drawn from Tormey et al. (in press).

In Georgia, that long tectonic inheritance becomes most legible, especially for geoheritage, in the late Cenozoic volcanic highlands of Samtskhe–Javakheti, where eruptive products, fault partitioning and river incision expose a coherent, dated volcanic stratigraphy that is physically and culturally inseparable from the landscape of Vardzia.

The Samtskhe–Javakheti volcanic highland is a large, high-elevation volcanic province in southern Georgia, occupying >4500 km^2 at mean elevations of ~1500–2000 m, with individual volcanic edifices exceeding 3000 m (e.g. Didi Abuli, 3305 m). It is bounded to the north by the Eocene Achara–Trialeti Range and related ridges, and it sits within the broader zone of late Cenozoic volcanism from Late Miocene (~9–7 Ma) to end-Pleistocene time, subaerial eruptions occurred contemporaneously with molasse trough development, and major volcanic highlands formed in Javakheti (Pasquare et al. 2011). Two main eruptive stages are distinguished, with absolute ages summarised as ~12–6 Ma and ~ 3.5–0.03 Ma, spanning calc-alkaline and sub-alkaline series. Within this broad framework, the Vardzia geoheritage work resolves the Samtskhe–Javakheti highland into a mulit-stage construction history that is particularly clear in the Vardzia section.

3.3.1 Stage 1: Upper Miocene Goderdzi Volcanic Formation and the Mtkvari Ignimbrite

The first stage is Upper Miocene volcanism of the Goderdzi Volcanic Formation, described as thick dacite–andesite tuffs and associated andesitic–basaltic lava flows totalling 700–1000 m. In the Vardzia section specifically, the Goderdzi Formation unconformably overlies Middle Eocene rocks and reaches ~900–1100 m total thickness, with an internal stratigraphy that is unusually useful for interpretation because it is both thick and well exposed. Locally, the formation is divided into a lower pyroclastic unit (200–240 m), a middle ignimbrite unit (60–120 m) made of the Mtkvari Ignimbrite, and an upper lithic breccia unit (700–800 m) containing additional ignimbrite layers and thick andesitic ash beds.

The second stage of the story is defined not only by rock type but by a specific deposit that functions as a national "time pin": the Mtkvari Ignimbrite Flow, which is both regionally extensive and precisely dated. In the Vardzia synthesis, this ignimbrite originates in the Niala Valley (remnants of an Upper Miocene caldera) and extends along the left bank of the Mtkvari River for ~35 km to Khertvisi. Its emplacement geometry is quantified: the ignimbrite moved south-to-north on a 2–3° palaeoslope, and its thickness increases down-flow from ~60 m near Karzameti, to ~90 m at Vardzia, to ~120 m near Saro, consistent with ignimbrite flow emplacement rather than ash-fall thinning.

Petrologically, the deposit is andesitic–dacitic (SiO_2 58–66%) with variable welding; isotopic parameters (εNd +3 to +4; $^{87}Sr/^{86}Sr$ 0.70341–0.70450) are interpreted as consistent with fractionation of mantle-derived melts. Zircon U–Pb dating (LA–ICP–MS) from multiple points along the flow yields a consistent mean age of 7.52 ± 0.21 Ma (Okrostsvaridze et al. 2023).

This ignimbrite story is strengthened by an independent, basin-scale archive. Okrostsvaridze et al. (2023) demonstrates that ash layers preserved in the Eastern Caucasus intermontane Neogene basin were deposited at ~7.5 Ma and that their age and geochemistry support a common source with ignimbrites of the Goderdzi Volcanic Complex, including the Mtkvari (Kura) Ignimbrite. From their distribution (up to 150–300 km from source) and the scale of the ignimbrite system, the authors conclude that both ignimbrites and ash layers were produced by large collapse caldera-forming eruption(s) in the Samtskhe–Javakheti volcanic highland, forming the Niala caldera, with the youngest andesitic flows in the caldera dated at ~6.4 Ma, implying a volcanic cycle of roughly 1 Ma (Okrostsvaridze et al. 2023).

The Vardzia narrative then introduces a structural "seam" that helps explain present-day landscape segmentation. The Samtskhe–Javakheti highland is divided into two blocks, western Samtskhe and eastern Javakheti, separated by a regional SW–NE fault zone along the Mtkvari valley. At the Vardzia section, differential GPS measurements indicate that the Javakheti block lies ~127 m lower than the Samtskhe block, based on the elevation of the upper boundary of the Mtkvari ignimbrite (1380 m asl in Javakheti vs 1417 m asl in Samtskhe). The Mtkvari ignimbrite thus functions as both a time marker and a displacement marker.

3.3.2 Stage 2: Early Pleistocene Flood and Plateau Basalts

After the Upper Miocene ignimbrite stage, the volcanic story shifts toward mafic plateau building. Following a pause of approximately 1.5–2 Ma (as summarised in the Vardzia synthesis), volcanism resumed in the Javakheti block as rapid eruption of flood and plateau basalts, regionally comparable to plateau basalts of NE Türkiye and northern Armenia. Published geochronology is summarised as Early Pleistocene, typically ~2.5–1.5 Ma across K–Ar and ^40Ar/^39Ar datasets (Okrostsvaridze et al. 2020). These basalts cap the Goderdzi Formation and build much of the present-day Javakheti plateau surface, into which rivers such as the Mtkvari and its tributaries have incised deep canyons.

3.3.3 Stage 3: Abul–Samsari Dacitic Ridge

The final major construction phase is a return to more evolved volcanism in the form of the Abul–Samsari intraplate volcanic ridge, described as the youngest and northernmost formation of the Samtskhe–Javakheti volcanic plateau. In the Vardzia synthesis, volcanism resumes around ~400 ka, and the ridge develops as a 42 km-long S–N belt reaching 18 km maximum width, hosting >20 edifices; Samsari volcano (3285 m) contains a 3 km caldera. The ridge lavas are reported as relatively homogeneous dacites with major-element ranges (SiO_2 64.5–66.55%, Al_2O_3 16.3–16.9%, Fe_2O_3 4.52–4.80%, CaO 4.60–5.32%, Na_2O 4.15–4.76%, K_2O 1.82–2.28%) consistent with prolonged fractionation (Okrostsvaridze et al. 2023).

A second, equally characteristic expression of the Georgian Lesser Caucasus belt is its metallogeny. The Adjara–Trialeti metallogenic province is described as exceptionally rich in ore deposits and occurrences and includes major copper, polymetallic, and gold districts; within this province, epithermal Au–Ag mineralization and porphyry-style Cu mineralization are reported to be associated with Eocene calc-alkaline magmatism (Moritz et al. 2016; Okrostsvaridze et al., 2018, 2021).

3.3.4 Geoheritage Implications: Vardzia and the Lesser Caucasus Volcanic KGA

Viewed through the Samtskhe–Javakheti highland and the Vardzia section, the Georgian Lesser Caucasus can be told as a coherent collision-era volcanic–tectonic story: an Upper Miocene caldera-forming ignimbrite province (Goderdzi Formation) anchored by a dated, mapped flow (Mtkvari Ignimbrite at 7.52 ± 0.21 Ma) that is independently recorded as ~7.5 Ma ash in intermontane basins; a fault-partitioned plateau whose block offsets can be quantified using the ignimbrite as a marker; a subsequent episode of mantle-derived flood basalts in Early Pleistocene time; and a final Mid–Upper Pleistocene ridge system of evolved dacitic volcanism with mapped geometry and internal temporal trends.

Within the KGA framework used in this volume, this area is expressed as the Lesser Caucasus Volcanic Highlands Key Geoheritage Area, with the Vardzia section as its clearest type example. Here, a thick Upper Miocene volcanic stratigraphy (900–1100 m Goderdzi Formation with measured subunits), a precisely dated, 35 km-long ignimbrite flow with quantified emplacement geometry and geochemistry, a measurable block-segmentation fault system, and younger plateau-basalt and ridge-building volcanism all meet in one landscape. This is precisely the kind of geology that functions as "stage" for geoheritage: deposits are dated, regionally continuous, morphologically expressive and directly tied to cultural landscapes—most notably at Vardzia, where the twelfth-century rock-cut city is literally carved into the Mtkvari Ignimbrite, synthesising geological and cultural heritage in a single iconic site.

4 Black Sea–Colchis Margin: The Lowland Stage

Western Georgia's Colchis Lowlands and Black Sea margin represent a distinct geomorphic and sedimentary environment: low-relief plains, large river systems, wetlands and coastal processes. Even where bedrock exposure is limited, this area matters for geoheritage because it expresses the mountain-to-coastal system: sediment eroded from the Greater and Lesser Caucasus is routed through alluvial fans and floodplains into deltas and coastal plains, where Quaternary sea-level changes and river avulsions have built the wide, flat Colchian surface.

This is also one of Georgia's most important climate and biodiversity archives. Thick successions of alluvial, deltaic and coastal deposits preserve evidence of past shoreline positions, wetland evolution and Holocene environmental change, while today's peatlands, back-swamps and floodplain forests support the Colchic World Heritage property and other protected areas. The same sedimentary framework underlies key cultural landscapes: ports and trading cities, irrigation systems, rice fields, tea plantations and villages organised along natural levees and former shorelines.

5 Karst, Caves and Mineral Waters: The Underworld Stage

Georgia's "underground wonders" are a prime example of a geoheritage theme that becomes immediately legible once the belt logic is established. In the Underground Wonders chapter draft, carbonate rocks are described as covering approximately 10% of Georgia's territory, predominantly associated with thick Lower Cretaceous sedimentary sequences along the southern slope of the Caucasus, where geological and hydrogeological conditions produce deep karst abysses and numerous cave systems. The same draft emphasises scale and global standing: the total number of discovered caves in Georgia exceeds 1500, with a cumulative surveyed length of approximately 280 km, and the Prometheus (Tskaltubo) Cave alone is reported as 15,500 m in length.

This is not only a scientific asset; it is a management and interpretation opportunity. Caves and karst systems are fragile environments that can anchor geopark education and sustainable geotourism if they are carefully integrated into protection and visitor management plans.

Sataplia provides a rare "integrated site cluster" that combines karst, tourism readiness and palaeontology. In the Springer synthesis of Georgia's geopark potential,

Sataplia is described as a rare combination of karst caves and dinosaur footprints preserved in Lower Cretaceous limestones, with the Sataplia cave located close to the footprint sites and benefitting from a long history of protection and visitor infrastructure. This cluster naturally supports geopark-style interpretation that links "surface life" (footprints, palaeoenvironments) to "underground process" (karst hydrology and speleogenesis).

Karst also provides a natural bridge to Georgia's sacred hydrology thread. In the Borjomi–Kharagauli geopark candidate narrative, the Borjomi mineral water springs are explicitly described as emerging from deep faults, tying taste and therapeutic culture directly to subsurface structure and uplift history. Similar connections between geology, mineral waters and cultural practice appear at Tskaltubo, Sairme and other spa regions across the country.

5.1 Geoheritage Implications: Imereti Karst and Mineral Waters KGA

Within the KGA framework, Georgia's major karst belts and mineral-water fields are expressed as an Imereti Karst and Mineral Waters Key Geoheritage Area, with strong links to adjacent KGAs. Here, karst caves, underground rivers, speleothems, mineral springs and fossil-bearing limestones form a single underworld stage that underpins both biodiversity and human use. For future geopark and georoute initiatives, sites such as Sataplia, Prometheus, Tskaltubo and Borjomi are not isolated attractions but parts of an integrated karst–hydrology system, where visitors can experience how water, rock and time work together to shape both subterranean landscapes and Georgian healing and hospitality traditions.

6 Dynamic Georgia: Active Processes and Living Geoheritage

Across Georgia, glacial sculpting, river incision, landslides and active slope processes remain interpretively powerful because they are visible to non-specialists and can be linked directly to hazards, livelihoods and stewardship, an emphasis developed in later chapters on peaks, hazards and geopark readiness.

The four-part framework explains where Georgia's geoheritage is. An "active overlay" explains why those same places remain dynamic, why exposures refresh, why hazards cluster, and why many Georgian landscapes feel simultaneously ancient and unfinished.

The territory of the Caucasus acquired its modern geomorphological appearance in the Quaternary period. During this period, deposits are unevenly distributed and are primarily represented by riverine and marine terraces, glacial moraines, volcanic lava flows, and pyroclasts. Also noteworthy are the extensive terigene-accumulative deposits of the foothills/forelands, which were formed in the Kavakasion and Small Caucasus areas. The Quaternary terraces of the Black Sea coasts can also be distinguished, which are located meters above the modern sea levels. The modern coasts of this sea took their present form 6000–4000 years ago (Tsagareli 1989).

Earthquakes and the hazard–heritage interface. Present-day seismicity of the Caucasus is defined by the northward movement of the Arabian plate toward the Eurasian plate. The convergence rate is estimated to be about 10–20 mm/yr., 2/3 of which is likely accommodated south of the Lesser Caucasus, and it completely vanishes at the Greater Caucasus, forming a complex structure of seismic faults with diffused seismicity (Philip et al., 1989). A smaller portion of the seismic energy is released in the form of earthquakes. Analysis of historical and instrumental seismological data shows that strong earthquakes with magnitudes up to 7.0–7.5 and macroseismic intensity of 9 (MSK scale) have occurred here during the Quaternary period. The recurring periodicity of such events is on the order of a thousand years.

The South Caucasus is within an area of high seismic activity, with the risk factor constantly present. The consequences were severe in the South Caucasus in the 20th and early 21st centuries, particularly in the 1980s and 1990s (Faravani, Spitak, Racha) and in 2002 (Tbilisi) (Okrostsvaridze et al. 2013).

Volcanism as an active connector. Volcanism from Late Miocene to Pleistocene is described as coeval with trough development, with volcanic highlands and plateaus formed in Javakheti and related regions. (Okrostsvaridze et al. 2023) ties eruptive sources to basin archives and supports interpretation of large explosive systems (such as the Mtkvari Ignimbrite and Niala caldera) that connect Georgia's belts through time threads visible in both source landscapes and basin deposits. In geoheritage terms, this means that active and recent volcanism is not only a local story in the Lesser Caucasus; it is also a connector between highlands and intermontane basins.

Geoheritage implications: a living stage. For geoheritage and geopark development, this "dynamic Georgia" overlay is not optional decoration. It turns static views of peaks, plateaus, caves and coasts into stories of ongoing change. Uplift and incision explain why geosites stay fresh; earthquakes, landslides and floods explain why hazards and heritage share the same spaces; volcanism and ash layers explain how belts are connected through time; faults and fluids explain why mineral waters and spas are where they are; and geology–climate interactions explain why rainforests cling to certain slopes and deserts to others.

In a national geoheritage framework, these active processes are the living component of Georgia's stage, processes that geoparks must interpret if they are to support both appreciation and wise stewardship of the country's remarkable landscapes.

7 Closing: What the Geological Stage Implies for Georgia's Geoheritage Programme

Seen through a belts-first lens, Georgia's geology does more than provide background context. It explains why the country's geoheritage looks the way it does, and it points directly to how a national programme can grow. Three conclusions follow from the Greater and Lesser Caucasus belts, the Inner Caucasian Microplate, the Black Sea–Colchis Lowlands and the karst–mineral water "underworld" described above, all aligned with the KGA–UGGp–World Heritage logic developed earlier in this volume.

First, Georgia's geoheritage is naturally networkable. The belts are continuous story platforms. The Greater Caucasus high-peak belt, the Inner Caucasian Microplate hinge, the Lesser Caucasus volcanic highlands and the Colchis coastal plain each connect multiple geosites into coherent territories. In KGA terms, these are the structural backbones on which candidate geoparks and national georoutes naturally sit: a Svaneti–Kazbegi high-peaks KGA, an Inner Caucasian Microplate KGA centred on Dzirula, a Lesser Caucasus volcanic highlands KGA anchored by Vardzia and Goderdzi, an Imereti karst and mineral waters KGA, and a Colchis coastal KGA. Georgia does not have to invent a network; its geology already provides one.

Second, Georgia's strongest geoheritage narratives are integrative. Across the belts, the most compelling sites braid geology with biodiversity, culture and livelihood. Mineral waters emerge from deep faults in Borjomi and other spa regions, linking structure, hydrology and therapeutic culture. Karst caves and natural freezers host palaeontological archives, unique microclimates and long histories of use. Volcanic ignimbrites host rock-cut monuments at Vardzia and other sites, tying caldera-forming eruptions to medieval engineering and spirituality. Mountain landscapes shape settlement patterns, identity and risk in Svaneti, Racha, Tusheti and beyond. A national geoheritage programme that foregrounds these intersections—rather than geology in isolation—will resonate more strongly with communities and decision-makers.

Third, active processes are part of the heritage, not just background hazard. Uplift, earthquakes, volcanism and slope processes make Georgia a living laboratory. They refresh exposures, reorganise river networks, trigger landslides and shape coasts. Good geopark practice does not treat these as inconvenient background risks; it treats them as interpretive opportunities linked to stewardship: risk-aware education on trails and at viewpoints, community discussions about living with earthquakes and floods, and geotourism that acknowledges both beauty and vulnerability. In this sense, "dynamic Georgia" is an essential layer of the geoheritage story.

Taken together, these three statements define what the geological stage offers to Georgia's emerging geoheritage programme. The belts and microplate explain where Key Geoheritage Areas lie and how they connect; the integrative narratives show how geology, life and culture come together at specific sites; and the active overlay reminds us that these landscapes are still being made. The chapters that follow draw on this stage—one KGA, one territory, one theme at a time—to show how Georgia can translate its remarkable geology into a future of celebrated and well-protected geoheritage.

References

Avagian A, Sosson M, Karakhanian A, Philip H, Rebai S, Rolland Y (2010) Recent tectonic stress evolution in the lesser Caucasus and adjacent regions, vol 340, Geological Society, London, Special Publications, pp 393–408. https://doi.org/10.1144/SP340.17

Cowgill E, Forte AM, Niemi N, Avdeev B, Tye A, Trexler C (2016) Relict basin closure and crustal shortening budgets during continental collision: an example from Caucasus sediment provenance. Tectonics 35:2918–2947. https://doi.org/10.1002/2016TC004295

Gamkrelidze I (1986) Geodynamic evolution of the Caucasus and adjacent areas in Alpine time. Tectonophysics 127:261–277

Gamkrelidze IP (1997) Terrane of the Caucasus and adjacent areas. Bull Georg Natl Acad Sci 155(3):75–81

Gamkrelidze IP, Shengelia DM (2005) The Precambrian-Palaeozoic regional metamorphism, magmatism and geodynamics of the Caucasus. Scientific World, Moscow, pp 1–458. (in Russian)

Gamkrelidze I, Okrostsvaridze A, Maisadze F, Basheleisvili L, Boichenko G, Skhirtladze I (2019) Main features of geological structure and geotourism potential of Georgia, the Caucasus. Mod Environ Sci Eng 5(5):422–442. https://doi.org/10.15341/mese(2333-2581)/05.05.2019/010

Gamkrelidze I, Shengelia D, Chchindaze G, Lee YH, Okrostsvaridze A, Beridze G, Vardanashvili K (2020) U-Pb LA-ICP-MS dating of zoned zircons from the greater Caucasus pre-Alpine crystalline basement: evidence for Cadomian and Variscan evolution. J Geol Carpathica 71(3):249–263. https://doi.org/10.31577/GeolCarp.71.3.4

Gamkrelidze I, Okrostsvaridze A, Koiava K, Maisadze F (2021) Geotourism potential of Georgia, the Caucasus. Springer, pp 1–140. https://doi.org/10.1007/978-3-030-62966-3

Gamkrelidze I, Okrostsvaridze A, Tormey D, Lee YH, Boichenko G, Gogoladze S (2023) Plutonic magmatism of the greater Caucasus

Svaneti segment: zircon U-Pb geochronology, Petrochemistry and geodynamic setting. Bul Georg Nat Acad Sci 17(1):7–21

Gudjabidze GE, Gamkrelidze IP, Abesadze GN et al (2003) Geological map of Georgia 1:500,000 scale. Georgian State Department of Geology and National Oil Company, Tbilisi, Georgia

Lin YC, Chung SL, AF Bingö, Yang L, Okrostsvaridze A, Yang L, Lee HY (2020) Diachronous initiation of post-collisional magmatism in the Arabia-Eurasia collision zone. Lithos 36:356–357. https://doi.org/10.1016/j.lithos.2020.105394

Mumladze T, Forte AM, Cowgill ES, Trexler CC, Nathan A. Niemi NA, Yıkılmaz MB, Kellogg LK (2015) Subducted, detached, and torn slabs beneath the greater Caucasus. GeoResJ 5: 36–46. http://creativecommons.org/licenses/by-nc-nd/3.0

Milanovsky EE, Khain VE (1963) Geological structure of the Caucasus. Issues of regional geology SSSR, vol 8, pp 1–357. (in Russian)

Moritz R, Melkonyan R, Selby D, Popkhadze N, Gugushvili V, Tayan R, Ramazanov V (2016) Metallogeny of the lesser Caucasus: from arc construction to postcollision evolution. Soc Econ Geol Spec Publ 19:157–192

Okrostsvaridze A, Tormey D (2011) Evolution of the Variscan orogenic plutonic magmatism: the greater Caucasus. J Nepal Geol Soc 43:45–52

Okrostsvaridze A, Tormey D (2013) Phanerozoic continental crust evolution of the inner Caucasian microplate. Episodes 36(1):17–23. https://doi.org/10.18814/epiiugs/2013/v36i1/005

Okrostsvaridze A, Elashvili M, Janelidze Z (2013) Quaternary geology, volcanism and seismicity of Georgia. In: Proceeding IGCP 610 project, plenary conference, Tbilisi, pp 108–110

Okrostsvaridze A, Chung SL, Chang YH, Boichenko G, Gogoladze S (2018) Zircons U-Pb geochronology of the ore-bearing plutons of Adjara-Trialeti folded zone, lesser Caucasus. Bul Georg Acad Scin 12(2):90–99

Okrostsvaridze A, Chung SL, Chang YH, Skhirtladze I (2020) Geology and zircon U–Pb geochronology of the Mtkvari pyroclastic flow and evaluation of destructive processes affecting Vardzia rock-cut city. Quaternary Int 540:137–145. https://doi.org/10.1016/j.quaint.2019.03.026

Okrostsvaridze A, Chang YH, Chung SL, Rabayrol F, Boichenco G, Gogoladze S (2021) Paleozoic xenoliths in Eocene plutons: evidence for the destruction of pre-Jurassic crystalline basement beneath Adjara-Trialeti belt, lesser Caucasus. Geol Carpath 72(4):299–315. https://doi.org/10.31577/GeolCarp.72.4.2

Okrostsvaridze A, Akimidze K, Bluashvili D, Boichenco G, Gabrielashvili R (2022) Thorium—the energy source of the future and its ore occurrences in Georgia. Universal, Tbilisi, pp 1–224. (in Georgian). https://doi.org/10.1016/j.quaint.2019.03.026

Okrostsvaridze A, Lee YH, Tormey D, Skhirtlade I (2023) U-Pb zircon chronological constraints for three stages of syn-orogenic plutonic magmatism of the greater Caucasus Svaneti segment, Georgia. Episodes 46(2):194–210. https://doi.org/10.18814/epiiugs/2022/022023

Okrostsvaridze A, Gamkrelidze I, Lee YH, Keskin M, Tormey D, Boichenko G, Gogoladze S, Makadze M (2024) U–Pb geochronology and geochemistry of volcanic ash layers in the Eastern Caucasus intermountain Neogene basin: Implications for their sources. Arab J Geosci 17:129. https://doi.org/10.1007/s12517-024-11914-7

Pasquare F, Tormey D, Vezzoli M, Okrostsvaridze A, Tutberidze B (2011) Mitigation the consequences of extreme events on strategic Fasilites: evaluation of volcanic and seismic risk affecting the Caspian oil and gas pipelines in the republic of Georgia. J Environ Manag 92(7):1774–1782

Philip H, Cisternus A, Gvishiani A, Gorshkov A (1989) The Caucasus: an actual example of the initial stages of a continental collision. Tectonophysics 161:1–21

Reilinger R, McClusky S, Vernant P, Lawrence S, Ergintav S (2006) GPS constraints on continental deformation in the Africa-Arabia Eurasia continental collision zone and implications for the dynamics of plate interactions. J Geophys Res 111:B05411. https://doi.org/10.1029/2005JB004051

Şengör AMC (1984) The Cimmeride orogenic system and the tectonics of Eurasia. Geol Soc Am Spec Pap 195:181–241

Somin ML (2011) Pre-Jurassic basement of the greater Caucasus: brief overview. Turkish J Earth Sci 20:3–65. https://doi.org/10.3906/yer-1008-6

Tsagareli AL (1989) Quaternary period of Georgia. Mecniereba, Tbilisi, pp 1–279. (In Russian)

Wang SJ, Lee YH, Chung SL, Okrostsvaridze A (2015) Cenozoic Mountain building process in the greater Caucasus, from subduction to collision, American Geophysical Union. Fall Meeting 2014, abstract #T21A-45

Zakariadze GS, Karpenko SF, Bazilev BA (1998) Petrology, geochemistry and Sm-Nd izotopic age of pre-late-Hercynian paleooceanic complex of Dzirula salient of Transcaucasian massif. J Petrol 6(4):422–444. (in Russian)

Zaridze G, Shengelia D (1978) Hercynian magmatism and metamorphism of the great Caucasus in the light of plate tectonics. Bull Soc Geol Fr 20(3):355–359. https://doi.org/10.2113/gssgfbull.S7-XX.3.355

Georgia as an Ideal Geotourism Destination, Supporting Aspiring Geoparks

Avtandil Okrostsvaridze, Mirian Makadze, and Rabi Gabrielashvili

Abstract

Georgia, located on the eastern coast of the Black Sea, encompasses the southern slope of the Greater Caucasus, the northern slope of the Lesser Caucasus, and the intervening Transcaucasian Depression. Despite its relatively small area (69,700 km2), the country hosts numerous geological monuments of global significance that record diverse stages and processes of Earth's history. Notable examples include the Katitsvera synclinal fold, the Chaukhi gabbro–diabase–pillow lava massif, the Kakliskure meander of the Alazani River, and many others. Many of these geosites are complemented by medieval fortresses and monasteries harmoniously integrated into the natural landscape, creating a distinctive and memorable visual character. Owing to its exceptional geological exposures, subtropical Black Sea coastline, ancient history, rich cultural heritage, favorable geographic position, and pleasant climate, Georgia represents an outstanding natural setting for the development of geotourism. The legendary voyage of the Argonauts in search of the "Golden Fleece," traditionally linked to the ancient Georgian kingdom of Colchis, further enhances the country's geotouristic appeal on a global scale. Moreover, several regions of Georgia combine geological and cultural heritage sites with communities preserving ancient traditions, providing a strong foundation for the establishment of geoparks. Such areas include Ushguli, the Kazbegi district, the Sataplia dinosaur footprints, the Dmanisi hominid site, the rock-hewn city of Vardzia, and the Gareja desert. Within the emerging IUCN Key Geoheritage Areas (KGA) framework, all of these sites can be regarded as strong KGA candidates. This chapter outlines Georgia's main potential geopark territories, highlights several outstanding geoheritage sites, and presents national georoutes that connect these areas into a coherent geotourism network.

Keywords

Georgia · Greater caucasus · Lesser caucasus · Geoheritage · Geoparks · Geotourism

1 Introduction

Like many nations around the world, the Georgian people have an enchanting legend about their homeland. According to this legend, when the gods were distributing territories among humans, the Georgians were reportedly celebrating and arrived late. When asked by the gods about their delay, they replied, "We were toasting to you, which is why we are late." The gods took pity on them and assigned to the Georgians the best land reserved for themselves. A visit to Georgia convinces one that this legend is not entirely unfounded, as the country is exceptionally captivating.

The geological and geographical foundations for Georgia's geotourism potential are described in detail in Chap. 1, which presents Georgia as a collision-zone country where high peaks, volcanic plateaus, deep gorges, karst systems and Black Sea coasts are brought into close proximity. Building on that overview, this chapter focuses specifically on how these landscapes translate into geotourism: where potential geopark territories are found, which geosites stand out as particularly distinctive, and how national georoutes can connect them into coherent visitor experiences.

A. Okrostsvaridze (✉) · R. Gabrielashvili
Institute of Earth Sciences, Ilia State University, Tbilisi, Georgia
e-mail: avtandil.okrostsvaridze@iliauni.edu.ge

M. Makadze
Faculty of Exact and Natural Sciences, Tbilisi State University, Tbilisi, Georgia

A. Okrostsvaridze and D. Tormey (eds.), *Geoheritage of Georgia*, Geoheritage, Geoparks and Geotourism,
https://doi.org/10.1007/978-3-032-27639-1_6

2 Key Geoheritage Areas as a Baseline for Georgia

The Key Geoheritage Area and UNESCO Global Geopark frameworks are described in Chaps. 2 and 3. In this chapter we focus on how geotourism fits within that national geoheritage picture. Within this framework, every site highlighted in the following sections can be regarded as a strong Key Geoheritage Area candidate. Among these KGAs, certain clusters stand out as particularly well suited for development as aspiring UNESCO Global Geoparks, because they combine high geoheritage value with existing tourism infrastructure, educational potential and engaged local communities. In the remainder of this chapter we therefore proceed in three steps: first, by outlining Georgia's main potential geopark territories; second, by presenting a selection of especially outstanding geoheritage sites; and third, by describing national georoutes that connect these KGAs into coherent, car-based itineraries across the country. Together, these elements illustrate how a future Georgian network of Key Geoheritage Areas, geoparks and related designations could underpin a robust, place-based geotourism strategy.

3 Potential Geoparks of Georgia

Building on the geological framework described in Chapter 5, we highlight here six territories whose geoheritage, cultural landscapes and existing infrastructure make them particularly strong candidates for future UNESCO Global Geoparks: Ushguli, Kazbegi, Sataplia–Tskaltubo, Dmanisi, Vardzia and Gareja (Fig. 1). To create a general overview of Georgia's potential geoparks, we will briefly describe each one separately.

Ushguli Potential Geopark. This geopark could combine the magnificent Paleozoic massif of Shkhara in the Upper Svaneti region and the ancient community of Ushguli, located to its south, at the headwaters of the Enguri River. Shkhara massif is exposed on the Main Range of the Greater Caucasus and creates a ~ 15 km long and ~ 5 km high beautiful ridge. The massif is predominantly composed of Lower Paleozoic biotite gneisses (489 ± 9 Ma), which are intersected by a granodiorite pluton of the Variscan generation (316.9 ± 8.8 Ma) (Okrostsvaridze et al. 2023). The geology and geoheritage of this area are described in Chap. 8.

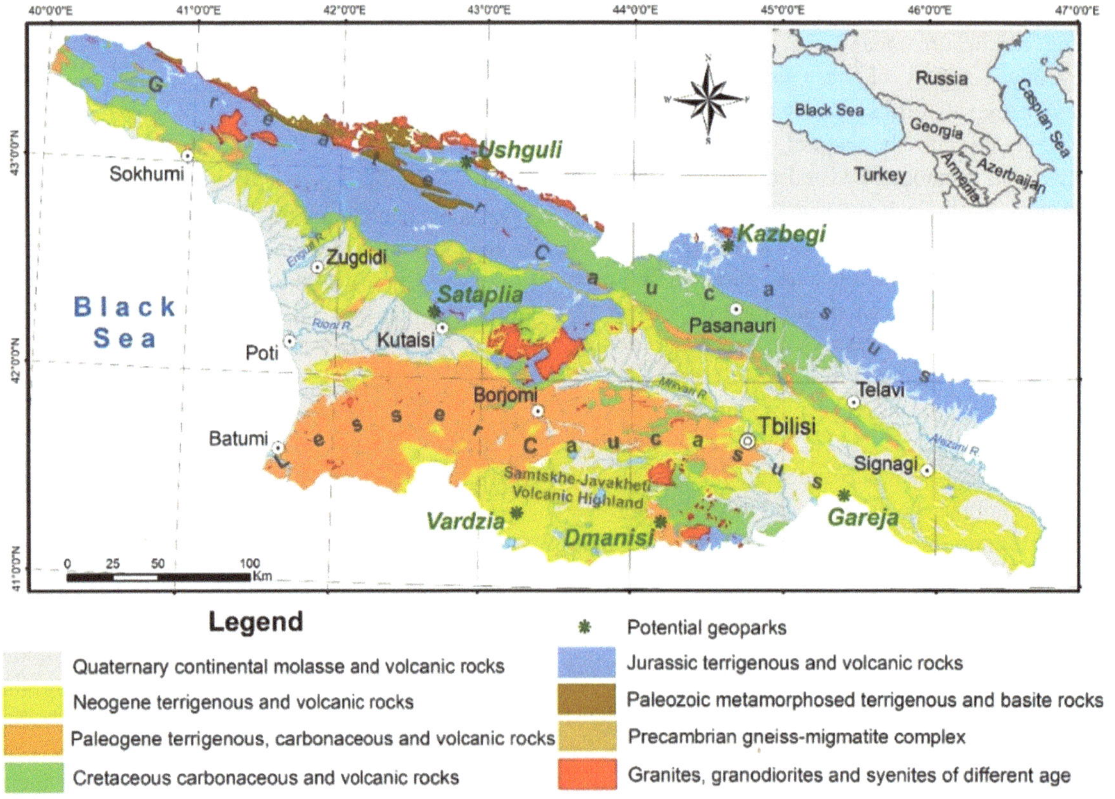

Fig. 1 Simplified Geological map of Georgia with potential geoparks sites. The geological framework is modified after Gudjabidze et al. (2003)

Kazbegi potential geopark. It is located on the Main Ridge of the Greater Caucasus, at 130 km from Tbilisi. In the small area of Kazbegi district has a long geological history (including Paleozoic to Pleistocene) and completely different genetic type rocks outcrop. Here crop out Paleozoic granitoid Dariali massif, Meso-Cenozoic sedimentary rocks, Pleistocene subaerial volcanic centers and lava flows. Quaternary stratovolcano Kazbegi (5053 m asl) is a dormant mount of the district and one of the major summits of the Greater Caucasus. It is covered by everlasting snow and glacier and with reference to its beauty is named 'bride' of the Greater Caucasus.

Sataplia Potential Geopark. It is located in Tskaltubo district, near the city of Kutaisi (~5 km), on 250 km from Tbilisi. In this area there is a rare combination of the important geological sites: karst cave and dinosaur footprints imprinted into the Lower Cretaceous limestones. These imprints are unique because there are two stratigraphic levels: lower imprinting of the predatory dinosaurs and ~ 2 m above—of herbivorous dinosaurs. From Sataplia dinosaur footprints on 100 m distance is located Sataplia karst of 600 m in length (Gamkrelidze et al. 2021).

Dmanisi Potential Geopark. Dmanisi is small town which is located at 90 km southwest of Tbilisi on the Quaternary Javakheti volcanic plateau, in the Mashavera river valley. The district offers two important geoheritige: Dmanisi hominid site and the Mashavera river Quaternary basalt flow of 30 km distance. Dmanisi hominid site discovered in 1991 (Gabunia and Vekua 1995), which is a world-class repository of archaeological and paleontological geoheritage. Situated in Georgia, this site is renowned for its exceptional findings, including some of the oldest hominin remains outside Africa, dating back to 1.85–1.78 Ma (Ferring 2011). The geology and geoheritage of this area are described in Chap. 4.

Gareja Potential Geopark. There is a semi-desert in Georgia, rich in both geoheritage and cultural heritage sites, known as the Gareja Desert. It is located 70 km southeast of Tbilisi and contains excellent geoheritage sites: colorful hills of sedimentary rocks, active mud volcanoes, volcanic ash layers of the Upper Miocene age (Okrostsvaridze et al. 2024), and a rich fossil fauna of terrestrial vertebrates, which together create the best conditions for establishing a geopark. In addition, there is the unique early-medieval David-Gareja monastery complex, which is carved into the sandstones and marls of the Miocene age.

Vardzia Potential Geopark. It is located in southern Georgia, on the Late Miocene-Quaternary Samtskhe-Javakheti volcanic plateau. The area includes for important geological sites: (1) The Upper Miocene Mtkvari Ignimbrite flow and post-caldera andesite Gumbati dome; (2) Quaternary continental flood basalts of the Javakheti Volcanic Plateau; (3) The Late Pleistocene intraplate Abul-Samsari volcanic ridge; (4) Goderdzi petrified wood site within Upper Miocene volcanic rocks (Gamkrelidze et al. 2021). The Vardzia area also houses the magnificent twelfth century Vardzia rock-cut city, carved into the Mtkvari Ignimbrite Flow. It represents a monumental creation, in which successfully synthesizing geological and cultural heritage. We propose that the Vardzia rock-cut city and surrounding unique geological and cultural heritage cites should be considered a promising region for aspiring an UNESCO global geopark in Georgia.

4 Outstanding Geoheritage of Georgia

As mentioned above, Georgia has a number of unique outcrops that clearly reflect various geological processes and represent important geoheritage monuments. They are formed in rocks of different ages and types, the number of which is large; however, we will only describe a few of them here.

Katitsvra fold. In the eastern part of the Shoda-Kedela Ridge (Upper Racha region, Oni Municipality), on the southern slope of the Greater Caucasus, the unique synclinal fold is exposed, which people named Katitsvra (the cat's beard in English). The fold extends approximately 3,000 m in length, with a height of about 600 m, and forms a magnificent double-peaked summit (42°39′2″ N, 43°37′49″ E; 3200 m esl) (Fig. 2). It is built up of Lower Cretaceous Neo-Tethyan marine limestone marls and sandstones. Most likely, the Katitsvra fold represents the geological heritage of global significance, formed by Alpine folding in the Greater Caucasus. Unfortunately, the access path to this unique geoheritage monument is quite difficult, which is why it is little known in Georgia.

Nijgori Tectonic Shear Zone. This zone is located in the Aspindza municipality of the Samtskhe district, on the left bank of the Mtkvari River, near the village of Nijgori (41°31′3″N, 43°15′32″E; 3,200 m asl). This unique tectonic structure has formed in non-carbonated clays, sandstones, and gravelites of the Oligocene coastal marine zone (Fig. 3).

As shown in the figure, a block within the shear zone has been uplifted and is currently positioned nearly vertically relative to the shear plane, while the adjacent rocks have undergone approximately 1,300° of rotation. Other tectonic structures are also clearly observable. Taken together, these features create a dynamic geological scenario, indicating that intensive collisional processes are occurring in the region. We consider the Nijgori shear zone to represent a unique geoheritage site with significant scientific and

Fig. 2 General view of the Katitsvera syncline fold from the southweast, Racha region. Photo © Ani Baidashvili

Fig. 3 General view of the Nijgori Tectonic Shear Zone from the northeast, Samtckhe region. Photo © Mirian Makadze

educational value, as well as considerable potential for geotourism.

The Katskhi Pillar. The Katskhi Pillar is a remarkable natural limestone monolith located in Zemo imereti, western Georgia, approximately 11 km from the city of Chiatura (42.2875°N 43.2156°E; 620 m esl). Unlike the Katitsvera Fold, this monument is accessible by a motor road, making it a popular tourist destination in Georgia. The structure rises about 40 m above the surrounding terrain and is situated near the valley of the Katskhura River. Morphologically, it represents a rectangular-shaped pillar with irregular sides (10.3 m × 16.5 m), while its summit platform covers an area of approximately 170 m^2 (Fig. 4). On the top of the Katskhi Pillar stand two churches—a larger and a smaller one—dated to the fifth and sixth centuries AD. In addition, the site includes three small monastic cells and a wine cellar, all enclosed within a stone wall (Katskhi Pillar 2012). In both form and function, the pillar closely resembles the Meteora pillars of Thessaly, Greece, with their cliff-top monastic complexes; however, the churches on the Katskhi Pillar are nearly one thousand years older than the Meteora monasteries.

Microscopic study of the Upper Cretaceous limestone forming the Katskhi Pillar shows that the rock mass is composed of a clastic limestone. It consists of carbonate clasts of variable size, a fine-grained carbonate matrix, and a small amount of terrigenous mineral grains. Sporadically, bioclastic carbonate fragments are also present. The

Fig. 4 General view of the Katskhi Pillar, Imereti region. Photo © Mirian Makadze

limestone contains small pores, micro-voids and joints, many of which are filled with calcite–barite mineral assemblages, indicating secondary diagenetic infilling and partial recrystallization processes within the carbonate framework.

Detailed structural surveys, supported by high-resolution terrestrial laser scanning and drone photogrammetry, have identified three dominant discontinuity sets: two sub-vertical and one sub-horizontal. This structural configuration governs the distribution of instabilities observed on the pillar's surface. Despite a high uniaxial compressive strength of approximately 150 MPa and a friction angle of about 46°, shallow rockfalls, wedge failures, and flexural toppling are common in the lower and middle portions of the monolith, while larger unstable blocks occur near the top, posing potential hazards for the built heritage and visitors (Margottini 2023).

Samshvilde. Another important example of Georgia's geoheritage is represented by the canyons of the Khrami and Chivchavi rivers and the Samshvilde Cape located between them (41°30′21″ N, 44°30′59″ E; 550–560 m a.s.l.). Both canyons are incised into Quaternary basaltic flows that overlie the southern periphery of the Paleozoic crystalline massif of the Khrami area. The Khrami Canyon exceeds 5 km in length, while the Chivchavi Canyon extends for about 3 km; both reach depths of up to 300 m. These conditions expose excellent geological sections, which together form a unique landscape (Fig. 5).

On the Samshvilde Cape which is naturally protected on three sides, a settlement already existed as early as the 4th millennium BC, as evidenced by habitation remains, caves, menhirs, and numerous burial grounds from that period. By the 8th–ninth centuries AD, Samshvilde had developed into a fortified city and served as the center of the Samshvilde Saeristavo (duchy) within the Kingdom of Kartli. The fortress city was the first to confront enemies invading from the south and therefore existed under conditions of constant warfare. In the second half of the seventeenth century, it was captured and destroyed by Iranian troops, after which it ceased to function (Sanadze et al. 2014).

Fig. 5 General view of the Samshvilde Cape (center) and the canyons of the Chivchavi (left) and Khrami (right) rivers. The ruins of the early–medieval fortress city are clearly visible on the Samshvilde Cape, Javakheti region. Photo © Georgian Travel Guide

Lakumurash Toba Lake. Among Georgia's more than 800 lakes, it is probably one of the most attractive. The lake is located on the Egrisi Ridge at an elevation of 2550 m asl (42°22′09″ N, 44°30′59″ E; surface area ~ 2000 m^2). It was formed within Bajocian-age volcanic rocks of predominantly andesitic composition, which acquired their exotic appearance as a result of Alpine orogeny (Fig. 6). There is no automobile road leading to the lake, and it can be reached only by a few mountain trails. The most optimal route begins from the townlet of Khaishi, although traversing these 8 km-long paths represents a rather challenging mountain adventure.

Bursamdzeli Mountain. In the Inner Kartli region, north of the city of Gori on the Greater Caucasus Southern Slope, lies the striking and scenic mount Bursamdzeli (Fig. 7). Although this mountain is not particularly high, (3661 m asl) it offers a commanding view from the Tiriponi Valley and the surrounding hills, providing a panoramic vista of the entire Inner Kartli region. Because of that, the mountain held cultic and divine significance for the region's inhabitants, with numerous historical legends associated with it. It serves as an excellent example of how geoheritage can shape and influence the cultural identity and consciousness of the local population.

Fig. 6 General view of Lakumurash Toba lake, Samegrelo region. Photo © Georgian Travel Guide

Fig. 7 A view of the captivating Bursamdzeli Mountain from the Tiriphoni Valley, Shida Kartli region. Photo © Travel Man

Bursamdzeli Mountain is composed of Lower Cretaceous limestones, sandstones, and argillaceous shales. The mountain itself forms a pyramid-like tectonic wedge, which is capped by permanent snow and glaciers.

Khornabuji Fortress. It is located in southeastern Georgia, within the strategically significant historical province of Dedoplistskaro, and represents an outstanding example of the integration between geoheritage and fortification architecture (Fig. 8). The fortress is constructed atop tectonic wedge of Upper Jurassic limestone that have undergone intense silicification. Although such tectonic slices are abundant in this region, their origin remains insufficiently explained and lacks firmly established geological evidence.

Fig. 8 Ruins of the citadel of Khornabuji Fortress built on a limestone tectonic wedge Kakheti region. Photo © Georgian Travel

Fig. 9 General view of the Church of Saint Elijah the Tishbite situated on a silicified limestone tectonic wedge, Kakheti region. Photo © Georgian Travel

Utilizing several of these naturally fortified and difficult-to-access geological features, Georgians built a powerful defensive complex along the Persian frontier in the fourth century. Khornabuji Fortress successfully protected this border for nearly 1,300 years. It was first plundered by Mongol forces in 1265, and was ultimately destroyed in the first half of the seventeenth century by the armies of Iran (Lolashvili 1971). Today, only ruins of the fortress have survived, and the site is legally protected by the state.

Church of Saint Elijah the Tishbite. This church is also located in the Dedoplistskaro Municipality and, similar to the Khornabuji Fortress, is built upon a tectonic slice of silicified Upper Jurassic limestone. Part of the structure is carved directly into the limestone blocks themselves, representing an exemplary case of the harmonious integration of geoheritage features with a monastic complex (Fig. 9).

The foundation of Saint Elijah's Church dates back to the sixth century. Unlike Khornabuji Fortress, it survived the waves of destruction that befell the region and remains active to this day. The distance between the church and Khornabuji Fortress is approximately 6.5 km, and according to a beautiful local legend, the two sites were once connected by a secret underground passage (e.g., Lolashvili 1971)—a feat that would have been practically impossible for that period.

The Kaklis Qure meander. It is formed by the Alazani River and is located within the Dedoplistskaro Municipality also**.** The mender is a landform distinguished by its unique geometry and exceptional beauty, representing an important element of geological heritage (Fig. 10). These circular meanders occupy an area of more than 200 hectares and, in the near geological future, are highly likely to be abandoned as the Alazani River shifts into a single straightened channel. In addition to its geological value, this area also hosts **unique biogeographical heritage.** Specifically, the only preserved stands in Georgia of endemic, wild walnut forests are located within this meander. Notably, the Georgian name *"Kaklis Qure"* translates as **"Walnut Bay,"** a reference to these native walnut woodlands that gave the area its name.

5 Discussion and Conclusion

From this review of Georgia's geoheritage monuments, it is evident that the country has exceptional geotourism potential and possesses all the prerequisites for establishing geoparks of global importance. Despite its relatively small territory, Georgia contains numerous geological sites of diverse genesis and age, in which striking tectonic, magmatic and geomorphological processes are clearly expressed. Many of these sites also host important cultural and biological values, forming integrated landscapes where geological history, biodiversity and human history are tightly interwoven.

As discussed in Chap. 2, the emerging IUCN framework of **Key Geoheritage Areas (KGAs)** provides a natural baseline for recognising this richness. Every site highlighted in this chapter would qualify as a strong KGA candidate. Together they form a dense national network of geoheritage localities that represent key themes of Earth history in the Caucasus: collisional tectonics, plutonism and volcanism, glacial and fluvial systems, karst and caves, desert and semi-desert environments, and outstanding fossil and archaeological records.

Within this KGA network, several clusters stand out as particularly suitable for development as aspiring UNESCO Global Geoparks, because they combine high geoheritage value with existing tourism infrastructure, educa-

Fig. 10 General view of the Kaklis Qure meander of the Alazani River, Kakheti region. Photo © Georgian Travel Guide

tional opportunities and engaged local communities. The diversity and abundance of both geoheritage and cultural heritage monuments, together with the presence of historic settlements, create all the necessary preconditions for the establishment of several geoparks of global importance. Of these, we consider the Vardzia rock-cut city and its surrounding ignimbrite plateau and river valleys to be the most promising area for aspiring to a UNESCO Global Geopark in Georgia, as it offers an outstanding combination of geological, cultural and landscape values already recognised in international discussions.

In addition to its diverse and rich geoheritage and cultural monuments, Georgia benefits from a favourable geographical location, a pleasant climate, and the historical identity of the legendary land of the "Golden Fleece." If the Argonauts' voyage is understood as one of the earliest geological expeditions, then Georgia's geoheritage story extends back to the very origins of exploration. Taken together, these factors, a dense KGA network, several strong candidate geopark territories, well-defined georoutes, and a powerful cultural narrative, define Georgia as an ideal destination for the development of international geotourism and as a country well placed to expand its role within the UNESCO Global Geoparks and broader geoheritage community.

References

Apollonius Rhodes (III BC) Argonautica: Book III. Editor Race H., 2008. Loeb Classical Library, pp 1–211

Casadevall TJ, Tormey D, Roberts J (2019) World Heritage Volcanoes: Classification, gap analysis, and recommendations for future listings. Gland, Switzerland: IUCN. pp viii + 68

Description of the Monuments of History and Culture of Georgia (2013) 1, Tbilisi

Ferring R, Oms O, Agusti J et al. (2011) Earliest human occupations at Dmanisi (Georgian Caucasus) dated to 1.85–1.78 Ma. Proc Natl Acad Sci USA 108(26):10432–10436. https://doi.org/10.1073/pnas.1106638108

Gabunia L, Vekua A (1995) A Plio-Pleistocene hominid from Dmanisi, East Georgia, Caucasus. Nature 373:509–512. https://doi.org/10.1038/373509a0

Gamkrelidze I, Okrostsvaridze A, Koiava K, Maisadze F (2021) Geotourism Potential of Georgia, The Caucasus. Springer, pp 1–140

Gamkrelidze I, Okrostsvaridze A, Tormey D, Boichenko G, Gogoladze S, Gabrielashvili (2024) Georgia's prospects to joining the UNESCO geoparks network and development of geotourism industry. Abstract of the 37th International Geological Congress, Busan, Republic of Korea

Golavito J (2014) Jason and the Argonauts through the Ages. "McFarland" Publ., pp 1–178

Gudjabidze G, Gamkrelidze I, Abesadze G et al (2003) Geological map of Georgia. Tbilisi, Georgia, Scale 1(500):000

Katskhi Pillar (2012) www. Dzeglebi.ge. (in Georgian)

Lolashvili I (1971) Georgian Pagan Cult Temples in Kambechovani. Tbilisi, pp 1–148

Margottini C, Spizzichino D, Sonnessa A, Puzzilli L. M (2015) Natural Hazard Affecting the Katskhi Pillar Monastery (Georgia). In: Lollino G, Giordan D, Marunteanu C, Christaras, B, Yoshinori I, Margottini C (eds) Eng Geol Soc Territ. Volume 8. Springer, Cham. https://doi.org/10.1007/978-3-319-09408-3_68

Meliksetian Kh (2018) The generation of collisional magmas on the example of Quaternary volcanism in the territory of Armenia. Abst. of doctoral dissertation. Yerevan, pp 1–38

Okrostsvaridze A, Gagnidze N, Bluashvili D (2014) Field investigation of the mythical "Gold Sands" of the ancient Colchis Kingdom and modern discussion on Argonauts mission. Episodes 37(2):10–17

Okrostsvaridze A, Chang YH, Chung SL, Rabayrol F, Boichenco G, Gogoladze S (2021) Paleozoic xenoliths in Eocene plutons: evidence for the destruction of Pre-Jurassic crystalline basement beneath Adjara-Trialeti belt, Lesser Caucasus. Geologica Carpathica 72(4): 299–315. https://doi.org/10.31577/GeolCarp.72.4.2

Okrostsvaridze A, Bluashvili D, Gogoladze S, Gabrielashvili R (2022) The discovery of U mineralization in the late Variscan plagiogranite vein of the Shkhara crystalline massif (Greater Caucasus, Georgia). Geologica Balcanica 51(1):3–14. https://doi.org/10.52321/GeolBalc.51.1.3

Okrostsvaridze A, Lee YH, Tormey D, Skhirtladze I (2023) U-Pb zircon chronological constraints for three stages of syn-orogenic plutonic magmatism of the Greater Caucasus Svaneti segment. Georgia. Episodes 46(2):195–209. https://doi.org/10.18814/epiiugs/2022/022023

Okrostsvaridze A, Gamkrelidze I, Lee YH, Keskin M, Tormey D, Boichenko G, Gogoladze S, Makadze M (2024) U-Pb geochronology and geochemistry of volcanic ash layers in the Eastern Caucasus intermountain Neogene basin: implications for their sources. Arab J Geosci 17:129. https://doi.org/10.1007/s12517-024-11914-7

Sanadze M, Berikashvili D, Gagoshidze G (2014) Samshvilde. Publishing house of georgian university, Tbilisi, pp 1–125. (in Georgian)

Placer Gold Geoheritage of Upper Svaneti (Greater Caucasus, Georgia): The Legend of the Argonauts' Expedition

Daniel Tormey, Avtandil Okrostsvaridze, and David Bluashvili

Abstract

One of the most important geoheritage features of Georgia is the placer gold of Upper Svaneti, particularly in the headwaters of the Enguri River. Numerous written accounts from antiquity describe gold-bearing sands in this region and the methods used to extract the metal. Modern geological exploration confirms substantial alluvial gold reserves in the Enguri River and its tributaries, and local inhabitants still recover gold using traditional panning techniques. According to ancient sources, the heroic expedition of the Argonauts, representatives of the Minoan and early Greek world, had as one of its main goals the capture of the "Golden Fleece" in the kingdom of Colchis. Historical, ethnographic, archaeological, and geological evidence now strongly supports the interpretation that this concept refers to the discovery of rich placer gold deposits and the mastery of their extraction, very likely in the Enguri basin of Svaneti. If so, the placer gold geoheritage of the Enguri River represents one of the earliest geological exploration sites in world history. On geological grounds, and taking into account regional cultural chronologies, the Argonauts' expedition must have taken place before about 1500 BC, prior to the catastrophic eruption of the Strongyle/Santorini volcano. We propose that this real story of early gold exploration was transformed into a captivating legend by the romantic Greek civilisation 800–700 years later, and was subsequently further veiled in mythic symbolism in Christian Europe. This chapter summarises the geological setting and placer deposits of Upper Svaneti, examines their relationship to the Argonauts' narrative, and evaluates the geoheritage, geotourism, and broader cultural significance of the Enguri placer gold system.

Keywords

Placer gold · Upper svaneti · Enguri River · Golden fleece · Argonauts expedition

1 Introduction

Ancient manuscript sources describe the expedition of the Argonauts as a real journey whose principal objective was to obtain the "Golden Fleece" (e.g., Homer, VIII–VII centuries BC; Herodotus, 440 BC; Herodorus of Heraclea, ca. 400 BC; Apollonius Rhodius, III century BC). The precise meaning of the "Golden Fleece" has been debated, but most scholars now agree that it refers to the discovery of gold deposits and the mastery of placer-gold extraction techniques, including the use of sheepskins in gold recovery. According to converging historical memory, ethnographic testimony, archaeological findings, and geological data, the Argonauts most likely targeted the gold-rich placers in the headwaters of the Enguri River in the Svaneti province of the ancient kingdom of Colchis (e.g., Okrostsvaridze et al. 2014; Okrostsvaridze 2017). On this interpretation, the gold placers of the Enguri River can be regarded as one of the first exploration geosites in world history.

The mobilisation and transport of eroded material from primary gold deposits by rivers is the fundamental process responsible for the formation of placer gold deposits. Coarser gold grains (larger than about 0.5 mm) are typically deposited relatively close to their source, whereas finer

D. Tormey (✉)
Catalyst Environmental Solutions, Santa Monica, USA
e-mail: dtormey@ce.solutions

IUCN World Commission on Protected Areas, Gland, Switzerland

A. Okrostsvaridze
Institute of Earth Sciences, Ilia State University, Tbilisi, Georgia

D. Bluashvili
Faculty of Geology and Mining, Georgian Technical University, Tbilisi, Georgia

A. Okrostsvaridze and D. Tormey (eds.), *Geoheritage of Georgia*, Geoheritage, Geoparks and Geotourism,
https://doi.org/10.1007/978-3-032-27639-1_7

grains can be carried by water currents for several kilometres downstream before settling. In mountainous regions with steep gradients and highly energetic rivers, this process can generate numerous, locally rich placer accumulations.

Upper Svaneti, the highest historical province of the Greater Caucasus, is crossed by many rivers that belong to the class of turbid, sediment-laden streams. Rivers of this type are capable of transporting large quantities of detritus along their valleys, including gold, scheelite, pyrite, chalcopyrite, and other ore minerals. As a result, placer gold deposits are abundant in Upper Svaneti, ranging in length from a few metres to several hundred metres. Because of the rugged high-mountain terrain their exact number is difficult to determine, but many placers are known, such as Latali, Yali, Tskhumari, Khudoni, and others (e.g., Geleishvili 1988). According to mythological, archaeological, and historical sources, local people began mining gold in prehistoric times and continue this activity today. Modern Svan gold miners know well where placer gold tends to accumulate and at what times of year extraction is most effective. The amounts of gold obtained demonstrate that there is a clear geological basis for the long-standing image of Svaneti as a land rich in gold.

This chapter describes the gold-bearing sands in the headwaters of the Enguri River, situated within the culturally and scenically outstanding landscape of Upper Svaneti. It presents the geological and geomorphological framework of the placer deposits, reviews historical and mythological sources linking them to the Argonauts' expedition, and evaluates their prehistoric, geoheritage, geotourism, and economic significance. In doing so, it positions the Enguri placer gold system as a key element in the emerging concept of Svaneti as a "Golden Fleece" geoheritage region, where ancient exploration, living mountain culture, and active alpine river processes intersect.

2 A Brief Description of the Ancient Kingdom of Colchis

According to ancient Greek sources, between the sixteenth and eleventh centuries BC there existed a powerful kingdom of Colchis on the eastern coast of the Black Sea, ruled at one time by King Aeëtes. During this period, long before similar developments in much of Europe, the people of Colchis mastered the technologies of ore mining and metal smelting. Early Colchis, together with the Iberian state union that formed around the same time in what is now eastern Georgia, is thought to have laid the foundations for the later formation of the Georgian nation (Fig. 1)

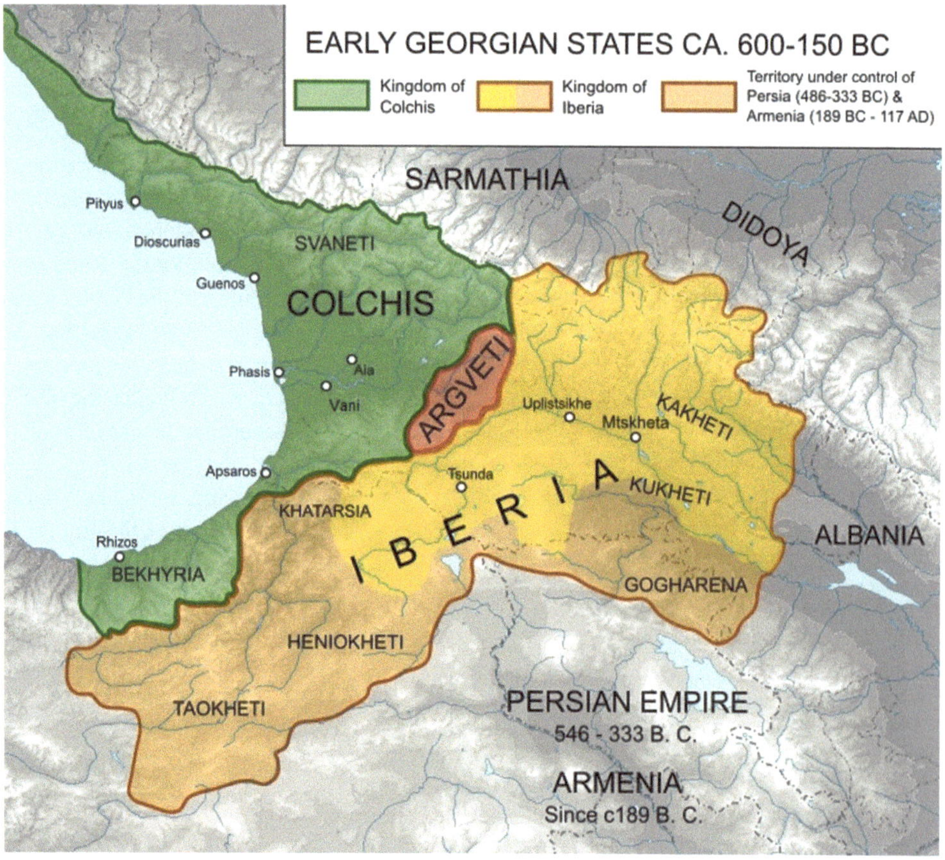

Fig. 1 Political map of the Eastern Black Sea region around in 600 to 150 BC (Braund 1994)

(e.g., Braund 1994). The mountainous hinterland of Colchis included the territory of present-day Upper Svaneti, where the gold-bearing rivers that later inspired the legend of the Golden Fleece originate.

It is worth noting that, internationally, the early kingdom of Colchis is best known because it is associated with the heroic voyage of the Argonauts, a masterpiece of Greek mythology, and as the homeland of Medea and the "Golden Fleece." The name Colchis is first mentioned in a literary source in Aeschylus's (525–456 BC) tragedy *Prometheus Bound* (also known in Georgian scholarship as "Amirani Chained"). In earlier literary sources, the western Georgian state union is referred to as "Aia." For example, Homer (eighth century BC) uses this name six times in the *Iliad* to describe a land "from where the sun rises." The territory of the ancient Colchis kingdom mainly included the western provinces of modern Georgia, Abkhazia, Svaneti, Samegrelo, Imereti, Guria, Racha, and Adjara, as well as the Sochi and Tuapse regions of modern Russia, and the Artvin and Trabzon provinces of modern Turkey.

3 Historical Svaneti Province

In Georgian, Svaneti (in ancient sources—*Svania*) is a historical province of the early and ancient Colchis kingdom, located on the southern slopes of the Caucasus Mountains near the Black Sea. The province has traditionally been inhabited mainly by the Svans, a geographical subgroup of Georgians who speak Georgian and the Svan language. Today the Svans live primarily in the headwaters of the Enguri, Tskhenistskali, and Kodori rivers of the Black Sea basin. The Enguri Valley is known as Upper Svaneti, the Tskhenistskali Valley as Lower Svaneti, and the Kodori Valley as Abkhazian Svaneti (Fig. 2).

Upper Svaneti is the historical centre of Svaneti and is one of the highest and most densely populated mountainous regions in the Caucasus. It currently has more than 23,000 residents, with the administrative centre in the town of Mestia (Fig. 3). The region is renowned for its ancient traditions, architectural masterpieces, and spectacular mountain landscapes. Of particular note are the multi-

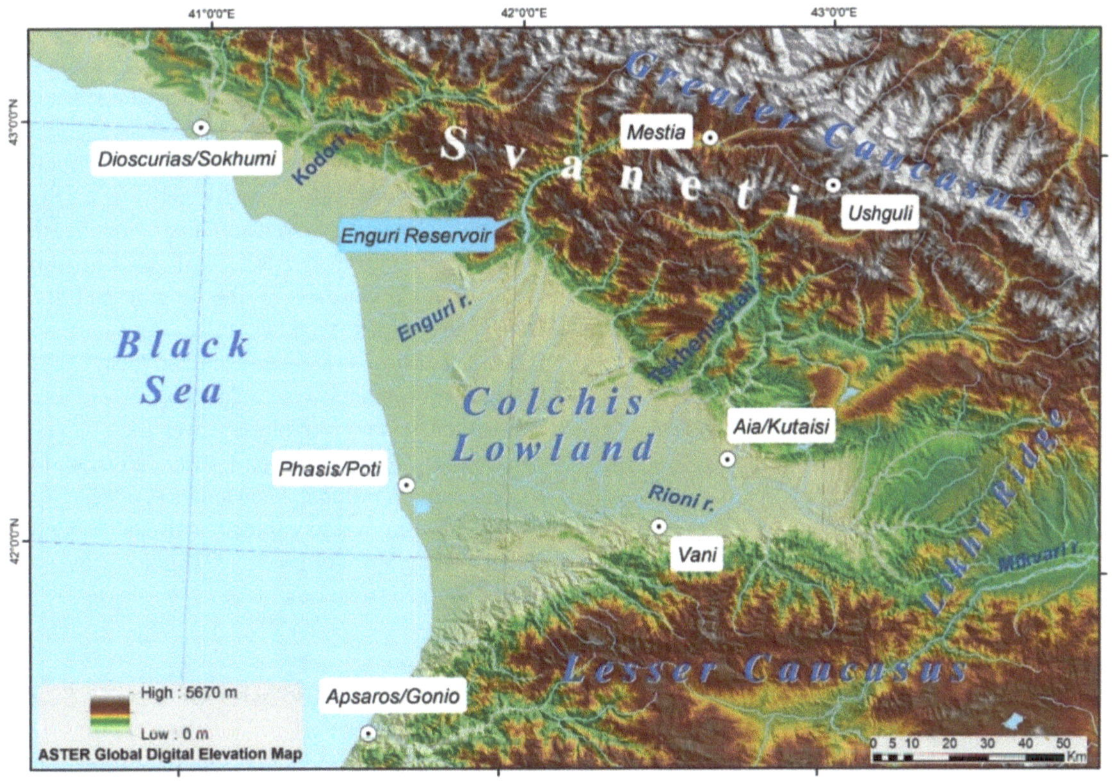

Fig. 2 ASTER Satellite imagesof modern-day Colchis Lowland and Svaneti Region

Fig. 3 Center of Mestia Town. The Greater Caucasus Ridge in the background. Photo © geotrp.ge

functional Svan stone towers of the 9th–twelfth centuries, inscribed on the UNESCO World Heritage List. These 3- or 4-storey square towers, which taper towards the top and are entered from the second floor, served both as dwellings and as protection against enemy attack and heavy snowfall. Written sources about the Svans and Svaneti date back to antiquity and consistently emphasise their bravery, the strength of their land, and their wealth in gold.

4 The History of Placer Gold Mining in Upper Svaneti

The first written information about the existence of gold in the river sands of Svaneti comes from the ancient geographer Strabo (64/63 BC—23/24 AD). In his *Geographica* (Book XI), he describes how the inhabitants of the mountainous hinterland of Colchis recover gold from rivers by placing sheepskins in the water so that wool fibres trap fine gold grains. When the fleece becomes saturated, the gold is shaken out and collected. This account provides a vivid early description of placer-gold extraction in the headwaters of rivers such as the Enguri and its tributaries, and strongly suggests that the technique was already well established by the first century BC.

After the ancient period, and almost until the eighteenth century AD, written information about gold mining in Upper Svaneti is sparse. Nevertheless, archaeological discoveries of gold artefacts, jewellery, and smithing tools, as well as remains of small forges and slag, indicate that placer gold continued to be exploited throughout the medieval period. These finds, together with local oral traditions, show that Svan communities maintained knowledge of gold-bearing river reaches and extraction methods over many generations, even when external written sources are nearly silent.

Historical sources indicate that from the 1960s onwards, placer gold mining in Upper Svaneti was carried out more systematically under Soviet administration. Seasonal mining brigades were formed from local residents, who worked along selected reaches of the Enguri and its tributaries using pans, simple wooden sluices, and mechanical equipment. According to archival data and local testimony, experienced Svan miners were able to obtain significant quantities of gold from river placers over the course of a single working season, which were then delivered to the state. In this way, traditional knowledge of placer locations and river behaviour was combined with state-organised exploration, linking ancient practices of gold recovery to modern geological investigation.

5 Placer Gold Deposits of Upper Svaneti

As noted in the introduction, the entire territory of Upper Svaneti is crossed by numerous steep mountain rivers. The region is undergoing very rapid uplift, and the rivers respond with intense erosion and sediment transport (e.g. Wang et al. 2014). Because of their high sediment loads, these streams are classified as turbid rivers. They are capable of mobilising large quantities of detritus from bedrock and unconsolidated

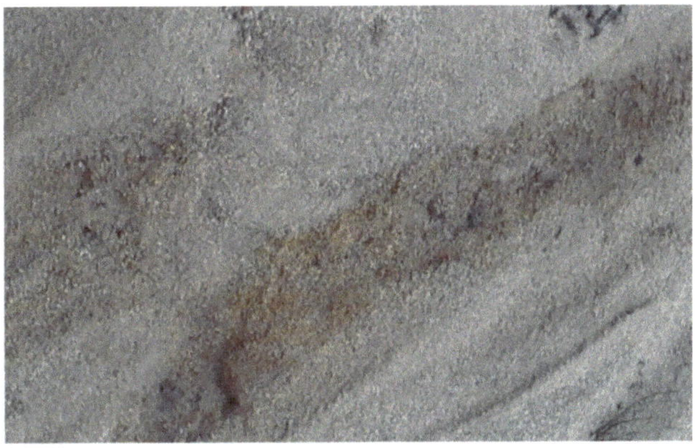

Fig. 4 The Mashrichala gold-bearing placer of the Nenskra River, under the water. Oxidized zones are potentially gold-bearing. Photo © A. Okrostsavridze

Fig. 5 A wooden pan for gold extraction from Upper Svaneti. Photo © A. Okrostsavridze

Fig. 6 Washed Lasili placer gold, Enguri River canyon. Here gold nuggets can be seen. Photo © A. Okrostsavridze

Fig. 7 Gold grains from the Mashrichala placer. Photo © A. Okrostsvaridze

deposits and transporting it through narrow, high-energy valleys. Along lower-gradient reaches, hydraulic sorting of this material leads to the formation of alluvial and eluvial placer deposits in which gold and other heavy minerals accumulate (Fig. 4).

Today, just as in the distant past, local inhabitants of Upper Svaneti still extract gold from these placer deposits. Using pans and simple wooden sluices, they wash gold-bearing sands along selected stretches of the Enguri and its tributaries (Figs. 5, 6). The work is usually carried out during the low-flow seasons, when river levels are favourable and the richest gravel bars are exposed. This traditional activity preserves techniques that probably date back many centuries and provides a living link to the ancient descriptions of Colchis as a land of gold.

The colour of Svaneti placer gold ranges from dark grey to light yellow, depending on the degree of oxidation and the presence of associated minerals. Gold grains vary in size from fine sand to multi-millimetre nuggets. Many particles are well rounded or flattened by transport, whereas others retain more angular shapes that indicate shorter travel distances. Microscopic examination shows a variety of textures and inclusions, while chemical analyses indicate that the fineness of Svaneti placer gold grains is high, generally ranging from 800 to 900. These characteristics are clearly visible in grains from the Mashrichala and Arshira placers, where both fine flakes and larger nuggets have been documented (Figs. 7, 8).

In 2008, a comprehensive study of placer gold deposits in Upper Svaneti was carried out, combining field mapping with the analysis of bedrock and unconsolidated sediments and the interpretation of satellite imagery and other remote-sensing data. This work focused on the Enguri basin and its tributaries and re-evaluated several known placers while identifying new prospective sites. The results demonstrated that gold has been repeatedly remobilised from primary sources in Mesozoic volcanogenic, sedimentary rocks and Paleozoic crystalline complexes into younger river and glaciofluvial deposits. On the basis of these findings, the area attracted interest from international exploration companies, including the "Golden Fleece Mining Corporation," which conducted further investigations (Okrostsvaridze and Bluashvili 2009).

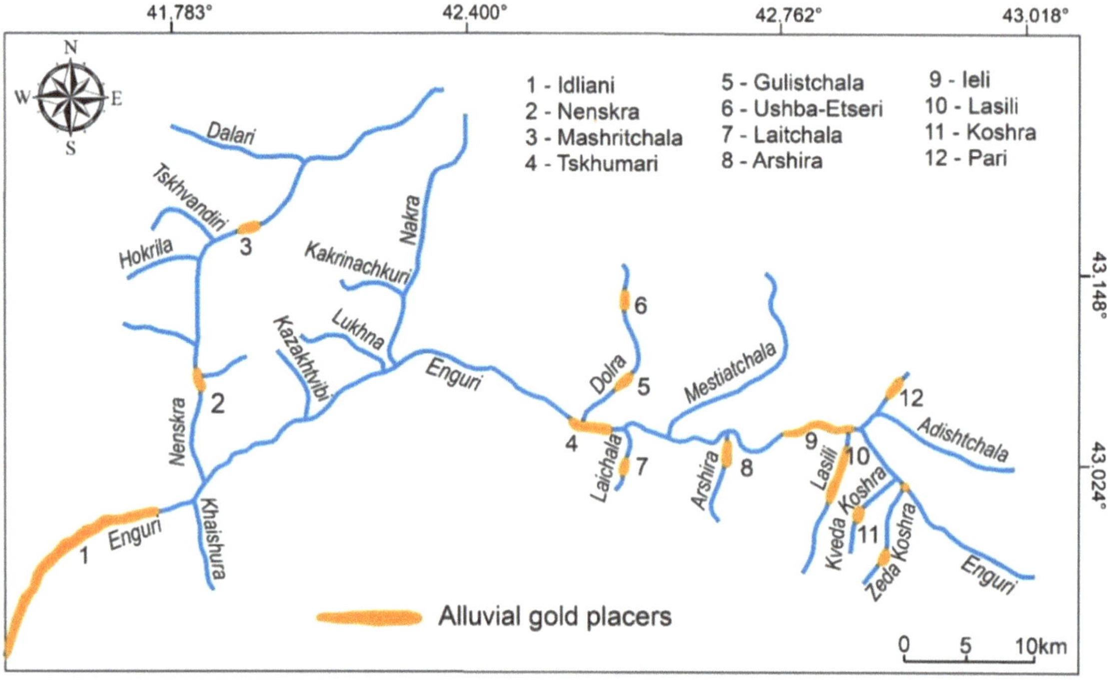

Fig. 9 Schematic map of placer gold geoheritage distribution in the headwaters of the Enguri River (Okrostsvaridze 2019)

Fig. 8 12.5 mm long gold grain from the Arshira placer. Photo © Rudi Gapel

The research confirmed that the headwaters of the Enguri River host a dense network of placer gold occurrences, many of which are closely associated with glaciofluvial terraces, confluence zones, and narrow canyon reaches (Fig. 9). This pattern strengthens the geological argument that Upper Svaneti, as part of the hinterland of ancient Colchis, could indeed have been one of the main targets of early expeditions seeking rich gold-bearing sands, an interpretation that aligns closely with the later literary tradition of the Argonauts' quest for the Golden Fleece.

6 The Argonauts' Expedition to the Ancient Colchis Kingdom: Myth or Reality

The story of the Argonauts' voyage to the ancient kingdom of Colchis has fascinated writers and artists for more than two and a half millennia. From early oral traditions through Homer and later classical authors to Apollonius of Rhodes' *Argonautica*, the narrative was repeatedly re-told and embellished, but its core elements remain constant: a journey by sea to a distant eastern land, a powerful king named Aeëtes and a quest to obtain the "Golden Fleece". Classical writers consistently describe Colchis as a wealthy, gold-bearing land on the eastern Black Sea coast, and assume that their audiences know the story.

When the various ancient accounts are compared, a consistent geographical core emerges. The Argonauts are described as sailing from the Aegean, through the Hellespont and the Sea of Marmara, along the Turkish coast of the Black Sea and on to the eastern shore near the mouth of a great river widely identified with the modern Rioni. Modern experimental voyages, most notably Tim Severin's reconstruction of a Bronze Age rowing vessel, have demonstrated that this route is entirely feasible with the shipbuilding and navigational skills of the time. In this sense, the Argonauts' expedition can be viewed as a mytholo-

Fig. 10 The Argonauts voyage route according to Herodorus of Heracles (Colavito 2014)

gised reflection of real journeys by Aegean seafarers to the gold-bearing hinterland of Colchis.

Geological, archaeological and ethnographic evidence from Georgia strengthens this interpretation. Rich placer gold deposits are found not only in the Rioni basin but also, and perhaps most importantly, in the headwaters of the Enguri River in Upper Svaneti, where traditional gold extraction using sheepskins persisted into the modern era. The landscape and river system of the Enguri basin match closely the setting implied by ancient sources: steep, gold-bearing catchments draining into a coastal plain accessible from the Black Sea. In this light, the Argonauts' voyage can be understood as an early exploration enterprise aimed at identifying and securing access to placer gold deposits in the mountainous hinterland of Colchis, with the Enguri basin as one of the most likely real-world targets.

When the various ancient accounts are compared, a consistent core emerges: the Argonauts sailed in a long, narrow rowing vessel along the coasts of the Aegean and Black Seas to reach the eastern shore of the Black Sea, in or near the territory of the ancient kingdom of Colchis. The overall distance and travel times implied by the story are compatible with what is known of contemporary shipbuilding and seafaring skills (Fig. 10).

7 Geoheritage Significance of the Enguri Placer Gold System

The Enguri River headwaters in Upper Svaneti present an exceptional concentration of geoheritage values within a relatively compact high-mountain setting. Geologically, the basin offers a clear, didactic example of how placer gold deposits form and evolve in an actively uplifting orogen. Gold-bearing primary sources in Mesozoic volcanogenic–sedimentary successions and Paleozoic crystalline rocks are exposed in steep catchments; intense physical weathering and erosion feed sediment into a highly energetic, turbid river system; and hydraulic sorting along channels and at confluences generates numerous alluvial and eluvial placer accumulations. The spatial relationship between source rocks, transport pathways, and placer deposits can be observed directly in the landscape and mapped using modern exploration techniques, making the Enguri basin a natural open-air laboratory for teaching source-to-sink processes.

The Enguri placer system is equally significant for its cultural and historical dimensions. From antiquity, the wider Colchis region was renowned as a land of gold, and numerous classical authors associated it with the quest of the Argonauts for the "Golden Fleece." Archaeological, ethnographic, and geological evidence together support the interpretation that this concept refers to the discovery and exploitation of rich placer gold deposits, very likely in the Enguri basin of Svaneti, and to the use of sheepskins in gold recovery. In this sense, the placer gold of Upper Svaneti can be regarded as one of the earliest geological exploration sites in world history, a rare case where an iconic Mediterranean myth can be traced back to a specific high-mountain river system and to techniques that are still remembered in local tradition.

The geoheritage value of the Enguri placers is further enhanced by their setting within the outstanding cultural landscape of Upper Svaneti, already recognised as a UNESCO World Heritage Site for its medieval stone towers, villages, and long continuity of mountain life. High-alpine geomorphology, active river processes, traditional Svan gold panning, and the enduring Argonauts narrative together create a tightly interwoven geo-cultural ensemble. Few places in the world combine an active placer-gold system, a well-documented resource history, and such a powerful mythological overlay in a single, largely intact landscape.

Taken together, these attributes give the Enguri placer gold system high scientific, educational, cultural, and symbolic value at national and international levels. It encapsulates key themes of Georgia's geoheritage, active tectonics, dynamic rivers, mineral resources, and deep cultural memory, and forms a natural cornerstone for any future geoheritage or geopark initiative in Svaneti, particularly concepts framed around the "Golden Fleece" as a bridge between geology and legend.

8 Geotourism and Geoheritage Potential of the Enguri Placer System

Upper Svaneti is already a major destination for cultural and mountain tourism, centred on the UNESCO World Heritage Site of the Upper Svaneti cultural landscape (Mestia, Ushguli and surrounding villages). The placer gold geoheritage of the Enguri River offers a powerful additional layer that can enrich existing visitor experiences without displacing the core cultural narrative. Along a relatively short reach of the Enguri and its tributaries, visitors can observe steep alpine valleys, glaciofluvial terraces and fans, active river channels, and sites where gold-bearing sands are still recognised and worked by local inhabitants. These features provide natural outdoor classrooms for explaining how high-mountain tectonics, Quaternary glaciation, and modern fluvial processes together create placer gold deposits.

The educational potential of this landscape is considerable. The Enguri basin can be used to illustrate several interconnected themes: "gold in the Earth and in the story", linking the physical concentration of heavy minerals in river sediments to the emergence of the Golden Fleece myth; "climate, glaciers, and gold", showing how past glaciations shaped valley profiles, sediment loads, and placer formation; and "sheepskin and science", connecting ethnographic accounts of fleece-based gold panning to modern sedimentology and mineral processing concepts. Guided walks along glaciofluvial gold terraces, interpretive stops at river confluences such as the Dolra and Nenskra junctions, and small exhibits on placer mechanics and sheepskin panning could serve school groups, university field courses, and general visitors alike.

There is also strong potential to integrate geoheritage interpretation into existing cultural circuits. In Mestia and Ushguli, visitor centres and museums could include concise panels or small displays explaining how the wealth of gold contributed to the historical reputation of Colchis and Svaneti, and how local communities learned to read the rivers and their sediments. A "Svaneti Gold Trail" could be developed as a themed walking route that links selected viewpoints, river terraces, and villages, combining stories of gold, towers, and mountain livelihoods. Cultural workshops on fleece washing, blacksmithing, and gold traditions, designed and led by local residents, would further reinforce the link between geology, traditional knowledge, and contemporary identity.

9 Discussion

A combination of mythological, archaeological, ethnographic, historical, and geological data strongly suggests that the story of the Argonauts preserves a real memory of voyages by Aegean seafarers to the eastern Black Sea in search of gold. Classical authors consistently describe Colchis as a wealthy, gold-bearing land and present the Argonauts' expedition as a factual event, set in a heroic past rather than in a purely imaginary realm. Archaeological discoveries in western Georgia and Svaneti, including gold artefacts, evidence of early mining and metalworking, and traces of long-distance contacts, demonstrate that by the second millennium BC the region already possessed advanced knowledge of ore extraction and metallurgy. Ethnographic records of fleece-based placer-gold recovery in the Caucasus, together with the geological reality of rich placer deposits in rivers such as the Enguri, provide a practical explanation for the "Golden Fleece" motif that is more convincing than purely symbolic interpretations.

Studies of Aegean civilisation show that the peoples of the Minoan and early Mycenaean world were accomplished sailors whose economy depended heavily on access to metals and other raw materials. The rapid development of their society, and later Greek colonisation of the Black Sea region, would not have been possible without systematic expeditions to areas rich in copper, gold, and other resources. In this light, the Argonauts' voyage can be viewed as an early exploratory enterprise by representatives of the Aegean civilisation, aimed at identifying and securing access to placer gold deposits in the hinterland of Colchis. The headwaters of the Enguri River in Upper Svaneti, where numerous placers are concentrated and where traditional gold extraction was still practised into the twentieth century, fit this role particularly well and correspond closely to the geographical and geological setting implied by ancient sources.

The timing of the earliest such expeditions can be constrained by independent historical and geological evidence. As discussed earlier, the Argonauts' expedition must pre-date the catastrophic Thera eruption of the Strongyle (Santorini) volcano, which took place around the mid-second millennium BC and had profound impacts on the Aegean world. This eruption disrupted maritime networks and contributed to major political and economic changes in the eastern Mediterranean. It is therefore reasonable to regard the Thera eruption as a lower time limit for the Argonauts' voyage: any large-scale, organised expedition of this kind is most likely to have occurred before this event, during the peak of Minoan and early Mycenaean maritime power. Later Greek authors, writing centuries afterwards, re-cast this earlier episode in a more romantic and moralised form, gradually clothing a real exploration story in a mythical veil.

From a geoheritage perspective, the Enguri placer system can thus be interpreted not only as a classic example of high-mountain placer formation, but also as one of the earliest documented exploration geosites in world history. Here, an exceptional convergence of geological processes and human initiative gave rise to a narrative that has shaped European cultural imagination for more than 2,500 years. The geoheritage significance and geotourism potential outlined in the preceding sections follow directly from this synthesis: the same landscape in which the Enguri River draws modern gold-panners is the likely stage on which Aegean explorers first encountered the gold-bearing rivers of the Colchis hinterland. Recognising and interpreting this deep connection between geology, early resource exploration, and enduring myth is central to understanding the Enguri River as a key element of Georgia's geoheritage and as a cornerstone for any future "Golden Fleece" geoheritage or geopark initiative in Svaneti.

10 Conclusion

The headwaters of the Enguri River in Upper Svaneti bring together an exceptional combination of geological processes, cultural history, and enduring myth. Geologically, the basin offers a clear and accessible example of placer-gold formation in an actively uplifting mountain range. Gold-bearing primary sources in Mesozoic volcanogenic–sedimentary rocks and Paleozoic crystalline complexes are exposed in steep catchments; intense physical weathering and erosion deliver sediment into a highly energetic, turbid river system; and hydraulic sorting along channels, terraces, and confluences generates numerous alluvial and eluvial placers. Modern field investigations, remote-sensing studies, and chemical analyses confirm that these placers are widespread, repeatedly replenished, and capable of yielding significant quantities of gold.

From antiquity to the present day, this geological reality has shaped human activity in the region. Ancient authors described Colchis as a land of gold and recorded techniques of placer extraction, including the use of sheepskins to trap fine gold grains in river currents. Archaeological finds, medieval artefacts, and ethnographic records show that Svan communities have mined placer gold for millennia, maintaining detailed knowledge of favourable seasons, river reaches, and extraction methods. Modern Svan gold-panners, washing Enguri gravels, thus continue practices that echo both early resource exploration and the classical accounts that inspired the legend of the Golden Fleece.

When viewed in this light, the story of the Argonauts can be interpreted as a mythologised reflection of real voyages by Aegean seafarers to the eastern Black Sea in search of rich gold-bearing rivers in the hinterland of Colchis. Constraints from the Santorini (Thera) eruption and the development of Aegean maritime power suggest that the earliest such expeditions most likely occurred before the mid-second millennium BC, during the florescence of Minoan and early Mycenaean civilisation. Later Greek and European authors re-cast these events in poetic form, layering symbolic and moral meanings onto an original core of exploration. The Enguri basin of Upper Svaneti, with its dense network of placer deposits and long-lived traditions of gold extraction, fits this reconstructed historical and geological setting particularly well.

As a geoheritage site, the Enguri placer gold system therefore possesses outstanding scientific, educational, cultural, and symbolic value. It provides a natural laboratory for demonstrating source-to-sink processes, an early example of organised exploration geoscience, and a rare instance where a major Mediterranean myth can be traced back to a specific high-mountain landscape. Its location within the UNESCO World Heritage cultural landscape of Upper Svaneti strengthens this significance, creating a tightly interwoven geo-cultural ensemble in which alpine geomorphology, river processes, traditional Svan life, and the Argonauts' narrative all intersect. Recognising and carefully managing this ensemble, through targeted geoconservation, sensitive geotourism, and locally driven interpretation, would not only safeguard a unique part of Georgia's geoheritage, but also provide a compelling cornerstone for any future "Svaneti Golden Fleece" geoheritage or geopark initiative.

References

Appian of Alexandria (II BC) Books XIII-XVII (Civil Wars). Translation: John Carter, 1996. Harmondsworth, Middlesex, England. Penguin books Ltd: pp 1–436

Apollonius Rhodius (III BC) Argonautica. Editor Race H (2008) Loeb Class. Library, pp 1–211

Balter M (2006) New Carbon Dates Support Revised History of Ancient Mediterranean. Science 312:508–519

Bennet J (2003) "Minoan civilization", Oxford Classical Dictionary, pp 1–985

Braund D (1994) Georgia in antiquity: a history of colchis and transcaucasian Iberia, 550 BC - AD 562: Oxford, USA. Oxford University Press, pp 1–378

Friedrich WL (2000) Fire in the Sea. Volcanism and the Natural History of Santorini. Cambridge University Press, Cambridge, pp 1–256

Geleishvili V (1988) Gold of Georgia. PhD Thesis. Tbilisi, pp 1–54

Golavito J (2014) Jason and the argonauts through the ages. http://www.jasoncolavito.com/jason-and-the-argonauts.html

Manning W, Ramsey C.B, Kutschera W et al. (2016) Chronology for the Aegean Bronze Age 1700–1400 BC. Science. Am Assoc Adv Sci, 312 (5773): 565–569

Okrostsvaridze A (1992) The first data on gold high concentration in the achapara and hokrila river sources. Bul Geor Nat Acad Scien. 5(1):127–131

Okrostsvaridze A, Bluashvili D (2009) The sakeni ore field-geological evidance and perspectives (greater caucasus): Bul. Georg Nat Acad Scien. 3(1):97–104

Okrostsvaridze A, Gagnidze N, Bluashvili D (2014) Field investigation of the mythical "Gold Sands" of the ancient Colchis Kingdom and modern discussion on Argonauts mission. Episodes 37(2):10–17

Okrostsvaridze A (2017) The Argonauts: A modern Investigation of the Mythical "Gold Sands." LAMBERT Academic Publishing, Saarbrucken, pp 1–92

Okrostsvaridze A (2019) Argonauts expedition: myth or reality? Geological Arguments. Publishing House of the Georgian Academy of Sciences, Tbilisi, pp 1–140

Okrostsvaridze A, Tormey D, Bluashvili D, Makadze (2024) The Argonauts Campaign to the Ancient Colchis Kingdom: One of the first documented geological exploration in the Human history. Abstract #359 of the 37th IGC, Busan, Republic of Korea

Strabo (I AD- I BC). Geography. Ed. Stefan Radt (2010) Vandenhoeck and Ruprecht, pp 1–321

Wang SJ, Lee YH, Okrostsvaridze A (2014) Cenozoic mountain building processes in the greater caucasus, from subduction to collision. American Geophysical Union, Fall Meeting, USA, San-Francisco, abs. #T21A-4565

Warren P. M (2006) The date of the thera eruption in relation to aegean-egyptian interconnections and the egyptian historical chronology. In Czerny E. et al. Stud Honor Manfred Betake. pp 305- 321. https://doi.org/10.1017/S0003598X00115388

Living Within Mtkvari Ignimbrites Geoheritage, Lesser Caucasus, Georgia

Avtandil Okrostsvaridze, Daniel Tormey, and Mirian Makadze

Abstract

Powerful subaerial volcanic activity in the Lesser Caucasus began in the Middle Miocene and led to the formation of the ~4,500 km^2 Samtskhe–Javakheti Upper Miocene–Quaternary volcanic highland in Georgia. The 35 km long, 60–120 m thick Mtkvari ignimbrite is the oldest component of this highland and was erupted from the Keltepe–Gumbati resurgent caldera ~7.51 Ma ago. The ignimbrite flow is divided into two parts by a regional fault running along the Mtkvari River in the Vardzia area: the western is located in the Samtskhe Block and the eastern one—in the Javakheti Block. Field observations show that the Javakheti Block has subsided by ~127 m relative to the Samtskhe Block over the past ~7.51 Ma. As a result, the main ignimbrite flow is exposed only on the left bank of the Mtkvari, whereas on the right bank it is submerged and does not crop out; here only thin (2–3 m) ignimbrite layers belonging to the upper part of the Goderdzi Volcanic Formation are exposed. These features make the Vardzia area a geoheritage site of global significance. In the ignimbrites of the Javakheti Block the Vanis-Kvabebi rock-cut monastic complex was founded in the eighth century, and in the ignimbrites of the Samtskhe Block the Vardzia rock-cut city was constructed at the end of the twelfth century. Both are now cultural heritage sites of outstanding importance. This chapter shows how a key geoheritage site of exceptional scientific significance can also host globally significant cultural heritage.

Keywords

Mtkvari ignimbrites · Vardzia rock-cut city · Vanis-Kvabebi

1 Introduction

In the Middle Ages, religious, residential, and defensive buildings in the volcanic provinces of the Byzantine Empire and the Kingdom of Georgia were often carved out into Miocene subaerial fine-grained tuffs and ignimbrites. It seems that the builders were good engineers: these rocks are easy to work, and their low iron and magnesium contents create a relatively warm interior environment. A good example of this is the well-known carved buildings that exist in the ignimbrites of the Cappadocia province of modern Turkey (e.g., Topal, Doyuran 1997; Le Pennec et al. 2005). Similarly, large numbers of carved buildings are in the Mtkvari ignibrite flow on the Samtskhe-Javakheti volcanic highland of Georgia (e.g., Spizzichino et al. 2017; Okrostsvaridze et al. 2020). However, unlike Cappadocia, these buildings are less well known, although the Vardzia rock-cut city is probably the largest of its kind in the world.

The Mtkvari Ignimbrites flow is exposed in the lower part of the Goderdzi Volcanic Formation of the Samtskhe-Javakheti volcanic highland and represents the product of the first major eruption of the Keltepe-Gumbati large resurgent caldera (Makadze et al. 2024). The flow of these ignibrites spilled only on the northern slope of this caldera, and here it is exposed by the deeply cut canyons of the Mtkvari and Uraveli rivers, although the best outcrop is in the exposure of the left slope of the Mtkvari River. This flow of ignimbrites, with a thickness of 60–120 m,

A. Okrostsvaridze (✉)
Institute of Earth Sciences, Ilia State University, Tbilisi, Georgia
e-mail: avtandil.okrostsvaridze@iliauni.edu.ge

D. Tormey
Catalyst Environmental Solutions, Santa Monica, USA

IUCN World Commission On Protected Areas, New Delhi, India

M. Makadze
Tbilisi State University, Faculty of Exact and Natural Sciences, Tbilisi, Georgia

A. Okrostsvaridze and D. Tormey (eds.), *Geoheritage of Georgia*, Geoheritage, Geoparks and Geotourism,
https://doi.org/10.1007/978-3-032-27639-1_8

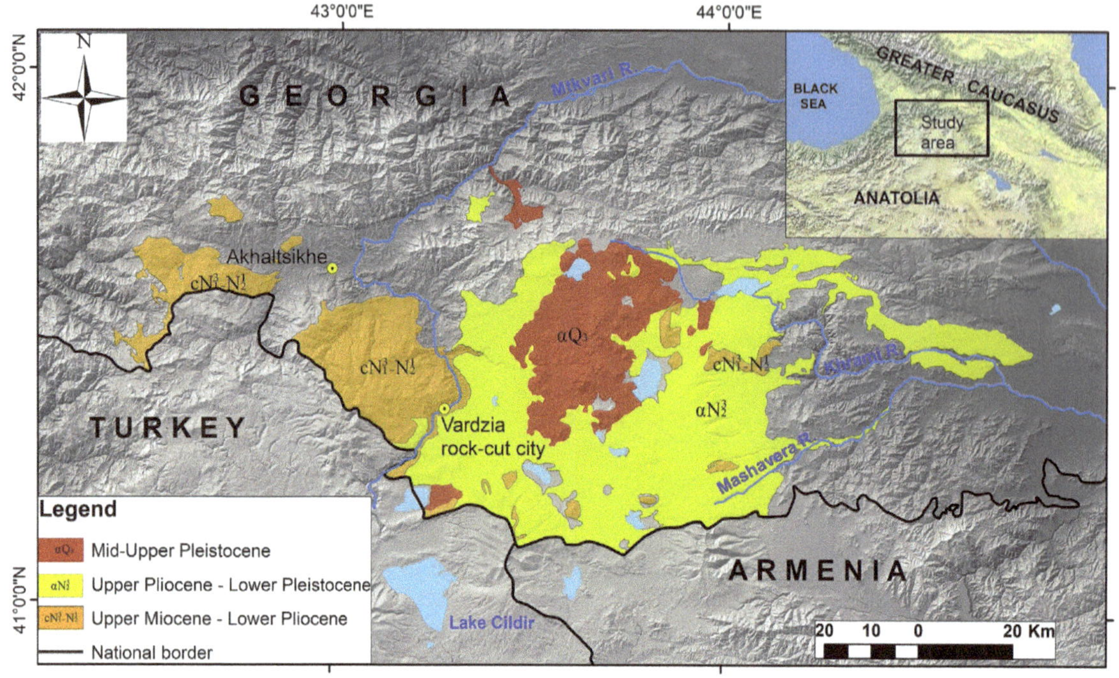

Fig. 1 Schematic geological map of the Upper Miocene–Quaternary Samtskhe–Javakheti volcanic highland (Okrostsvaridze et al. 2020)

follows a distance of 35 km, from the Karzameti Fortress to the Khertvisi Fortress. Throughout this distance, numerous defensive, residential and religious structures were carved into in these rocks in almost every favorable location during the Middle Ages and represented a unified space of urbanization. It can be said that this system of buildings, as well as the ignibrite flow of Mtkvari, played an important role in protecting the historical province of Samtskhe-Javakheti from destruction and preserving its ethnicity in the Middle Ages.

In this chapter, we focus on one civilian rock-cut city of the Vardzia and one rock-cut monastery complex of the Vanis-Qvabebi. Both of them are located in the central part of Mtkvari Ignimbrites Flow and present excellent examples of how a key geoheritage sites, of outstanding scientific significance, can be transformed into a globally significant cultural heritages. Tormey et al. (2026) discuss the geopark potential of Vardzia and its surroundings in greater detail.

2 Brief Geology of the Samtskhe-Javakheti Volcanic Highland

The Caucasus mountain-building area expresses the collision between the still convergent Arabian Plate and the Scythian Platform of the Eurasian continent. Post-collisional powerful subaerial volcanic activity in Eastern Anatolia and the western Lesser Caucasus started from Middle Miocene, which continued until Holocene. This volcanic activity shaped large volcanic highlands, present in the Georgian, Turkish and Armenian territories, covering an area of approximately 30,000 km^2. Part of the highland, located in Georgia, is known as the Samtskhe-Javakheti and covers approximately 4500 km^2 (Fig. 1).

The Samtskhe-Javakheti volcanic highland, is located in southern Georgia in Lesser Caucasus, north of the Erzurum-Kars volcanic highland (NE Anatolia) and discordantly covers Middle Eocene tuff-breccias and the Upper Cretaceous sandstones. The mean elevation of the highland is about 1200–2200 m above the sea level; individual volcanic edifices, however, reach up to 3000 m (Gumbati—2996 m, Didi Abuli—3305 m). The highland has a long history of the investigation and has been studied by many generations of geologists (e.g. Skhirtladze 1958; Ustiev and Jigauri 1971; Tutberidze 2011; Okrostsvaridze et al. 2020; Lin et al. 2020; Okrostsvaridze et al. 2024).

Recent studies have shown that the Samtskhe-Javakheti volcanic highland is divided into two tectonic blocks: western (Samtskhe) and eastern (Javakheti). These blocks are separated by a regional fault running SW-NE along the Mtkvari river Canyon. Differential GPS measurements indicate a vertical displacement along this fault, particularly evident at the Vardzia section, where the Javakheti Block is currently ~127 m lower than the Samtskhe Block (Okrostsvaridze et al. 2020).

In Samtskhe-Javakheti volcanic highland three main stages of magmatic activity are observed: (1) Upper

Miocene-Lower Pliocene, when huge, 700–1100 m thick dacite-andesitic volcanic tuffs and andesitic-basaltic lava flows were formed (Goderdzi Volcanic Formation); (2) Upper Pliocene-Lower Pleistocene, when continental flood basalts were formed with thicknesses ranging from 120 to 170 m and (3) Mid-Upper Pleistocene—associated with the formation of the Abul-Samsari intraplate volcanic range (Okrostsvaridze et al. 2020).

2.1 Goderdzi Volcanic Formation

The Upper Miocene-Lower Pliocene Goderdzi Volcanic Formation is the oldest and most widespread volcanic unit within the Samtskhe-Javakheti volcanic highland. It unconformably overlies Middle Eocene tuff-breccias and the Cretaceous sandstones and reaches a total thickness of 900–1000 m in the Vardzia area (Fig. 2) (Skhirtladze 1958).

The Goderdzi formation in the Vardzia area, is divided into three parts: the lower part with a thickness of 200–250 m is made up of pyroclastic material, in which mainly dark, weakly cemented coarse-grained clasts of hypersthene and two-pyroxene-bearing andesitic and andesitic-dacitic composition dominate. At some parts of this coarse-grained tuff formation one can notice thin-layered basaltic-andesitic and pyroxene-bearing andesitic lava-flows.

This section is overlain by an ignimbritic flow with a thickness ranging from 60 to 120 m, which is well exposed in the deeply eroded valleys of the Mtkvari (Fig. 2) and Uraveli rivers.

The Mtkvari Ignimbritic flow is overlain by a third major unit, consisting of 600–700 m of coarse-grained lithic breccia of two-pyroxene andesitic, pyroxene-hornblende andesitic-dacitic and hornblende-biotite dacitic composition. In the upper part of the stratigraphic section, subordinate andesitic and andesitic-dacitic lava flows are observed, along with thin ignimbrite layers (1–3 m thick) and volcanic ash beds of andesitic composition, also ranging from 1 to 3 m in thickness. In this part of the Goderdzi Formation, volcanic breccias are notably abundant and characterized by large clast sizes. In specific locations, such as near the village of Chachkari and the Vahanis Caves, individual fragments reach dimensions of 1.2–1.5 m.

The Goderdzi volcanic formation exhibits several characteristics indicative of caldera-forming eruptions, including: large volumes of eruptive material, rhythmic alternation of different volcanic rock types, substantial lithic breccias (sometimes exceeding 1 m in size), thick volcanic ash layers (up to 3 m), and widespread pyroclastic flow deposits. We propose that one such eruption was responsible for the formation of the Mtkvari ignimbrites (Okrostsvaridze, Popkhadze 2016).

2.1.1 Mtkvari Ignimbrites Flow

The Upper Miocene Mtkvari Ignimbrite flow extends continuously for approximately 35 km along the left bank of the Mtkvari River, from the Niala caldera to the Khertvisi

Fig. 2 General view of the Goderdzi volcanic formation along the Mtkvari River canyon.White-colored rocks ignimbrites flow is the rock-cut city of Vardzia. Photo © Giorgi Boichenko

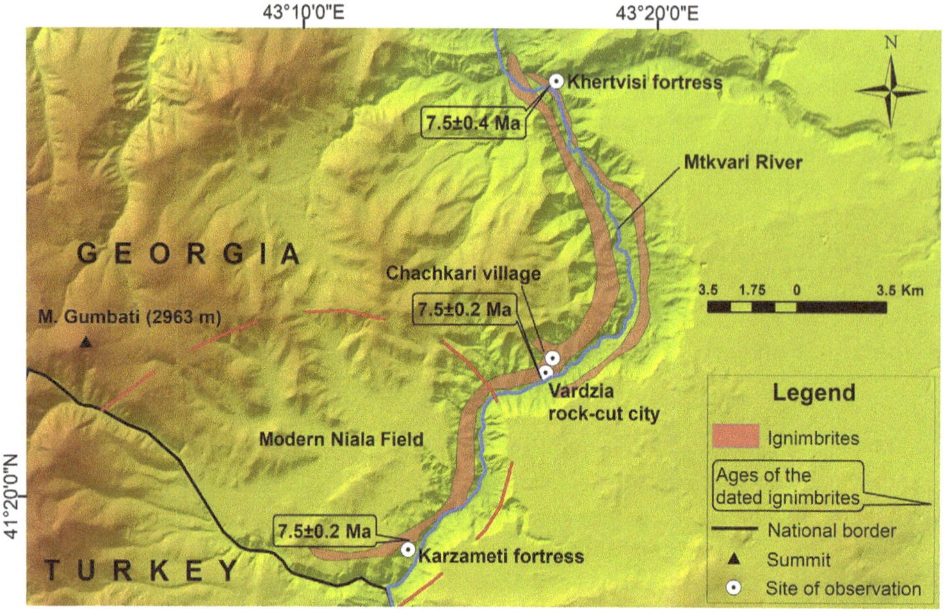

Fig. 3 Exposed part of the Mtkvari Ignimbrite flow with zircon U–Pb ages (Okrostsvaridze et al. 2020)

Fig. 4 Mtkvari Ignimbrite flow in the Vardzia area. The break-off zone of the Vardzia Block is visible on the left; Vardzia rock-cut city is situated on the right. Photo © Giorgi Kirkitadze

Fortress (Fig. 3). The flow moved from south to north with a slope ranging between 2° and 3°, and its thickness gradually increased in the direction of flow. For instance, near the Karzameti Fortress, the ignimbrite thickness reaches up to 60 m, increasing to 100 m at the Vardzia rock-cut city (Fig. 4), and attaining a maximum of 120 m near the village of Saro. This pattern of increasing thickness from the volcanic center outward is characteristic of ignimbrite flows, whereas the thickness of ash-fall tuffs typically decreases with distance from the source (Branney, Kokelaar 2002).

The base of the Mtkvari ignimbrites conforms to the underlying topography, while the upper boundary forms an ideally planar surface, a characteristic feature of flow structures. Additionally, the internal structure of the ignimbrites exhibits features indicative of turbulent flow, providing further evidence that the Mtkvari ignimbrites represent a pyroclastic flow deposit.

It is noteworthy that along the entire length of the flow surface, a 30–50 cm thick, fine-grained, white-colored layer is developed, the origin of which remains a subject of debate. According to one interpretation, it represents a post-eruptive volcanic ash deposit; however, it is also possible that it formed as a result of ignimbrite cooling and degassing.

The width of the Mtkvari Ignimbrite flow also remains a subject of debate. Some researchers suggest that the flow followed the course of the present-day Mtkvari River valley. However, field observations indicate that the current canyon morphology of the Mtkvari developed at a later stage, whereas the ignimbrite flow was significantly broader and extended across a depression on the northern flank of the volcanic structure. Supporting evidence includes extensive ignimbrite exposures in the valleys of the Mtkvari's left tributaries, located at higher hypsometric levels—for instance, in the vicinity of the village of Chachkari.

The structural characteristics of the ignimbrites exposed within the Mtkvari Canyon offer valuable information regarding the tectonic history of the region. Notably, a thick ignimbrite flow is exposed exclusively along the left slope of the canyon, while no corresponding deposit is observed on the right slope. Instead, only thin ignimbrite layers belonging to the uppermost (third) unit of the Goderdzi Volcanic Formation are exposed on the right flank. This asymmetrical distribution has long presented a geological

Fig. 5 Geometric correlation of ignimbrite flow outcrops on the right and left banks of the Mtkvari River, near the Vardzia rock-cut city. Photo © Giorgi Boichenko

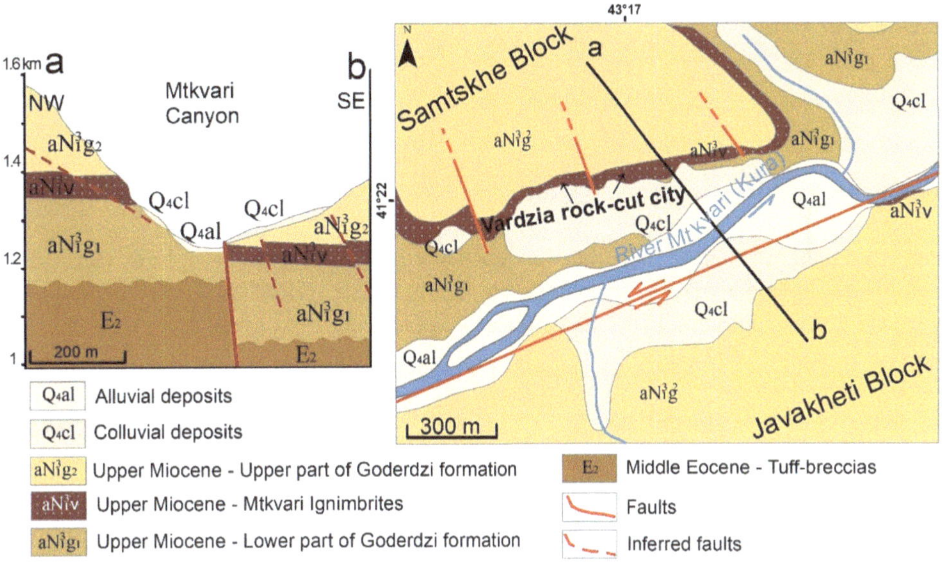

Fig. 6 Geological map of the Goderdzi Volcanic Formation in the vicinity of the Vardzia rock-cut city, with accompanying geological cross-section along the (**a**), (**b**) line

conundrum, for which a definitive explanation had remained elusive. However, recent field investigations have revealed an analogous ignimbrite deposit on the right bank of the Mtkvari River, situated directly adjacent to the present-day riverbed (Fig. 5) (Okrostsvaridze et al. 2020).

Analysis of the geological map and cross-section of the Vardzia area clearly reveals that the Mtkvari ignimbrite flow is distributed at different hypsometric levels on the right and left banks of th Mtkvari River (Fig. 6), with a vertical offset of approximately 127 m. This hypsometric asymmetry indicates the presence of a regionally significant fault, which follows the course of the Mtkvari River and marks the boundary between the Javakheti Block to the east and the Samtskhe Block to the west (Okrostsvaridze et al. 2016).

The internal structure of the Mtkvari Ignimbrite is of particular importance for the construction of rock-cut structures within it. The ignimbrite is primarily composed of volcanic ash and fine breccia fragments (0.5–3 cm), with occasional larger clasts reaching 5–15 cm in size. The material is unsorted, likely a result of the turbulent emplacement of the ignimbrite flow. The lithic clasts are predominantly composed of hypersthene–hornblende andesite–dacite, with rare biotite inclusions. The porphyroclastic component accounts for approximately 40–50% of the total volume and consists mainly of broken or euhedral plagioclase crystals (0.5–1 mm). Smaller amounts of hornblende, biotite, hypersthene, and augite are also present. Accessory minerals, including magnetite, apatite, and zircon, occur as isolated grains. The cementing matrix is composed chiefly of fine grained volcanic ash with abundant glass shards (Ustiev and Jigauri 1971).

The Mtkvari pyroclastic flow consists petrographically of andesitic–dacitic ignimbrites that range from welded to weakly welded and non-welded varieties, with SiO_2 content varying between 58 and 66%. The color of the ignimbrites correlates with the degree of welding. In partic-

Fig. 7 Ruins of a rock-cut dwelling carved into Upper Miocene ignimbrites on the right bank of the Mtkvari River. The historical background of the structure remains unknown. Photo © A. Okrostsavridze

ular, the southern segments near the Karzameti Fortress are highly welded and exhibit a pink coloration. Progressing northward, the ignimbrites become weakly welded and display a whitish-pink hue. Further to the north, near the Khertvisi Fortress, the ignimbrites are white and non-welded (Okrostsvaridze et al. 2020).

Isotopic parameters of the Mtkvari Ignimbrite (εNd values ranging from +3 to +4 and $^{87}Sr/^{86}Sr$ ratios between 0.70341 and 0.70450) indicate that the flow originated through fractional crystallization of mantle-derived melts (Lin et al. 2020). Zircon crystals from the ignimbrite were dated using the U–Pb method with LA-ICP-MS instrumentation. Samples were collected from three key locations: the proximal section near the Karzameti Fortress, the medial section near the Vardzia rock-cut city, and the distal section near the Khertvisi Fortress. The obtained ages were consistent across all sites, yielding a weighted mean age of 7.52 ± 0.21 Ma (Okrostsvaridze et al. 2020).

Due to their petrographic characteristics described above, the Mtkvari Ignimbrites are relatively soft rocks, which makes them easy to work and suitable for carving various types of structures. Moreover, the structural homogeneity of the ignimbrites allows such work to be carried out at virtually any level within the flow.

For this reason, along the 35 km extent of the Mtkvari Ignimbrite flow, residential, religious, and defensive structures were carved into the rock during the early Middle Ages at nearly every suitable location (Fig. 7). Among the civil constructions, the most remarkable is the rock-cut city of Vardzia, while the Vanis (Vahani) Qvabebi monastery complex stands out among the religious edifices. Notably, both complexes also served defensive purposes, as the ignimbrites are exposed in steep cliff formations along the Mtkvari River Canyon.

3 Vardzia Rock-Cut City

The rock-cut city of Vardzia, carved into the Upper Miocene ignimbrites of the Mtkvari Valley, is arguably one of the most significant monuments of Georgia's cultural heritage. Its construction took place between 1156 and 1202, during the reigns of King George III (1156–1184) and Queen Tamar (1184–1213) (Metreveli 1979). The fortress-city included a water supply system, secret tunnels, residential quarters, pharmacies, libraries, wine cellars, council chambers, and various other multifunctional structures (Fig. 8). The city is a terraced structure, and currently, 19 levels are preserved, containing over 630 rooms. Among these spaces are 28 wine cellars, 12 pharmacies, 15 churches, 6 halls and 22 warehouse facilities (Fig. 9). In addition to these, the complex includes nearly 640 other rock-cut chambers of various functions (Gafrindashvili 1976).

From the late twelfth century onward, Vardzia functioned for more than three centuries as a fortress-city located along a major trade route. Throughout this period, it played a significant role in the defense of Georgia's southern borders and in maintaining the territorial integrity of the kingdom.

In addition to its political and economic significance, Vardzia also served as one of Georgia's major educational and cultural centers. Although it may not have featured lavishly decorated halls and chambers, the wall paintings

Fig. 8 General view of the Vardzia rock-cut city. Photo © Guram Kharshiladze

Fig. 9 View of the central part of Vardzia rock-cut city. On the right—the entrance to the Church of the Assumption of the Virgin Mary. Photo © Guram Kharshiladze

Fig. 10 Fresco of King George III and Queen Tamar on the wall of the Church of the Assumption of the Virgin Mary in Vardzia. Photo © A. Okrostsvaridze

within its chapels represent an invaluable treasure of Georgian cultural and historical heritage. Of particular importance is the mural cycle of the Church of the Assumption of the Virgin, completed between 1184 and 1186. In addition to scenes depicting the life of Christ, the frescoes include portraits of King George III and Queen Tamar (Fig. 10).

In 1551, the Vardzia complex was raided by the Persian army, and following the conquest of the region by the Ottoman Empire in 1578, it was ultimately destroyed. After this period, the fortress-city ceased to function and, for nearly three centuries, was used by Turkish shepherds as shelter for their flocks. Following the restoration of Georgia's independence at the end of the twentieth century, efforts to rehabilitate Vardzia gradually began. In 2007, the Vardzia complex, together with the Khertvisi medieval fortress, was designated by UNESCO as a World Heritage Site.

3.1 Living Water: Ancient Hydraulics of Vardzia

For a rock-cut city in a semi-arid canyon, water was as critical as walls or weapons. Vardzia was supplied through a 3.5 km gravity system cut into volcanic rocks, bringing spring water from the highlands into the heart of the complex (Gafrindashvili 1976) (Fig. 11). At Vardzia the

Fig. 11 Fragments of the water supply tunnel of the Vardzia rock-cut city. Photo © A. Okrostsvaridze

solution was not a surface aqueduct visible to enemies, but a largely hidden hydraulic system cut directly into volcanic tuffs and ignimbrites. Springs emerging high on the Erusheti Ridge above and behind the cliff were captured and directed into the heart of the complex, turning an exposed canyon wall into a self-reliant mountain monastery.

Historical sources and field observations indicate that Vardzia's main supply arrived via a gravity channel approximately 3.5 km in length, of which about one third was driven as a tunnel through volcanic tuffs and ignimbrites. The line originated near the village of Zeda Vardzia and is estimated to have delivered on the order of 150,000–170,000 L of water per day to the fortress-city. Within the complex itself, the channel entered a finely bedded volcanic unit above the main ignimbrite, where it could be cut as a relatively stable, gently sloping gallery. From there water was dropped into cisterns and small distribution channels that fed cells, refectories, bakeries, and other key spaces. A perennial spring within the cliff still flows today, a reminder that this hydraulic intelligence was grounded in the rock's own fracture and permeability pattern.

The system did more than provide drinking water. Lateral channels and rock-cut pipes directed overflow towards terraces and small gardens below the main galleries, allowing the cultivation of vines, fruit trees, and vegetables on narrow ledges otherwise too dry and exposed for agriculture. At the same time, discreet drains and weep holes bled excess moisture back to the cliff face, helping to regulate humidity in the interior. This buffering effect was important for the preservation of frescoes and plaster, and likely reduced dust, making conditions more favourable for monastic life and liturgy. The 1283 earthquake that removed a large part of Vardzia's façade exposed a section of the water tunnel on the southern margin of the complex, revealing a passage nearly two metres high that may have served both as aqueduct and as a concealed evacuation route.

In engineering terms, Vardzia's waterworks form part of a wider Eurasian tradition of rock-cut hydraulics, comparable in principle to the covered channels and pressure-breaking devices of Petra in Jordan. Here, however, the system is adapted to the specific properties of the Mtkvari Ignimbrite and its overlying ash and breccia layers, exploiting their workability and structural behaviour. As such, the ancient water infrastructure is itself a key element of Vardzia's geoheritage: a demonstration of how geological structure, hydrology, and human ingenuity were combined to make a vertical volcanic cliff both habitable and resilient.

4 Vanis-Kvabebi Rock-Cut Monastery Complex

As previously described in the geological overview, only the upper part of the Goderdzi Volcanic Complex is exposed on the right slope of the Mtkvari River, comprising thin ignimbrite layers ranging from 2 to 3.5 m in thickness. Despite the limited rock mass, numerous ritual and residential structures were carved into these ignimbrite layers during the Early Middle Ages. The most prominent among them is the Vanis-Kvabebi rock-cut monastery complex (Fig. 12), located approximately 3 km northeast of the Vardzia rock-cut city This site stands as one of the most significant monuments of Georgia's cultural heritage. It was founded in the eighth century AD and served as a center of educational and scholarly activity for nearly eight centuries. It is highly probable that the existence of this monastic center later influenced the establishment of the rock-cut city of Vardzia.

At present, approximately 200 rock-cut structures of various functions are preserved in the Vanis-Kvabebi complex, carved into thin ignimbrite layers and volcanic tuffs. These include monastic cells, storage facilities, shelters, tunnels, three water supply channels, and six churches (Fig. 13). All of these structures are arranged across 16 vertical levels on steep, inaccessible cliffs composed of pyroclastic rocks, volcanic ash, and ignimbrites.

Despite numerous challenges, this monastic complex remained active from the eighth century until the mid-sixteenth century AD. Like Vardzia, the Vanis-Kvabebi Monastery Complex was raided by the Persian army in 1551 and ultimately destroyed by the Ottomans in 1578. Since

Fig. 12 General view of the Vanis-Kvabebi Rock-Cut Monastery Complex. Photo © A. Okrostsvaridze

Fig. 13 View of the central segment of the Vanis-Kvabebi Monastic Complex. At the center is the Church of St. George. Photo © A. Okrostsvaridze

the late twentieth century, monastic life has gradually been restored at the site.

5 Geological Hazards of Ignimbrites Dwellings

In the Mtkvari Ignimbrite flow, the inhabitants faced not only invasions by foreign powers but also the impacts of active geological processes. As these rocks are composed of weakly welded ignimbrites, they are highly susceptible to erosion and weathering, resulting in the gradual degradation of the rock-cut structures (e.g., Spizzichino et al. 2017). In addition to these destructive processes, the Vardzia area is traversed by a regional fault zone (see Fig. 5), making the region prone to earthquakes of varying magnitudes.

According to historical sources, the first major earthquake in the region occurred on April 1, 1089. This event caused the collapse of the central part of the Vanis-Kvabebi complex, including the Church of St. George. Nevertheless, the monastery was subsequently restored and resumed its regular functions. The second major earthquake in the region, which occurred in 1283 and also damaged part of Vardzia, severely affected the Vanis-Kvabebi Monastery Complex as well. Despite this, it was once again restored and continued to operate.

At present, the Vanis-Kvabebi Monastic Complex faces several geological challenges. The first involves the influx of water during periods of heavy rainfall, which leads to the washing away of the frontal sections of the complex. This phenomenon is particularly damaging along a major fracture in the central part of the site, where large volumes of water accumulate in the courtyard. The second geological issue is the sudden detachment and fall of large clasts (volcanic bombs) from the overhanging cornices composed of pyroclastic rocks. This poses a serious hazard not only to residents but also to visitors. As a result, access to the complex has been restricted since 2022.

The Vardzia rock-cut city is currently exposed to even greater geological hazards than the Vanis-Kvabebi complex (Spizzichino et al. 2017). As in Vanis-Kvabebi, the ignimbrites at Vardzia are subject to intense weathering and erosion, which progressively reduces the structural stability of the carved architecture. In addition to these processes, the 900-m-long tectonic block into which the city is carved has detached from the surrounding bedrock and is moving downslope under gravitational influence toward the Mtkvari River valley. At present, this block has subsided by 30.6 m relative to the adjacent rock mass and continues

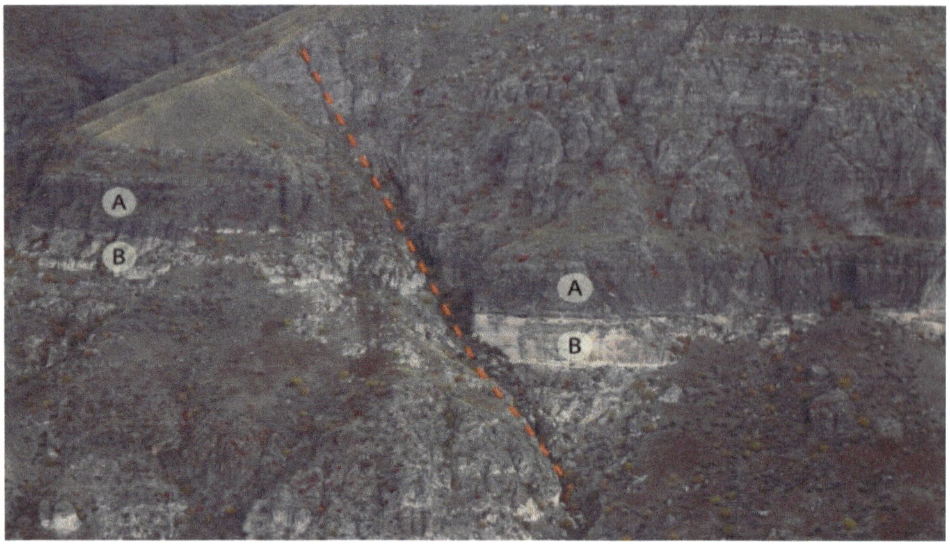

Fig. 14 View of the southern boundary of the Vardzia tectonic block. **A**—Pyroclastic rocks; **B**—Mtkvari ignimbrite flow (Okrostsvaridze et al. 2020)

to exhibit active displacement (Fig. 13). Additionally, the Vardzia Block is subdivided into several smaller blocks, which further weakens its structural stability (Fig. 14).

In 1283, a major earthquake struck the region, severely damaging the Vanis-Kvabebi Monastic Complex and collapsing nearly one-third of Vardzia's façade. Interestingly, this collapse exposed a segment of the water supply tunnel on the southern edge of the Vardzia complex, with a preserved height of nearly 2 m (see Fig. 11). It appears that the tunnel may have served not only for water conveyance but also as a secret passage—wide enough to allow the passage of a horse. Notably, the collapse surface closely follows the alignment of the tunnel, suggesting that water infiltration may have weakened the underlying rock mass, ultimately triggering failure along this plane.

Despite this major natural disaster, both the rock-cut city of Vardzia and the Vanis Monastic Complex were restored by the end of the thirteenth century and continued functioning even under Mongol rule. However, in 1551, both sites were raided and looted by Persian forces, and in 1578, they were ultimately destroyed by the armies of the Ottoman Empire. Today, both sites are protected by the state and are considered one of the most popular tourist destinations in Georgia.

6 Conclusion

Thus, the globally significant cultural heritage sites, the 12th-century rock-cut city of Vardzia and the early medieval Vanis-Kvabebi Monastic Complex, are carved into multiple layers of ignimbrites. These volcanic rocks were deposited between approximately 7.5 and 6.8 million years ago as a result of large-scale eruptions from the Keltepe–Gumbati caldera system. In light of various geological factors, these sites are not only of outstanding cultural value but also represent geological heritage sites of global importance and considerable scientific interest.

It is evident that the builders of these structures possessed advanced engineering knowledge—ignimbrites are not only easily workable rocks but also provide favorable conditions for habitation. Moreover, the locations of the structures were selected with a comprehensive consideration of geological, geographical, and architectural factors. Undoubtedly, these remarkable cultural heritage monuments could not have been created without the presence of thick and extensive ignimbrite layers. Today, these structures serve as outstanding examples of how geological heritage sites of significant scientific value can be transformed into cultural heritage sites of global importance.

Carved into ignimbrites, both of these monumental structures served the Georgian state for centuries, fulfilling defensive, educational, and religious functions, and are currently under state protection. However, due to the relatively soft nature of ignimbrites, these rocks are highly vulnerable to weathering and erosion, which leads to the gradual degradation of the carved architecture. The situation is further aggravated by the fact that the region lies within a tectonically active zone, posing a persistent seismic hazard to both the Vardzia rock-cut city and the Vanis-Kvabebi Monastic Complex. Therefore, considering all the aforementioned factors, it is essential to implement continuous engineering and geotechnical monitoring, along with appropriate protective interventions, to mitigate destructive processes and ensure the long-term preservation of these unique heritage sites. In doing so, Georgia can continue to present the Vardzia–Vanis-Kvabebi ignimbrite corridor as a flagship example of how geological and cultural heritage together support education, identity, and sustainable geotourism.

References

Gamkrelidze I, Okrostsvaridze A, Maisadze F, Basheleisvili L, Boichenko G, Skhirtladze I (2019) Main features of geological structure and geotourism potential of georgia, the caucasus. J Modern Environ Sci Eng 5(5):422–442

Gamkrelidze I, Okrostsvaridze A, Koiava K, Maisadze F (2020) Geotourism potential of Georgia. The Caucasus. Springer International Publishing, pp 1–140. https://doi.org/10.1007/978-3-030-629 66-3

Gamkrelidze I, Okrostsvaridze A, Tormey ID, Boichenko G, Gogoladze S, Gabrielashvili (2024) Geoheritage of georgia (The Caucasus) and potential for aspiring UNESCO Geoparks. Abstract #360, 37th International Geological Congress, Busan, Republic of Korea

Gaprindashvili G (1976) Vardzia. Publishing House "Avrora", Sankt-petersburg, pp 1–215 (in Russian)

Gudjabidze GE, Gamkrelidze IP, Abesadze GN, Buadze VI, Djanelidze TV, Dudauri OZ, Kandelaki DN, Nadareshvili GS, Papava DI, Pruidze MP, Topchishvili MV (2003) Scale 1:500,000 Geological Map of Georgia. Tbilisi

Keskin M, Hanilçi N, Yılmaz I, Okrostsvaridze A, Şişman F, Koral H (2024) New findings regarding the volcano-stratigraphy and volcanic activity history of the Erzurum-Kars (NE Anatolia, Türkiye) and Samtskhe-Javakheti (SW Georgia) Plateaus. Abstact of the 76th Geological Congress of Türkiye with International Participation

*Le Pennec J*A, Temel LA, Froger J, Sen S, Gourgaud A, JL, Bourdier J, (2005) Stratigraphy and age of the Cappadocia ignimbrites, Turkey: reconciling field constraints with paleontologic, radiochronologic, geochemical and paleomagnetic data. J Volcanol Geotherm Res. 141(1):45–64. https://doi.org/10.1016/j.jvolgeores.2004.09.004

Lin YC, Chung SL, Bingöl AF, Okrostsvaridze A, Yang L, Lin TH, Lee HY (2020) Diachronous initiation of post-collisional magmatism in the Arabia-Eurasia Collision Zone, Litoth 356–357. https://doi.org/10.1016/j.lithos.2020.105394

Makadze M, Okrostsvaridze A, Keskin M, Pophkhadze N (2024) Structure and composition of the Late Miocene Goderdzi Volcanic Complex, Lesser Caucasus, Georgia: implication for resurgent caldera existence. Abstract #1511 of the 37th International Geological Congress, Busan, Republic of Korea

Metreveli RV (1979) The History of Georgia: a dictionary. Publishing House "Ganatleba", Tbilisi, pp 1–351. (in Georgian)

Okrostsvaridze A, Elashvili M, Popkhadze N, Kirkitadze G (2016) New data on the geological structure of the Vardzia cave city. Georgia Bull Georgian Natl Acad Sci 10(3):98–105

Okrostsvaridze A (2017) Late cenozoic samtskhe-javakheti volcanic highland, georgia: the result of mantle plumes activity. Abstarct of 19th EGU General Assembly

Okrostsvaridze A, Chung SL, Chang YH, Skhirtladze I (2020) Geology and Zircon U-Pb geochronology of the mtkvari pyroclastic flow and evaluation of destructive processes affecting vardzia rock-cut city. Quaternary Geology 540:137–145. https://doi.org/10.1016/j.quaint.2019.03.026

Okrostsvaridze A, Gamkrelidze I, Yuan Hsi Lee YH, Keskin M, Tormey D, Boichenko D, Gogoladze S, Makadze M (2024) U-Pb Geochronology and geochemistry of volcanic ash layers in the eastern caucasus intermountain neogene basin: implications for their sources. Arab J Geosci 17:129. https://doi.org/10.1007/s12517-024-11914-7

Skhirtladze NI (1958) Postpaleogene effusive volcanism in georgia. Publishing house of the georgian academy of sciences, pp1–159 (in Russian)

Spizzichino D, Boldini D, Frodella W, Elashvili M, Margottini C (2017) Landslide risk analysis and mitigation for the ancient rock-cut city of Vardzia (Georgia). In: Proceedings of 2017 IPL symposium, UNESCO, Paris, pp 1–8

Topal T, Doyuran V (1997) Engineering geological properties and durability assessment of the Cappadocian tuff. Eng Geol 47(1–2):175–187

Tormey D, Okrostsvaridze A, Makadze M, Boichenko G, Gogoladze S, Gabrieloashvili R (2026) Rock-cut city of Vardzia and its surroundings (Georgia): a potential UNESCO Global Geopark in the Lesser Caucasus. Geoheritage 18:156. https://doi.org/10.1007/s12371-026-01366-8

Tutberidze B (2004) Geology and Petrology of Alpine late orogenic magmatism of the central part of Caucasian Segment. Tbilisi University Publisher, Tbilisi, pp 1–362 (in Russian)

Ustiev EK, Jigauri DI (1971) The welded tuffs from Vardzia formation (southern Georgia). Izvestia AN SSSR 4:3–16 (in Russian)

The Majestic Summits of the Greater Caucasus: Mountain Geoheritage of Georgia

Avtandil Okrostsvaridze, Lasha Sukhishvili, and Rabi Gabrielashvili

Abstract

Georgia encompasses extensive parts of the Greater and Lesser Caucasus ranges, making the country remarkably mountainous. The summits of these mountainous are not only a defining feature of the landscape but also a source of natural wealth, an inseparable part of daily life, and consequently a constant sense of national pride. The Lesser Caucasus has few high summits in Georgian territory, with only three exceeding 3,000 m and at the same time all of which represent volcanic cones. In contrast, the Greater Caucasus is abundant in towering summits. Georgia is home to six summits exceeding 5,000 m, 34 summits rising above 4,000 m, and an untold number surpassing 3,000 m—so numerous that their full identification has yet to be completed. Beyond their sheer number, these summits captivate with their unique allure, with Ushba ranked among the world's ten most beautiful summits. As in many cultures worldwide, Georgia's most inaccessible summits inspire deep reverence, often regarded as the realm of the gods. In this context, the ancient Greek legend is particularly noteworthy—the tale of the gods chaining Prometheus to one of the Caucasus' unreachable peaks as punishment. Today, a wealth of mythological, geographical, and tourist information surrounds these summits, yet geological data remains remarkably scarce. This chapter provides a brief overview almost all of Georgia's summits above 3000 m. However, the geological structure and origin history of outstanding summits, which represent a truly significant geological heritage of Georgia, are discussed in detail. We consider such summits: Shkhara, Kazbegi/Mkinvartsveri, Gistola, Tetlundi, Ushba, Shkhelda, Ailama, Burchula, Laila and Chaukhi. They have played a significant role in shaping the identity and consciousness of the Georgian people and still have a influence in their spiritual life.

Keywords

The greater caucasus · Georgia · Mountain geoheritage · Shkhara · Tetlundi · Ushba

1 Introduction

Majestic summits, covered with snow and glaciers, are probably the most unexplored places on Earth, which is why they continuously inspire admiration and curiosity in people. Their conquerors are portrayed in society as heroes who demonstrate the maximum of human potential.

In the collective consciousness of all ethnic groups within our civilization, peaks and mountains of exceptional beauty occupy a special place, often regarded as the realm of the gods or as sacred mountains. or instance, in Greek mythology, the gods reside on Mount Olympus, while for the Japanese, the young volcano Fuji is considered a sacred mountain. A distinct topic of discussion is the deep respect and reverence the Tibetan people hold for the mountains and peaks that surround them. The Caucasus region is no exception in this regard, with its impassable mountains also regarded as the realm of the gods. For example, according to one version of the myth of Prometheus, the gods chained him to the peak of Mkinvartveri, which is one of the most prominent peaks of the Greater Caucasus. Undoubtedly, due to a multitude of factors, mountains will continue to be a subject of unwavering admiration, intrigue, and research for humanity.

According to data from the International Mountaineering and Climbing Federation in 2023 (UIAA Report 2024), there are only 14 peaks on Earth that exceed 8,000 m, all of which are located in the Himalayas and Karakoram Mountains. There are numerous peaks exceeding 7,000 m, also

A. Okrostsvaridze (✉) · L. Sukhishvili · R. Gabrielashvili
Institute of Earth Sciences, Ilia State University, Tbilisi, Georgia
e-mail: avtandil.okrostsvaridze@iliauni.edu.ge

A. Okrostsvaridze and D. Tormey (eds.), *Geoheritage of Georgia*, Geoheritage, Geoparks and Geotourism,
https://doi.org/10.1007/978-3-032-27639-1_9

found in the Himalayas and Karakoram, with others in the Pamirs, Hindu Kush, Tian Shan, and Annapurna ranges. Additionally, there are over 1,000 peaks above 6,000 m worldwide, and the number of peaks surpassing 5,000 m is even greater. All of these peaks were formed as a result of active magmatic or geotectonic (or both) processes in the Earth's crust, and therefore they represent an important geoheritage of our planet.

2 The Georgian Summits

The Caucasus Mountains were formed by the collision of remnants of the Gondwana megacontinent with the southern edge of the Eurasian continent. This process began in the Late Paleozoic, around 340–320 million years ago, and continues to the present day. Today, this mountain range stretches approximately 1,200 km in a SE-NW direction, from the Caspian Sea to the Black Sea. It consists of two nearly parallel mountain systems: the Greater Caucasus to the north and the Lesser Caucasus to the south. This mountain range features several peaks of magmatic or geotectonic origin, which we have briefly discussed in this chapter.

If we consider summits as geoheritage, then the small country of Georgia (69,700 km^2, nine times smaller than France) should be regarded as one of the richest in the world in this respect, as it is home to a number of summits of global significance. Among them, it is worth noting that Ushba, located in the Greater Caucasus, holds a place of honor among the 10 most beautiful peaks in the world. Within the Georgian Greater Caucasus, at least 34 peaks exceed 4,000 m in elevation, with the highest concentration in Upper Svaneti. Their geological foundations range from Paleozoic crystalline massifs to younger volcanic centres such as Kazbegi, and together they provide the high-altitude core of Georgia's mountain geoheritage (Fig. 1).

The Lesser Caucasus runs south of the Greater Caucasus at a distance of approximately 100–200 km and represents a mountain system of complex structure. In contrast to the Greater Caucasus, it has fewer peaks, which are relatively lower in elevation, and all of them are of volcanic origin. Within the territory of Georgia in the Javakheti volcanic plateau of the Lesser Caucasus, there are three stunning cone volcanoes of Holocene age, each exceeding 3,000 m in height: Didi Abuli (3301 m esl), Samsari (3285 m esl) and Emlikli (3054 m esl). In addition to the above, the Erusheti Highland hosts the Mount Gumbati (2,996 m), which dates to the Upper Miocene and represents a post-caldera andesitic dome.

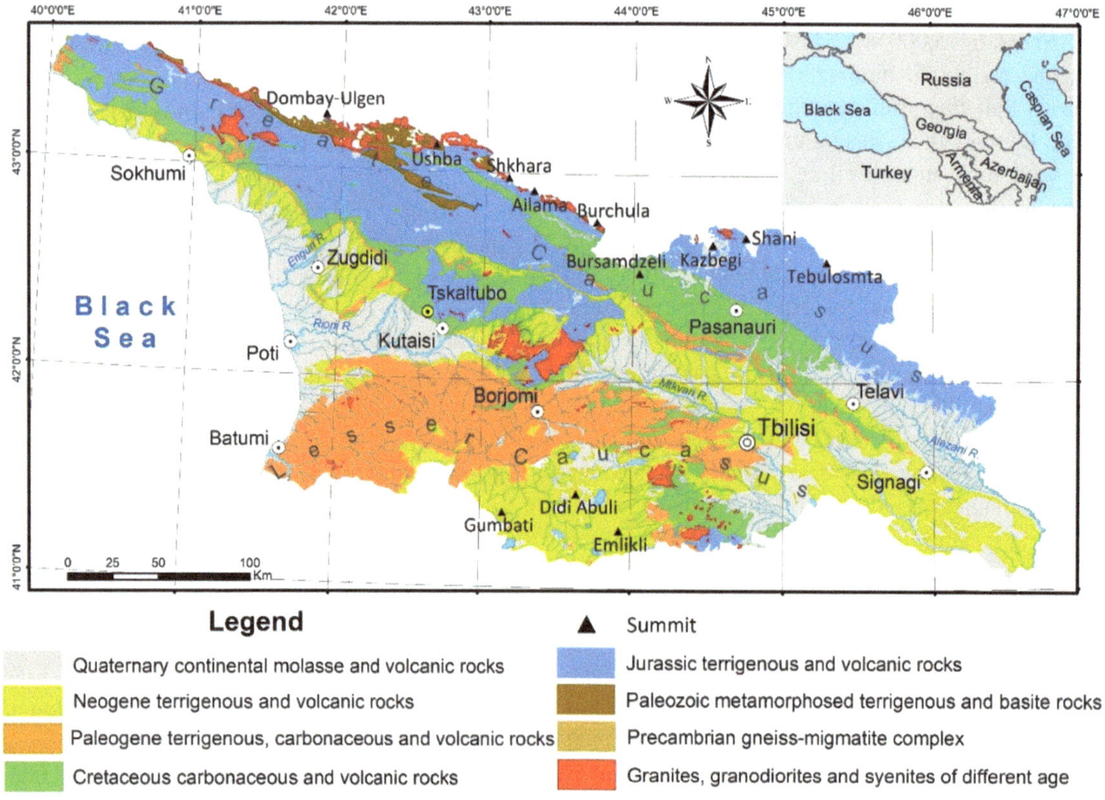

Fig. 1 Georgian significant summits on the simplified geological map of Georgia. The geological framework is modified after Gudjabidze et al. (2003)

There is no doubt that the summits represent the wealth and pride of Georgia, which is why there is an abundance of geographical and tourist information about them. However, data on the geological history of their formation remains quite scarce. Therefore, the main goal of this chapter is to provide a geological characterization of outstanding summits of the Greater Caucasus as an important geological landmark. Since each of them has a unique and fascinating geological history of formation and at the same time they are built of different types and ages of rocks.

3 The Summits of the Georgian Greater Caucasus

The Greater Caucasus and especially its the Main Range is particularly rich in summits, which rapidly took on its current form over the last 6–5 Ma (e.g. Wang et al. 2015). This range consists of two main age and structural components: ancient Paleozoic crystalline basement rocks (gneisses, migmatites, granitoids) and a younger Meso-Cenozoic sedimentary cover (shales, sandstones, limestones). During the Alpine collisional processes, the crystalline basement was fractured into more than 10 tectonic blocks, which are represented in the modern relief as crystalline massifs. These fractured solid blocks, as a result of Alpine contraction processes, gradually moved southward over the Meso-Cenozoic sedimentary cover, a process that continues to this day (e.g., Okrostsvaridze and Tormey 2011). As a result of this process, the Earth's crust thickened in the Greater Caucasus, and most of the peaks now represent the highest points of the crystalline basement massifs. However, this does not apply to peaks of young volcanic origin, such as Kazbegi in Georgia (5054 m esl) and Elbrus in Russian (5642 m esl), which were formed much later, within the last half-million years, on these massifs as a result of recent magmatic activity (Fig. 2).

From the east, the Greater Caucasus in Georgia begins with Kakheti's Cavcasioni, which does not have outstanding summits, but it is a very beautiful sight in the background of Alazani Valley (Fig. 3). We believe that this ridge represents the important geoheritage of Georgia.

The first summits exceeding 4,000 m in the eastern part of the Georgian Greater Caucasus are Diklosmta (4,285 m asl, 42°29′18″N, 45°46′31″E) and Tebulosmta (4452 m esl., 42°34′24″N, 45°18′43″E). They are composed of highly fractured Lower Jurassic shales and sandstones, ~90–170 Ma, intruded by Middle Jurassic diabase dykes dating to around 168–165 Ma.

To the west of these summits stands one of the most striking peaks of the Caucasus mountains, yang volcano, Mkinvartsveri/Kazbegi (5,054 m asl; 42°41′57″N, 44°31′06″E). Rising 700–1,000 m above the surrounding mountains, it dominates the landscape. The peak has been called Kazbegi since the late eighteenth century, but its historical Georgian name is Mkinvartsveri (Batonishvili 1745), meaning "Glacier Peak". Covered in white snow year-round, it is also known as the "Bride of the Greater

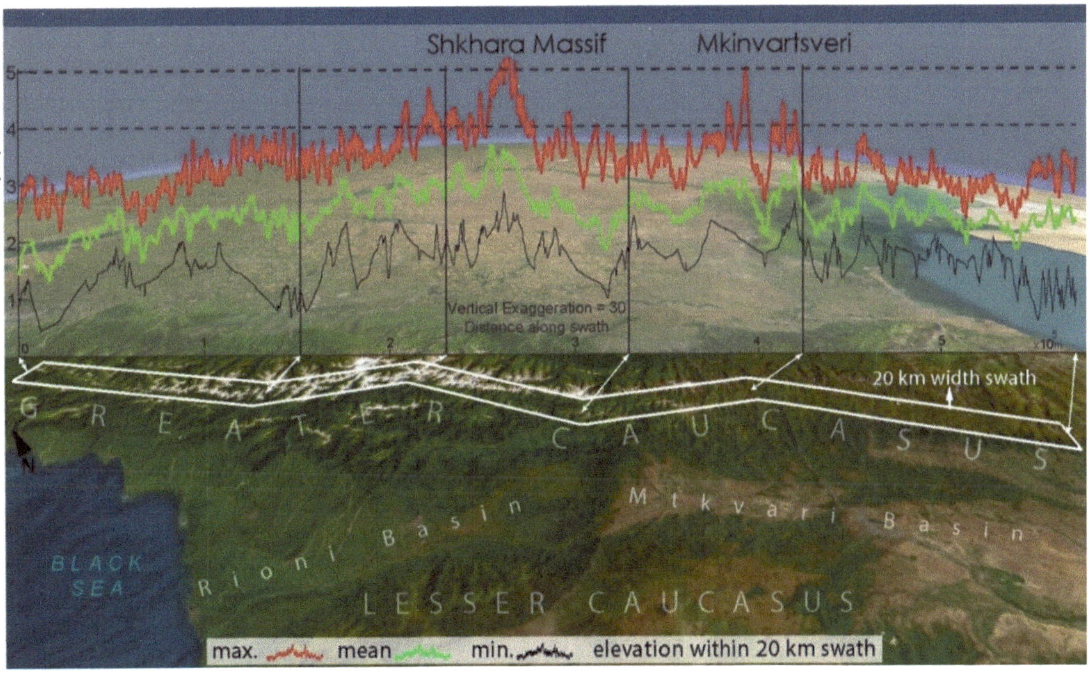

Fig. 2 The diagram-model of the hypsometric distribution of the Greater Caucasus Georgian summits for a 20-km width segment

Fig. 3 The greater caucasus of Kakheti (Kakheti's Cavcasioni) from the Gombori Ridge. Alazani valley in the middle. Photo © Georgian Travel Guide

Fig. 4 View of mount Kazbegi from the southeast. In the foreground on the left is the medieval gergeti trinity Church. Photo © Andrew Stanilard

Caucasus.“ The view of this peak is particularly striking, enhanced by a 14th-century orthodox church built a top one of the andesite lava flows to its south (Fig. 4).

Kazbegi is a Late Pleistocene stratovolcano of mainly andesitic composition. It began to form around 40 000 years ago, at the base of the Paleozoic Dariali massif, of which granodiorites have been dated to 314 ± 1.7 Ma ago using the zircon U–Pb method (Gamkrelidze et al. 2021). This basement is currently located at approximately 1,800–2,000 m above sea level, with the volcanic structure forming a massive dome about 3 km high on top of it. According to modern studies, the magmatic source of the Kazbegi volcano is likely associated with the northernmost activity of the Afar-Anatolian-Caucasian mantle asthenospheric flow (e.g., Hua et al. 2023).

The volcanic structure of Kazbegi is complex, with several distinct magmatic activity periods identified in its formation. According to the K–Ar dating (Chernishev et al. 1999) at first stage which occurred 455,000 ± 40 Ka ago predominantly andesitic lava with small amount of pyroclastic material was erupted. At the first stage of its action the Kazbegi volcano a small amount of dacitic pyroclastic material is also observed. According to this study, the Kazbegi volcano was most active between 235,000 and 185,000 years ago, during which its main structure formed. During this period, volcanic activity was primarily characterized by andesitic deposits, with dacite and basaltic flows occurring in limited quantities. During the third period (50,000 ± 20 ka), a limited amount of andesitic lava flows formed, accompanied by the eruption of pyroclastic material.

Wood remains from the youngest Goritsikhe andesite flow of the third period have been dated using the radiocarbon method, yielding an age of 6,000 years (Burchuladze et al. 1976). Zircons from the same flow were dated using the U–Pb method (LA-ICP-MS study), yielding an age of 32 ± 1.6 ka (Gamkrelidze et al. 2021). According to the Global Volcanism Program Report 2021, the last known eruption of the Kazbegi volcano occurred in 750 BC (Kazbek 2021).

Fig. 5 A general view of the Chaukhi massif, from the southwest, looking from the village of Juta. Photo © Giorgi Tsiklauri

Thus, based on geological data, isotopic dating, and historical sources, Kazbegi is a relatively young volcano, from which we should continue to expect surprises.

Southeast of the young volcano of Kazbegi lies the Middle Jurassic subvolcanic massif of Chaukhi, which forms the magnificent summit. It is rising over 3,000 m, is exposed in the axial part of the Eastern Greater Caucasus, between the villages of Roshka and Juta. This massif features seven peaks, reaching an elevation of 3,842 m. It is located only 100 km from Tbilisi and therefore represents a "paradise" for both local and international geotourists as well as novice mountaineers (Fig. 5).

Geologically, the Chaukhi massif represents a fragment of the Middle Jurassic Neotethys oceanic crust, currently located over 3,000 m above sea level. This massif contains all three main components of oceanic crust: gabbro-sheeted dykes, thick diabasic intrusions, and pillow lavas. Notably, most of the peaks in the Chaukhi massif are composed of thick diabasic intrusions, with the host rocks consisting of pillow lavas. The thickness of the diabasic intrusions ranges from 15 to 40 m, while the diameters of the pillow lavas vary from 25 to 80 cm. We carried out a chemical analysis of the Chaukhi massif's rocks at the MSLALAB laboratory in Canada using LA-ICP-MS (Okrostsvaridze et al., 2024). According to this study, the SiO_2 content of these rocks' ranges from 35.7% to 44, 7% Al_2O_3—18.42–22.745%, MgO - 6.6–11.18%, Na_2O—2.1–2.6% and KO_2—0.14–0.22%. These values correspond to the typical chemical composition of oceanic crust, providing solid grounds to classify the Chaukhi massif as a well-preserved fragment of a Middle Jurassic, Tethyan oceanic rift. Thus, we believe that the Chaukhi massif, with its charm and unique geological structure, represents a geological heritage of global importance, possessing both geotouristic and educational values.

West of Kazbegi mount region in the Rioni River headwaters, located the Zemo Racha region. This area falls under the Oni municipality, which borders the North Ossetian Republic of the Russian Federation to the north. Here, on the Main Ridge of the Greater Caucasus the pre-Mesozoic Adaykhokh crystalline massif is exposed. The length of the massif is about 40 km, the width is 7–8 km and the average height is 4 km above sea level. It is thrust over on the Lower Jurassic clay shales, which is why it is very uplifted. The Adaykhokh massif is composed of Lower Paleozoic biotite crystalline shales, gneisses, and micaschists, which are intruded by Upper Paleozoic biotite granitic plutons (Okrostsvaridze 2007).

Within Georgia, the Adaykhokhi Crystalline Massif contains several peaks that exceed 4,000 m in elevation but do not surpass 4,500 m above sea level. These peaks, listed from east to west, are: Chanchakhi (4462 m esl), Buba (4419 m esl), Burchula (4364 m esl) and Laboda (4313 m esl). Among them, the most outstanding in terms of geoheritage is Burchula Peak, of which there is an excellent historical photograph in the National Archives of Georgia (Fig. 6). It was taken in 1890 by the famous Russian traveler and photographer Dimitri Ermakov. As evident from this photograph, Burchula Peak was at that time surrounded by extensive glaciers, but unfortunately, only half of these glaciers remain today.

To the west, the Upper Racha region is bordered by Lower Svaneti, which includes the Tskhenistskali River basin. In this area, along the Main Ridge of the Greater Caucasus, four peaks rise above 4,000 m: Tetri Utshonbi

Fig. 6 View of Burchula pik from the southern perspective. Photograph by Dmitry Ermakov, 1890. The photo is stored in the National Archives of Georgia

(4,049 m asl), Shavi Utshonbi (4,113 m asl), Tsurungala (4,250 m asl), and Ailama (4,547 m asl). All of these peaks are bordered to the north by the Kabardino-Balkaria Republic of the Russian Federation.

Ailama Peak (42°57′26″N, 43°10′46″E) the tallest and most impressive among the listed peaks. It is situated directly along the Main Thrust of the Greater Caucasus and is composed of Late Paleozoic biotite granodiorite intrusive. A glacier on the southern slope of Ailama feeds the Khoruldashi River, which flows into the Tskhenistskali River. At the base of Mount Ailama, in the valley of the Koruldashi River, during the Soviet period there was a big mountaineering camp called "Ailama," where over 200 mountaineers trained simultaneously.

West of Lower Svaneti is the bordered Upper Svaneti region, which extends along the Main Range of the Greater Caucasus for approximately 125 km. As Upper Svaneti is home to some of Georgia's most significant summits, we have also included a geological map of this region here.

In the Upper Svaneti, at the headwaters of the Enguri River, the Shkhara Paleozoic crystalline massif is exposed. It forms the highest mountain in the Greater Caucasus (Fig. 8), with an average elevation of approximately 5,000 m asl and a length of 15 km. This massif, located in the south along the Main Thrust of the Greater Caucasus is thrusted onto Lower Jurassic clay-shales, that feature is distinctly visible in the relief. In the main fault zone, this massif forms a great wall of gneisses and granites, with an average height of 2200–2300 m. This wall is especially striking in spring when blanketed in snow. During this period, the Jurassic

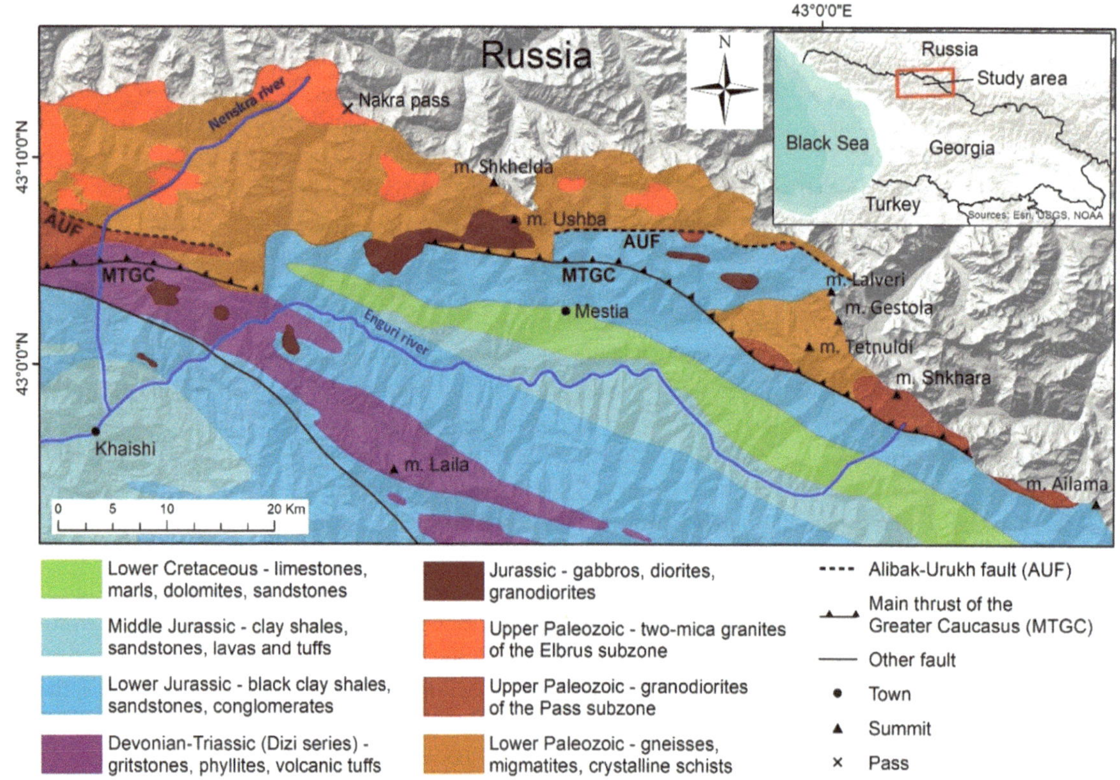

Fig. 7 Geological map of Upper Svaneti, showing the outstanding summits of the region. (Modified after Gudjabidze et al. 2003)

Fig. 8 View of the southern slope of the Shkhara Massif in spring, as seen from the village of Ushguli. In the foreground, the Svan early medieval tower stands on the left, while the Church of the Virgin Mary is on the right. Photo © Rudi Geipel

sedimentary rocks are green in color, and the crystalline massif is covered with white snow (Figs. 7 and 8).

The Shkhara Massif is composed of Lower Paleozoic crystalline shales, gneisses, and migmatites (gneiss-migmatite complex). These rocks were intruded by plutons of biotite granodiorites during the Upper Paleozoic, coinciding with the Variscan orogeny.

The petrology and isotopic ages of this massif, and all plutonic rocks of Upper Svaneti in general, have been studied in detail in the work of Okrostsvaridze et al. (2023). According to this study, the U–Pb isotopic age of zircons from the biotite gneisses of the Shkhara Massif is 488.5 $\pm$ 8.5 Ma, while the intruding biotite granodiorites date to 316.9 $\pm$ 8.8 Ma. Most of the peaks of the Shkhara Massif are composed of these biotite granodiorites, which, during the Variscan compression and folding, intruded the gneiss-migmatite complex and occupied the upper part of the structure. These processes took place at a depth of approximately 30–35 km within the Earth's crust. Their exhumation and exposure at the modern surface resulted from Late Alpine geodynamic processes over the past ~ 6–5 Ma (Wang et al. 2015).

The 15 km long Shkhara Massif is naturally divided into southern, central, and western sections, each containing multiple peaks. Interestingly, in the local Svan dialect, *Shkhara* means "nine" or "nine-peaked (headed)," aligning with the number of peaks in the massif. The highest peak in the central part of the Shkhara Massif is Main Shkhara (5,203 m; 42°59′58″ N, 43°06′42″ E), which is the third highest in the Greater Caucasus after Elbrus (5,642 m) and Dykh-Tau (5,205 m).

One of the most notable ascents of Shkhara massif was the traverse completed in 1940 by a group of Georgian alpinists, including Aleksandre and Alyosha Japaridze, Goji Zurabiani, and Gabriel and Beknu Khergiani. During this traverse, they conquered 20 peaks along the Ailama, Namkvani, and Shkhara ridges. The most recent winter ascent of Main Shkhara via the southern wall was achieved in 2018 by Georgian climbers Archil Badriashvili and Giorgi Tepnadze. The International Federation of Mountaineering and Climbing rated this ascent as the highest difficulty category, "6B" (Japaridze 2010).

West of the Shkhari massif, on the Georgia–Kabardino-Balkaria border, stands another prominent peak, Gestola (4860 m asl; 43°02′54″N, 43°01′28″E). It is composed of Lower Paleozoic crystalline schists, gneisses, and migmatites. Gestola is situated within the Alibek–Urukh regional fault zone, which is reflected in the landscape by its imposing northern wall (Fig. 9).

Located 3.3 km south of Mount Gestola is one of Georgia's most beautiful peaks, Tetnuldi (4852 m asl; 43°01′50″ N, 42°59′35″ E), which is permanently covered with snow. It is noteworthy that Tetnuldi, like Mount Ushba, lies entirely within the borders of Georgia and does not share a border with the Russian Federation. This peak dominates the landscape and overlooks the villages of the historical Mulakhi and Adishi communities (Fig. 10).

The summit of Tetnuldi represents a part of the Sgimazuki tectonic block, whose southern tectonic zona runs along the its southern slope, while the northern fault plane is of this tectonic block expressed on the northern slope of Gestola summit (see Fig. 10).

Fig. 10 Tetnuldi from the southeast, showing the tectonic displacement plane and tectonic wall clearly. Photo © Mykola Sakovych

Fig. 9 The summit of Gistola from the northern side, where the tectonic wall is clearly visible. Behind of Gestola, the peak of Tetnuldi is visible. Photo © Mikhail Golubev

The Sgimazuki massif, unlike the Shkhara massif, which is primarily composed of biotite gneisses and granitoids, is mainly represented by hornblende-plagiogneisses and their intersecting biotite plagiogranite intrusions. Zircon LA-ICP-MS U–Pb dating was conducted on both the plagiogneisses and plagiogranites, as well as the alaskitic veins of the Sgimazuki massif According to this study, the isotopic age of the plagiogneisses in the Sgimazuki Massif corresponds to 454.4 ± 9.3 Ma, the plagiogranites to 324.9 ± 3.5 Ma, and the alaskites to 309.2 ± 7.5 Ma (Okrostsvaridze et al. 2023).

From a mountaineering perspective, Tetnuldi is one of the most captivating peaks in the Greater Caucasus, distinguished by its beauty, moderately challenging routes, and convenient accessibility. The first recorded ascent was made in 1887 by a group of English alpinists led by Douglas Freshfield. In 1930, the first officially documented Georgian ascent was carried out by a team of Georgian alpinists under the leadership of the renowned female mountaineer Aleksandra Japaridze (Japaridze 2010). Today, the peak remains a popular destination for sport climbing, attracting both local and international alpinist groups.

23 km west of Mount Tetnuldi at the headwaters of the Dolra River, stands one of the most magnificent peaks of the Greater Caucasus Mount Ushba (43°07′35″N, 42°39′38″E). Ushba has two peaks: the southern peak (4,710 m asl) and the northern peak (4,690 m asl), separated by a ~200-m-deep saddle. Renowned for its striking beauty and grandeur, Ushba is often referred to as the "Crown of the Greater Caucasus. Notably, it is among the peaks that lie entirely within Georgian territory. Like Tetnuldi, the Ushba summit standing out from the surrounding structure and dominating the landscape (Fig. 11).

Compared to the Shkhara and Tetnuldi massifs, the Ushba Massif has a distinct geological composition. It represents a Middle Jurassic lens-shaped pluton, covering an area of approximately 35 km^2, which is emplaced along the Main Thrust of the Greater Caucasus. To the north, this pluton intersects with older Paleozoic crystalline schists, while to the south, it contacts Late Carboniferous, weakly metamorphosed sandstones and phyllites. The pluton is primarily composed of diorites, granodiorites, and syenites, and is cut by a system of plagiogranitic veins of varying thickness. The twin peaks of Ushba directly represents the frontal part of this pluton, which has intruded the host rocks to the highest elevation. According to zircon LA-ICP-MS

Fig. 11 View of the mount Ushba from SE, from Hatcvali area. Photo © A. Okrostsvatidze

Fig. 12 Magnificent mount Shkhelda, viewed from the south. Photo © Archil Badriashvili

U–Pb dating, the isotopic age of this pluton corresponds to 177 ± 1.8 Ma (Okrostsvaridze et al. 2023).

The first successful ascent of Ushba was made in 1888 by British climbers John Cockin and Ulrich Almer, who reached its northern peak, which presents fewer technical challenges. The first ascent of the southern summit was accomplished in 1903 by a joint German-Swiss expedition led by Adolf Schulze. In 1934, a team of Georgian climbers—siblings Alexandra and Alyosha Japaridze, along with Iagor Kazalikashvili and Gio Niguriani—became the first to reach the northern peak. Notably, Alexandra Japaridze was the first woman to set foot on Ushba. Many of the legendary Georgian climber Mikheil Khergiani's greatest achievements, earning him the title "Tiger of the Cliffs," are associated with Ushba. Among them was the first ascent of its 900-m, negatively inclined rock face (Japaridze 2010).

In about 3 km northwest of Ushba, on the Main Ridge of the Greater Caucasus along the Georgia-Russia border, lies another magnificent rocky massif—Shkhelda (43°08′54"N, 42°08′42"E) (Fig. 12). This massif consists of five peaks, the highest reaching 4,310 m asl, while the lowest stands at 4,129 m. Shkhelda is entirely composed of Lower Paleozoic crystalline schists, gneisses, and migmatites, which are intruded by small Upper Paleozoic biotite-bearing granitoid bodies. The massif is located within the Alibek-Urukh regional fault zone, causing it to stand out in the landscape and undergo fragmentation into small blocks. Tragically, in 2024, renowned Georgian mountaineer Archil Badriashvili, a recipient of the prestigious Piolet d'Or (Golden Ice Axe), lost his life while descending from Shkhelda, as a result of a lightning strike.

In the Upper Svaneti region, Svaneti Ridge extends for approximately 90 km, running almost parallel to the Main Ridge of the Greater Caucasus from the south. It is primarily composed of terrigenous and volcanogenic-terrigenous rocks, which were continuously deposited from the Devonian to the Triassic period (e.g., Somin, 2011). There is a magnificent peak on this ridge, Laila/Lahili (4,008 m esl) 42°55′14″N, 42°33′24″E), permanently covered with snow and ice. It undoubtedly represents an important geoheritage and that is why many local heroic legends are associated with it. Located just 20 km from the town of Mestia, peak Laila's striking beauty makes it a popular destination for both Georgian and international mountaineers and geotourists (Fig. 13).

4 Concluding Remarks

At the end of this chapter, it should be noted that the stunning summits described above, the Greater and Lesser Caucasus mountains, have played a significant role in the formation and development of Georgian state. The ranges have served as a natural barrier, protecting the country from nomadic tribes and invaders from the north and from the south. Furthermore, the Caucasus mountain system traps cold air masses from the north, and hot air masses from the south, which has a substantial positive impact on the climate of Georgia.

Additionally, the Greater Caucasus contains 725 glaciers within Georgian territory, covering a total area of approximately 340 km^2 (Maruashvili 1964). It is noteworthy that all these glaciers are primarily located around summits higher than 4000 m, with the existence of these high mountains playing a decisive role in their formation. As we can see, the summits and mountains of the Caucasus are not only a proud, natural, and captivating heritage of Georgia, but they also hold vital significance for the existence and well-being of the country.

Fig. 13 A panoramic view of the Svaneti Ridge, from the north. Summit Laila pik visible in the central part. This photograph was taken of 1890, by the renowned Italian traveler Vittorio De Sella. The photo is stored in the National Archives of Georgia

References

Batonishvili V (1745) The geography of Georgia, Book 2, Karti. Published by Mose Janashvili, 1895, Tbilisi, pp 1–148

Burchuladze A, Janelidze C, Togonidze G (1976) Application of radiocarbon method to resolution some of the Pleistocene and Holocene paleogeography of Georgia. The actual problems of contemporary geochronology. Nauka, Moscow, pp 238–243

Chernishev I, Arakelianc M, Lebedev V, Bubnov S, Golcman I (1999) K-Ar geochronology of eruption of the newest volcanic centres of Kazbegi area of the Greater Caucasus. Russian J Earth Sci 1:61–72. https://doi.org/10.2205/1999ES000024

Gamkrelidze I, Okrostsvaridze A, Maisadze F, Basheleisvili L, Boichenko G, Skhirtladze I (2019) Main features of geological structure and geotourism potential of Georgia, the Caucasus. J Modern EnvironSci Eng 5(5):422–442. https://doi.org/10. 15341/mese(2333–2581)/05.05.2019/010

Gamkrelidze I, Okrostsvaridze A, Koiava K, Maisadze F (2021) Geotourism potential of georgia, the caucasus. Springer, pp. 1–81 https://doi.org/10.1007/978-3-030-62966-3_1.

Gushchenko II (1979) Eruptions of volcanoes of the world: a catalog. Moscow: Nauka Pub, Acad Sci USSR Far Eastern Sci Center, pp 1–474. (in Russian)

Gudjabidze GE, Gamkrelidze IP, Abesadze GN et al (2003) Geological map of georgia: georgian state department of geology and national oil company. Tbilisi, Georgia, Scale 1(500):000

Hua J, Fischer KM, Gazel E, Parmentier EM, Hirth G (2023) Long-distance asthenospheric transport of plume-influenced mantle from Afar to Anatolia. Geochemistry, Geophysics, Geosystems, 24, e2022GC010605. https://doi.org/10.1029/2022GC010605

Japaridze I (2010) History of georgian mountaineering, 1923–1991. Tbilisi, pp 1–245 (in Georgian)

"Kazbek", Global Volcanism Program Report, Smithsonian Institution. Retrieved 2020

Maruashvili L (1964) Physical Geography of Georgia. Publishing House "Codna". Tbilisi, pp. 1–355. (in Georgian)

Okrostsvaridze A (2007) Hercynian granitoid magmatizm of the greater caucasus. Nergebi, Tbilisi, pp 1–223

Okrostsvaridze A, Boichenko G, Gogoladze S, Gabrielashvi R, Baratashvili Z (2024) Middle Jurassic chaukhi gabbro-diabase-pillow lava massif: an impressive geoheritage of the greater caucasus, Georgia. In: Abstract #983 of the 37th international geological congress Busan, Republic of Korea

Okrostsvaridze A, Lee YH, Tormey D, Skhirtlade I(2023) U-Pb zircon chronological constraints for three stages of syn-orogenic plutonic magmatism of the Greater Caucasus Svaneti segment, Georgia. Episodes 46 (2): 194–210. https://doi.org/10.18814/epiiugs/2022/022023

Okrostsvaridze A, Tormey D (2011) Evolution of the Variscan orogenic plutonic magmatizm: The Greater Caucasus. Journal of Nepal Geological Society 43:45–52

Report of the International Mountaineering and Climbing Federation (UIAA), 16 May of 2024. https://www.theuiaa.org/uiaa-releases-2023-annual-report

Pasquare F, Tormay DR, Vezzoli LM, Okrostsvaridze A, Tutberidze B (2011) Mitigating the consequences of extreme events on strategic facilities: evaluation of volcanic and seismic risk affecting the caspian oil and gas pipelines in the Republic of Georgia. J Environ Manage 92(7):1774–1782. https://doi.org/10.1016/j.jenvman.2011.02.003

Somin ML (2011) Pre-Jurassic basement of the greater caucasus: brief overview. Turkish Journal of Earth Sciences 20:3–65

UIAA Report (2024) https://www.theuiaa.org member-meeting-sep-2024

Wang SJ, Lee YH, Chung SL, Okrostsvaridze A (2015) Cenozoic mountain building process in the greater caucasus, from subduction to collision, American Geophysical Union, Fall Meeting 2014, abstract #T21A-45

Magnificent Glaciers of the Greater Caucasus in Georgia

George Kordzakhia, Larisa Shengelia, and Genadi Tvauri

Abstract

As of the 2010s, more than 600 glaciers were identified in Georgia, with a total area of about 350 km^2. By 2020 this number had fallen to just over 400, and cumulative glacial area had decreased substantially relative to the catalogue baseline. Analysis of glacier inventories and satellite remote sensing data shows marked regional contrasts: in eastern Georgia (Tergi and adjacent basins), glacier area has shrunk by a little more than half, whereas in western Georgia (Enguri, Kodori and Rioni basins) the total glacial area has decreased by roughly forty percent. In both regions, glacier fragmentation, the disappearance of small glaciers and the accelerated retreat of large valley glaciers are evident, particularly since the late twentieth century. In this chapter we summarise the current distribution, types and characteristics of Georgian glaciers, outline their cultural and touristic significance, and analyse recent changes in glacier number and area using a combination of historical catalogues and high-resolution satellite imagery. We distinguish the dynamics of glacial basin degradation in eastern and western Georgia and discuss the implications of these trends for water resources, natural hazards and climate-change adaptation. At the same time, we emphasise that Georgia's glacial basins are key elements of the country's geoheritage: in the IUCN Key Geoheritage Areas (KGA) framework introduced in Chap. 2, they represent important examples of "glacial and periglacial systems" and form high-mountain cores of potential geopark territories such as Upper Svaneti and Kazbegi–Keli. Recognising their dual role as climate indicators and geoheritage assets is essential for integrating geoscientific knowledge, geotourism and conservation in the Georgian Caucasus.

Keywords

Georgian glaciers · Greater caucasus · Remote sensing · Climate change

1 Introduction

Nestled within the heart of the Greater Caucasus, Georgia's high-mountain landscape offers a striking blend of grandeur and fragility. From the jagged ridges of Svaneti and Racha to the volcanic peaks around Kazbegi, glaciers carve valleys, feed rivers, and shape some of the country's most iconic views. They are visible markers of the active collision between the Arabian and Eurasian plates, storing winter snow and releasing meltwater through the dry seasons to sustain downstream ecosystems, agriculture and hydropower.

While glaciers worldwide are admired for their visual splendour, in Georgia they also hold deep cultural and historical significance. In Upper Svaneti, medieval towers rise above glacier-fed valleys, while in the Kazbegi region the silhouette of Gergeti Trinity Church stands against the ice of Mkinvartsveri (Kazbegi) in one of the most recognisable vistas of the Caucasus. Legends such as that of Prometheus (Amirani in Georgian tradition), chained to the icy slopes of Kazbegi, and numerous local tales of sacred peaks and dangerous crevasses underline the place of glaciers in Georgian imagination and identity. At the same time, these high-mountain ice masses are among the most sensitive indicators of regional climate change, responding rapidly to shifts in temperature and precipitation.

G. Kordzakhia · L. Shengelia (✉)
Institute of Hydrometeorology of the Georgian Technical University, Tbilisi, Georgia
e-mail: l.shengelia@gtu.ge

G. Tvauri
Ivane Javakhishvili Tbilisi State University, Elevter Andronikashvili Institute of Physics, Tbilisi, Georgia
e-mail: genadi.tvauri@tsu.ge

A. Okrostsvaridze and D. Tormey (eds.), *Geoheritage of Georgia*, Geoheritage, Geoparks and Geotourism,
https://doi.org/10.1007/978-3-032-27639-1_10

In this chapter we provide an overview of the distribution, types and characteristics of Georgian glaciers, examine their cultural and touristic roles, and analyse recent changes in glacier number and area using historical catalogues (Maruashvili et al. 1975) and satellite remote sensing. We distinguish the dynamics of glacial basin degradation in eastern and western Georgia and discuss implications for water resources, natural hazards and climate-change adaptation. At the same time, we emphasise that the major glacial basins, particularly those of the Enguri and Tergi, are key elements of Georgia's geoheritage. In the IUCN Key Geoheritage Areas (KGA) framework introduced in Chap. 2, they represent important examples of "glacial and periglacial systems" and form high-mountain cores of potential geopark territories such as Upper Svaneti and Kazbegi–Keli.

2 Setting of Georgian Glaciers

Georgia's glaciers are predominantly located in the Greater Caucasus and are distributed across twelve glacial basins. Six basins lie in western Georgia (Bzipi, Kelasuri, Kodori, Enguri, Khobistskali and Rioni) and six in eastern Georgia (Tergi, Aragvi, Liakhvi, Assa, Arguni and Pirikiti Alazani). Large glaciers (with areas exceeding 2 km^2) are concentrated in four of these basins, three in the west and one in the east (Tergi). The overall distribution of glaciers in the Caucasus and the location of Georgian glacial basins are shown in Fig. 1, based on the Global Land Ice Measurements from Space (GLIMS) database.

In western Georgia (Enguri, Kodori and Rioni glacial basins), the combination of high precipitation from the Black Sea and rugged topography results in larger and thicker glaciers than in the eastern part of the country. Examples include the Lekhziri, Tviberi and Chalaati glaciers in the Enguri basin, which descend from high cirques and ridges into deeply incised valleys.

In contrast, eastern Georgia (Tergi glacial basin, Kazbegi region) has a generally drier climate and therefore hosts smaller glaciers on average than the western basins, yet the landscapes are equally dramatic. Gergeti Glacier on Mount Mkinvartsveri (Kazbegi) is one of the most iconic examples, forming a striking icefield above the Tergi valley and the Gergeti Trinity Church (Fig. 2).

This high mountain system features peaks exceeding 5000 m, including Mount Shkhara (5193 m) and Mount Kazbegi (Mkinvartsveri) (5047 m), both adorned with iconic glaciers (Fig. 3). Figure 4 presents the Asmashi–Tviberi Glacier as of 2016 and the area of the Toti Glacier's inflow near the Tviberi Glacier, illustrating the complexity of glacier networks in the upper Enguri basin. These examples highlight the main centres of modern glaciation in Georgia: in the west, the Shkhara–Tetnuldi–Ushba region of Upper Svaneti and the highest parts of the Kodori and Rioni basins, and in the east, Mount Mkinvartsveri and adjacent ridges. Together, these basins define the core of Georgia's glacial geoheritage and represent typical examples of "glacial and periglacial systems" in the sense of

Fig. 1 Location of Caucasus glaciers (red color) from the GLIMS database

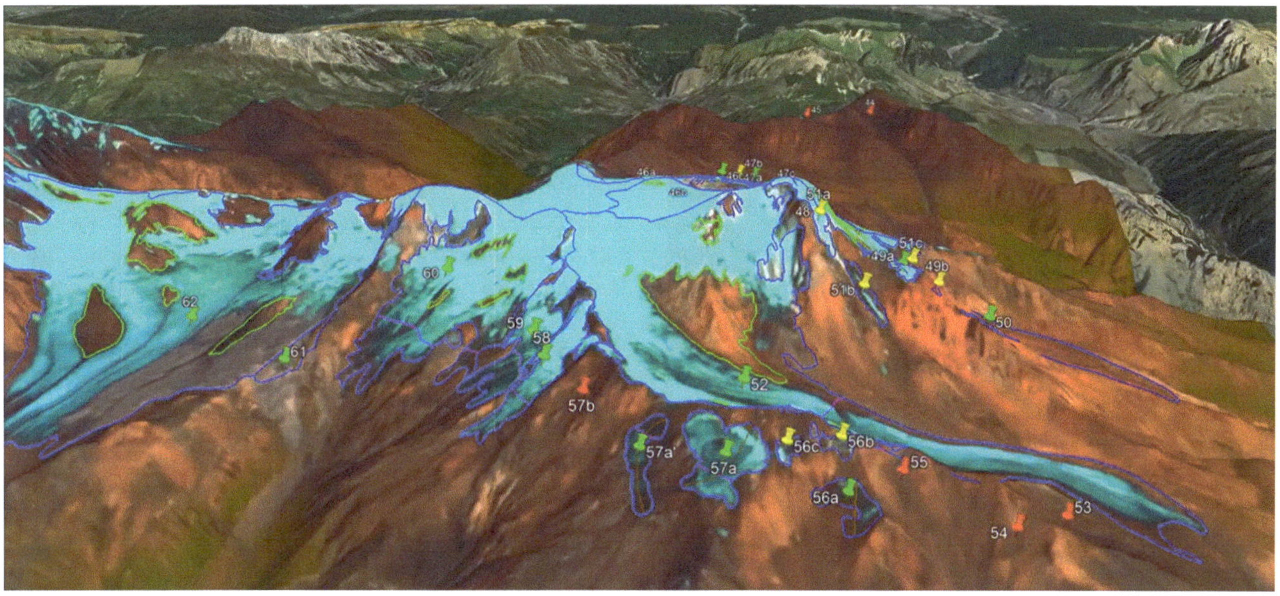

Fig. 2 Mkinvartsveri (Kazbegi) glaciers (East Georgia) derived from Landsat 8 OLI TIRS imagery, September 19, 2020

Fig. 3 Gergeti Glacier (Mount Kazbegi) r. Terek (Tergi) Glacial Basin (materials of the National Environmental Agency's (NEA) glaciological expedition)

the Key Geoheritage Areas (KGA) framework introduced in Chap. 2.

Here and below, three colours of pins are used on maps to mark glaciers in the different basins. Green pins indicate glaciers with an area of 0.1 km^2 or more, yellow pins indicate glaciers with an area of less than 0.1 km^2 or snowfields, and red pins indicate glaciers that have disappeared since the time of the earlier catalogues. The pins also show each glacier's catalogue number, allowing direct comparison between historical and modern inventories.

3 Glaciers Types and Characteristics

Glaciers in Georgia vary widely in type and size depending on local topography and climate. The main forms include valley glaciers that flow down mountain valleys (for example Tviberi Glacier in the Enguri basin), cirque glaciers occupying bowl-shaped hollows at the heads of valleys, and smaller hanging glaciers clinging to steep slopes and feeding icefalls and waterfalls. Together they create the dramatic ice–rock landscapes characteristic of the Greater Caucasus, with deep crevassing, proglacial lakes and strong seasonal contrasts in snow and vegetation (Fig. 5). These

Fig. 4 Asmashi-Tviberi Glacier, as of 2016 (Materials of the NEA's glaciological expedition)

Fig. 5 The area of the Toti Glacier's inflow near the Tviberi Glacier. 03.08.2016 (materials of the NEA's glaciological expedition)

visual features are important not only for glaciological research but also for geotourism, as they frame many of Georgia's most iconic mountain views.

4 Cultural and Historical Significance

Georgia's glaciers are not only natural wonders but are also woven into the fabric of its cultural identity. For centuries, mountain communities have lived in close visual and practical relationship with snow and ice, and many of the country's most recognisable cultural landscapes are framed by glaciated peaks and valleys.

Medieval architecture and glaciers. In Upper Svaneti, medieval stone towers and fortified houses rise above glacier-fed valleys, with views towards Shkhara, Ushba and other ice-clad summits. In the Kazbegi region, the silhouette of the Gergeti Trinity Church against the backdrop of Mkinvartsveri (Kazbegi) and the Gergeti Glacier has become an iconic image of the Georgian Caucasus. In these places, sacred buildings and traditional settlements are visually and symbolically anchored in glaciated mountain scenery.

Legends and myths. Glaciers and high peaks also play a central role in Georgian myth and folklore. The best-known example is the legend of Prometheus, known as Amirani in Georgian tradition, who was chained to the icy slopes of Mount Mkinvartsveri as punishment for bringing fire to humanity. Numerous local tales speak of sacred summits, dangerous crevasses and glacier spirits, underlining the deep symbolic role of snow and ice in the spiritual imagination of mountain communities.

Alpine traditions and livelihoods. Herding, haymaking, mountaineering and pilgrimages to glacier viewpoints have long shaped life in the highlands. Knowledge of snow conditions, avalanche paths and glacier melt patterns has historically been essential for safe travel and seasonal movement of livestock. In recent decades, guided trekking and climbing have added a new layer, bringing visitors into direct contact with glaciated landscapes that locals have known for generations.

These cultural connections add depth to the magnificence of Georgia's glaciers: they are not merely physical features, but integral elements of the country's heritage landscape. In the context of this book, they can be regarded as key components of Georgia's geoheritage. Within the IUCN Key Geoheritage Areas (KGA) framework introduced in Chap. 2, the main glacial basins of Upper Svaneti and Kazbegi in particular represent areas where glacial and periglacial systems are closely intertwined with long-lived cultural traditions, making them strong candidates for recognition as KGAs and as high-mountain cores of future geopark territories.

5 Glacier Tourism and Landscape Aesthetics

The combination of glaciers and cultural heritage has made Georgia a compelling destination for nature-based and adventure tourism. Glaciated valleys, sharp peaks and glacier-fed rivers create dramatic scenery, while traditional villages, towers and churches provide a strong cultural foreground. For many visitors, the first encounter with Georgian glaciers is not on a map or in a scientific paper, but as a powerful visual impression: ice and rock rising above stone towers, mountain churches and winding roads.

In Upper Svaneti, easily accessible glaciers such as Chalaati, Tviberi and the glaciers beneath Shkhara are among the main attractions for hikers and trekkers starting from Mestia and Ushguli. Short walks to glacier tongues, longer treks to high passes and viewpoints, and multi-day routes between villages all allow visitors to experience glaciated landscapes from different distances and angles. In clear weather, the contrast between white ice, dark rock and the stone architecture of Svan towers is particularly striking.

In the Kazbegi region, the view of the Gergeti Glacier and the upper slopes of Mkinvartsveri (Kazbegi) from the Gergeti Trinity Church and nearby ridges has become one of the most photographed scenes in the Caucasus. The Georgian Military Road offers numerous roadside viewpoints into glaciated cirques and valleys, while day hikes to Gergeti Glacier and other high-altitude locations provide close-up encounters with ice, moraines and meltwater streams. Here, glacier vistas are tightly interwoven with places of worship, war memorials and historic crossing points between north and south.

Other regions, such as Racha and Lechkhumi, also contain small hanging and cirque glaciers in relatively quiet mountain settings, with high potential for low-impact hiking, birdwatching and landscape photography. Although less developed for tourism, these areas offer opportunities to diversify glacier-related travel beyond the well-known hubs of Svaneti and Kazbegi.

In the context of potential geoparks, the Enguri and Tergi glacial landscapes form the high-mountain cores of the proposed Upper Svaneti and Kazbegi–Keli geopark territories, where glacier viewpoints, trails and cultural sites can be integrated into coherent geotourism products.

6 Glacial Retreat and Change

Over the past decades, Georgian glaciers have undergone marked retreat and fragmentation. As of the 2010s, more than 600 glaciers were identified in the country, with a total area of about 350 km^2. By 2020 the number of mapped glaciers had fallen to just over 400, and the cumulative glacierised area had decreased relative to the catalogue baseline. This reduction reflects both the disappearance of many small glaciers and strong area loss from larger valley and cirque glaciers (Fig. 6).

Analysis of satellite remote sensing data and updated glacier inventories shows that retreat rates have accelerated since the late twentieth century, in line with regional warming trends. In eastern Georgia (Tergi and adjacent basins), glacier area has diminished by a little more than half compared to the historical catalogue, while in western Georgia (Enguri, Kodori and Rioni basins) the total glacierised area has decreased by roughly forty percent. In both regions, previously continuous glacier bodies have broken into smaller units, and numerous new proglacial and supraglacial lakes have formed in front of retreating ice tongues. These changes alter meltwater timing and volume, increase the potential for glacier-related hazards, and transform the visual character of high-mountain landscapes.

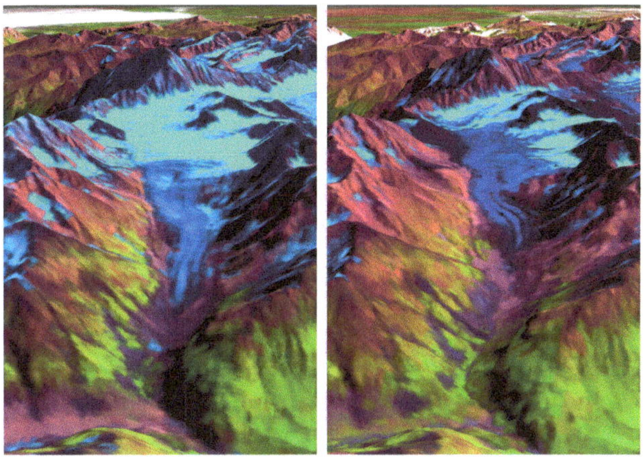
a b

Fig. 6 Comparative satellite images showing Kvitlodi Glacier retreat: a. represents 1987 and b—2024

Despite this retreat, glaciers remain among Georgia's most remarkable natural assets. They continue to play a crucial role in regulating river flow, supporting mountain ecosystems and agriculture, and providing the scenic backdrop for cultural landscapes and geotourism. Understanding the rates and patterns of glacier change is therefore essential not only for climate and hydrological studies, but also for planning adaptation measures and for safeguarding the geoheritage and aesthetic values of Georgia's high-mountain regions.

Glacier degradation also poses a significant threat to Georgia's sustainable development, making the monitoring of glacier retreat a national priority. Conducting large-scale, repeated ground-based observations (GBO) is costly and logistically challenging; therefore, satellite remote sensing (SRS) offers an efficient alternative, delivering comprehensive spatial and temporal coverage with the necessary level of detail and accuracy. For consistency, only glaciers with termini free of debris are selected for corresponding studies to ensure reliable identification of tongue positions. Methodological validation has shown that high-resolution SRS, combined with established best practices, provides accurate results (Conrad et al. 2015; Gorelick et al. 2017). These approaches are integrated with the authors' methods developed through previous work (Kordzakhia et al. 2016, 2019; Kordzakhia 2024) and form the basis for the detailed analyses of glacial basin degradation presented in the following sections.

6.1 Dynamics of Glacial Basin Degradation in Eastern Georgia

According to the catalogue, eastern Georgia counted 132 glaciers with a total area of 82.8 km^2. By 2010, the area had decreased to 63.1 km^2 (−23.8%), by 2015 to 48.9 km^2 (−22.5%), and by 2020 to 39.9 km^2 (−18.4%). Overall, glacier area declined by 42.9 km^2 (−51.8%) relative to the catalogue baseline. Glaciers number is not a representative characteristic, but it shows the process variability. Glacier numbers also declined sharply: from 132 in the catalogue to 114 in 2010 (−13.6%), 82 in 2015 (−28.1%), and 54 in 2020 (−34.1%). Thus, by 2020, glacier numbers had fallen by 78 (−59.1%). These results indicate that, in eastern Georgia, both glacier area and number decreased more than twofold over the study period.

6.2 Dynamics of Glacial Basin Degradation in Western Georgia

According to the glacier catalogue, western Georgia contained 409 glaciers with a combined area of 465.7 km^2. By 2010, the total glaciered area had decreased to 384.4 km^2, and by 2015 and 2020 to 318.7 km^2 and 274.9 km^2, respectively. This corresponds to a cumulative area loss of 190.8 km^2 (41.0%) relative to the catalogue baseline. The reduction occurred in distinct phases: a loss of 81.3 km^2 (17.5%) during the initial six decades, followed by an additional 47.5 km^2 (13.7%) over the subsequent five-year period. Glacier number exhibited marked temporal variability. While the catalogue recorded 409 glaciers, the number increased to 510 in 2010 (a 19% rise), mainly as a consequence of large and medium glacier fragmentation into smaller units. After 2010, glacier numbers declined, reaching 432 by 2015 and 375 by 2020, reflecting a net decrease of 34 glaciers (8.3%) compared with the catalogue inventory. These quantitative indicators demonstrate pronounced glacier degradation and fragmentation in western Georgia. However, the overall rate of glacier degradation remains lower than that observed in eastern Georgia, suggesting regional variability in response to climate forcing and local topo climatic conditions.

7 Conclusion

Modern studies of Georgian glaciers have allowed us to draw several key conclusions:

1. Comparison of glacier retreat trends derived from the SRS and GBO confirms the consistency of these datasets. This provides confidence in the reliability of the quantitative estimates based on the SRS and validates the analytical framework used in the study.
2. Glaciers across Georgia have experienced significant degradation in recent decades, but the extent and dynamics of degradation vary regionally. The glaciers of western Georgia, despite having significantly reduced area (41%) compared to the catalogue baseline, exhibit slower degradation and a temporary increase in glacier number due to fragmentation processes. In contrast, glaciers in eastern Georgia exhibit even higher relative retreat rates, reflecting their smaller size, greater continentality, and increased sensitivity to temperature increases and reduced accumulation.
3. The dynamics of the retreat of large glaciers over the past 60 years show a clear acceleration, especially in the second half of the observation period. This indicates that large glaciers' retreat is nonlinear and can be effectively approximated by a polynomial curve, namely a parabola.
4. Regional differences in retreat of large glaciers are evident. In the Enguri basin (e.g., Northern Liadeshti, Kvishi, Adishi), the acceleration of retreat has been observed since the early 2000s, which coincides with an increase in summer air temperature of up to 0.6 °C at the Mestia meteorological station. In the Tergi Basin (e.g., Gergeti, East Suatisi) even stronger warming of up to 1.1 °C at the Stepantsminda station contributed to similarly rapid retreat.
5. The consistent occurrence of nonlinear, accelerated retreat trends across all studied glaciers demonstrates the high sensitivity of the Georgian cryosphere to modern climate variability. The acceleration since 2000 is closely consistent with regional and global warming trends and confirms global estimates of accelerated glacier mass loss. These results highlight the need for continued monitoring and provide a scientific basis for climate adaptation planning, particularly in glacier-fed river basins and in areas vulnerable to glacier-related hazards such as icefalls and glacial lake outburst floods (GLOF).
6. Projections based on a business-as-usual (BaU) climate scenario indicate that glacial basins containing large glaciers will likely experience complete melting during the next century.
7. Glacial basins without large glaciers are likely to lose their remaining ice even earlier, with most of them experiencing complete melting within the current century.
8. The projected melting timelines indicate a distinct spatial and temporal variability in glacier response to climatic forcing. Large Glaciers located at lower altitudes and with smaller volumes (e.g. Kvishi, Kirtisho) are expected to disappear within the first half of the twenty-first century. In contrast, larger and higher elevation glaciers (e.g. Northern Liadeshti, Adishi, Shkhara, Boko, East Suatisi, Gergeti) are projected to persist until the

late twenty-second century. Variations in glacier geometry, altitude, and basin-specific microclimatic conditions primarily drive such differences. These findings underscore the importance of considering individual glacier characteristics, rather than generalized regional averages when assessing the long-term impacts of climate change on the Georgian cryosphere.

Beyond their critical roles in regional hydrology and climate regulation, Georgia's glaciers and their associated landforms also represent some of the country's most important geoheritage landscapes. Their rapid degradation therefore threatens not only water resources and natural hazards, but also the loss of iconic glacial valleys that underpin Georgian identity and mountain tourism. Recognising the main glacial basins as Key Geoheritage Areas (KGAs), and integrating them into future geopark, protected area and climate adaptation planning, can help ensure that their scientific, educational and cultural values are considered alongside the physical impacts of climate change in the Georgian Caucasus.

Acknowledgements This research was carried out with the support of the Shota Rustaveli National Science Foundation (Project FR-21-1996: Research on the Glacier Degradation in Georgia Over Recent Decades and the Creation of the "Electronic Atlas of Georgian Glaciers").

References

Conrad O, Bechtel B, Bock M, Dietrich H., Fischer E, Gerlitz L, Wehberg J, Wichmann V, Böhner J (2015) System for Automated Geoscientific Analyses 2.1.4. pp 1991–2007. https://doi:10.5194/gmd-8-1991-2015

Gorelick N, Hancher M, Dixon M, Ilyushchenko S, Thau D, Moore R (2017) Google earth engine: planetary-scale geospatial analysis for everyone. Remote Sens Environ 202:18–17. https://doi.org/10.1016/j.rse.2017.06.031

Kordzakhia G (2024) Convention on climate change georgia's fifth national communication of georgia to the the united nations framework convention on climate change. 4.5. Impact of Climate Change on Georgian Glaciers. Tbilisi, pp 299–319 https://unfccc.int/sites/default/files/resource/5NC_Eng.pdf

Kordzakhia G, Shengelia L, Tvauri G, Dzadzamia M (2016) Research into glacier variation dynamics in East Georgia under the impact of modern climate change. In: IGCP 610 fourth plenary conference proceedings, Tbilisi, Georgia, pp 96–100. https://www.caucasus-mt.net/research-into-glacier-variatio.html

Kordzakhia G, Shengelia LD, Tvauri GA, Dzadzamia MS (2019) The climate change impact on the glaciers of Georgia. World Science 4(44):29–34

Maruashvili L, Kurdgelaidze G, Lashkhi T, Inashvili S (1975) Glacier catalog. SSSR 9 (1) Gidrometeoizdat:1–86. (In Russian)

Underground Wonders of Georgia: Karst Caves and the Myth of Amirani

Kukuri Tsikarishvili, Zaza Lezhava, and Lasha Asanidze

Abstract

Humanity has been connected to the underground world (caves) shrouded in eternal darkness since ancient times. In this regard, Georgia is one of the most remarkable countries on Earth. Caves are formed in soluble limestone rocks, which cover a significant portion of Georgia's territory (over 10%). These areas exhibit a variety of surface (karst sinkholes, poljes, uvalas, ponors etc.) and underground (abysses, caves and shafts) karst features. Karst is particularly well-developed in western Georgia, where it stretches continuously for 325 km in length and between 2–3 and 30–35 km in width, from the Psou River to the vicinity of Lake Ertso. This area includes the northern hilly part of the Colchis lowland and the adjacent southern slopes of the western Caucasus. The vertical spread of karst begins from the Gagra coastal zone and extends up to an altitude of 2757.6 m above sea level (Speleologists' Peak in the Arabika Massif). These limestone formations and hydrogeological factors account for the presence of deep karst abysses and numerous caves in Georgia's limestone massifs. It is noteworthy that, due to the abundance of these magnificent natural phenomena and the promising opportunities for their exploration, Georgia holds a significant position globally in this field. Nowadays, the total number of caves discovered in Georgia exceeds 1500, with a cumulative surveyed length of approximately 280 km. These karst systems and their associated caves form an important part of Georgia's geoheritage, linking geological processes with archaeological and cultural history.

K. Tsikarishvili (✉) · Z. Lezhava · L. Asanidze
Tbilisi State University, Vakhushti Bagrationi Institute of Geography, Tbilisi, Georgia
e-mail: kukuritsikarishvili@gmail.com

Keywords

Karst · Limestones · Caves · Speleothems · Georgia

1 Introduction

As is widely recognized, karst caves develop in easily soluble rocks. In Georgia, such rock formations (mostly carbonate rocks) cover approximately 10% of the country's territory, predominantly associated with the thick sedimentary sequences of the Lower Cretaceous period along the southern slope of the Caucasus. Geological conditions, along with abundant precipitation, significant hypsometric variations in the landscape, and other contributing factors, have resulted in the formation of deep karst abysses and numerous cave systems. The karst caves in Georgia are primarily formed along the southern slope of the Caucasus in western Georgia. This region extends approximately 325 km in length and 3–35 km in width, spanning from the Psou River to the Ertso Lake (Kipiani et al. 1966). The vertical distribution of karst formations is equally remarkable, ranging from the Black Sea coastline to an altitude of 2757 m above sea level. These exceptional geological features form a chain of underground wonders. The diversity and abundance of karst formations in Georgia position this region as one of the most significant karst landscapes globally. The continuity and accessibility of this western Caucasus karst belt make it an ideal natural laboratory for the study and interpretation of underground landscapes, and a foundation for geotourism and education in Georgia.

2 Georgian Karst Caves

To date, the total number of caves discovered in Georgia exceeds 1500 (Asanidze et al. 2019), with a cumulative surveyed length of approximately 280 km (Fig. 1). Among

A. Okrostsvaridze and D. Tormey (eds.), *Geoheritage of Georgia*, Geoheritage, Geoparks and Geotourism,
https://doi.org/10.1007/978-3-032-27639-1_11

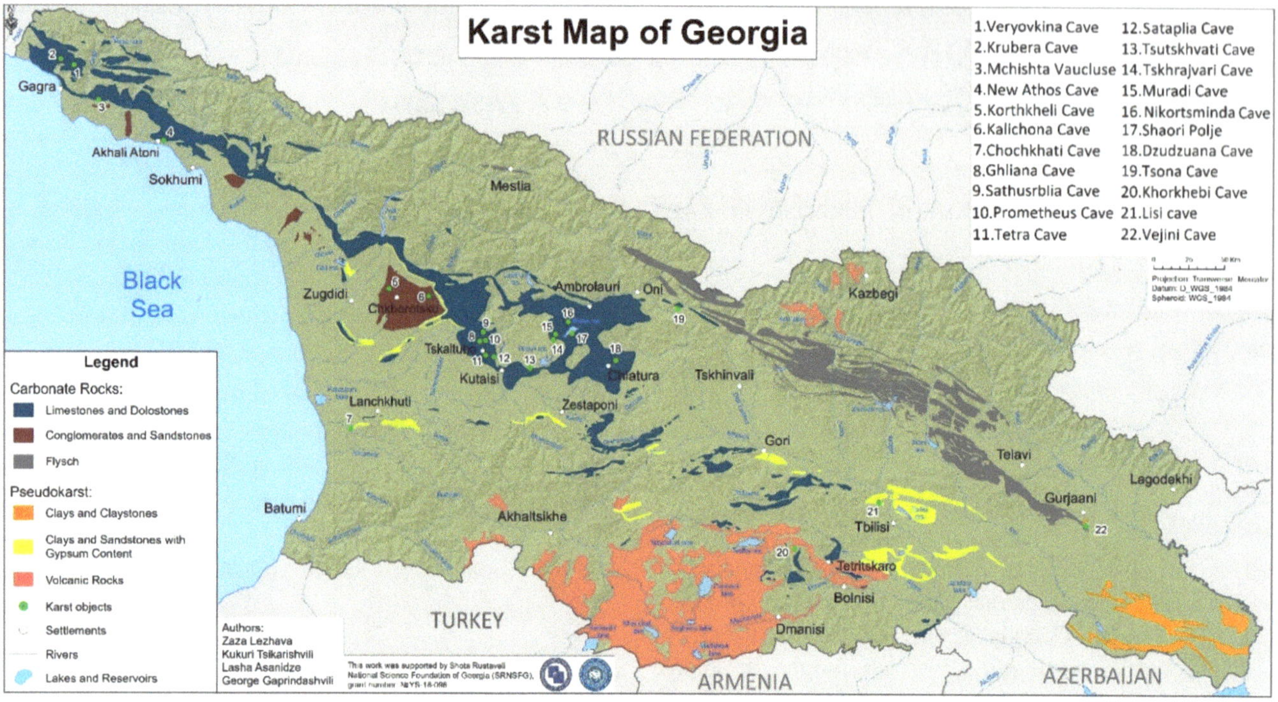

Fig. 1 Distribution of karst and pseudokarst features in Georgia: the areas marked with green dots on the map represent both regional and national speleological features. However, it should be noted that the actual number of such features in Georgia is likely higher than those shown on the map (Asanidze et al. 2019)

the subhorizontal cave systems, the largest is the Mchishta Cave, located along the Bzipi-Khipsta watershed ridge, with an estimated length exceeding 30,000 m (Tatashidze et al. 2003). Other notable examples include the Prometheus (Tskaltubo) Cave in Tskaltubo Municipality, measuring 15,500 m in length, and the New Athos Cave system, which extends 3285 m (Tatashidze et al. 2003).

Evidence suggests that caves in Georgia were utilized by early human populations for habitation, shelter, and, later, for ritual purposes. Recent archaeological investigations have uncovered remains of human settlements in these caves, spanning from the Stone Age to the late feudal period. These archaeological traces demonstrate how geology, ecology, and human history intersect, strengthening the case for Georgia's caves as living geoheritage sites rather than static geological formations.

The karst caves of Georgia provide a unique chronological record encompassing all stages of prehistory. Their study enables the reconstruction of the timeline of early human presence in Eurasia, including the processes of human settlement and all stages of cultural evolution. This, in turn, facilitates a deeper understanding of the origins and development of early human societies (Lezhava et al. 2019, 2020).

It is noteworthy that during periods of frequent enemy invasions, Georgian kings safeguarded state documents, valuable manuscripts, and national treasures in inaccessible natural caves. For example, historical sources indicate that in the early thirteenth century, when Queen Tamar's daughter Rusudani (1194–1245) was defending Georgia against Mongol invaders, she concealed the kingdom's treasures in the Khvamli Limestone Massif, which is located between the Tskhenistskali and Rioni rivers and rises to an elevation of 2000 m above sea level. The massif overlooks the Black Sea and the Colchis Plain from the north and is characterized by towering cliffs. According to one interpretation of the Prometheus myth, the legendary figure was chained to the cliffs of this massif (Fig. 2). Khvamli thus illustrates how natural landforms become repositories of both myth and memory, embodying the union of geological and cultural heritage that defines Georgia's broader geoheritage narrative.

Speleological investigations in Georgia began with the discovery of the Sataplia Cave in 1925 by local resident Petre Chabukiani within the Nature Reserve of the same name, near city of Kutaisi. The cave expands for 900 m, with a remarkable "domed hall" located 100 m from its entrance. This hall features a ceiling height of 10 m and a width ranging from up to 12 m. Diverse stalagmites and stalactites are developed in the cave (Fig. 3). Its easily accessible setting near Kutaisi allows Sataplia to serve as a model for how scientific research, conservation, and public education can be integrated within a geopark framework.

Fig. 2 The southern slope of the Khvamli limestone massif

Fig. 3 A giant stalagmite of the Sataplia "domed hall"

It is noteworthy that more than 200 well-preserved dinosaur footprints were found near the cave, which further increased interest in the Sataplia Cave. Consequently, today the Sataplia Nature Reserve stands as one of Georgia's premier geotourism destinations.

Among the caves cataloged in Georgia, the New Athos Cave System in Abkhazia (north-western part of Georgia)

Fig. 4 Petrified "waterfall" of 20 m in height in the New Athos Cave (1988)

was the first to be developed in accordance with international tourism standards. The system boasts a total of 1125 m of tourist trails and is recognized as one of Europe's most remarkable subterranean attractions. Annually, the New Athos Cave welcomes more than 800,000 visitors, quickly establishing itself as a popular destination. The cave is celebrated for its striking karst features, including astonishing stalagmite columns of human height, calcite-coated surfaces, petrified waterfalls (Fig. 4), intricately carved chandeliers, and other unique formations (Fig. 5) (Tintilozov 1983).

Among the caves in Georgia, the cave system in the Tskaltubo region stands out. Discovered in 1984 by researchers from the Institute of Geography of the Georgian Academy of Sciences, it is one of the largest and most intriguing cave systems in the country, with a total length of 15,500 m. Officially opened to the public in 2012, it

Fig. 5 Speleothems in New Athos Cave (1988)

Fig. 7 Stalagmite "Bermukha" in the Prometheus Cave

Fig. 6 Eccentrical shape of calcite deposits in the Prometheus Cave

Fig. 8 Spherical examples of pool speleothems in Muradi Cave (Racha Limestone Massif)

was modernized to meet international standards and is now widely known as the Prometheus Cave.

Within the halls of the Prometheus Cave, stalagmites create an elegant calcite shapes. Notable features include "castle towers" constructed from calcite, columns adorned with intricate patterns (Fig. 6). Notably, another natural wonder located in the cave's final section is the stalagmite known as Bermukha (Fig. 7).

Another fascinating and unique discovery is the Muradi Cave (Asanidze et al. 2017a), located in the Racha Limestone Massif. The Muradi Cave is remarkable for its diversity of formations (Asanidze et al. 2017b), including stalactites, stalagmites, stalagnates, helictites, travertines, calcite "curtains," and petrified waterfalls and "forests and flowers". The unique and unparalleled beauty of the Muradi Cave is accentuated by eccentric, ball-shaped formations composed of oolites, known as "Raphaelo Balls", which have no analogues in the Caucasus region (Fig. 8). Muradi Cave demonstrates the ongoing discovery potential of Georgia's karst systems, where new formations and mineral assemblages continue to expand our understanding of subterranean geoheritage.

From a scientific perspective, the multi-level karst system of Tsutskhvati, situated in the Tkibuli Municipality, is particularly significant. Its 75-m vertical profile comprises 13 levels, representing an extensive geological history. Of special interest is the fifth level, known as the "Bronze Cave," where 12–13-meter-thick stratified deposits contain artifacts dating back tens of thousands of years (Maruashvili 1978). The combination of archaeological artifacts and geomorphic structure at Tsutskhvati provides a textbook example of the integrated natural and cultural values central to UNESCO Global Geopark principles.

This brief overview highlights only a fraction of Georgia's karst wonders. The true scope is far more extensive and diverse, making it impossible to comprehensively describe them within the constraints of this book. However, it is worth noting that of the 118 karst caves worldwide with depths exceeding 1000 m, nine are located in Georgia (http://www.caverbob.com/wdeep.htm). Collectively, Georgia's caves, whether scientific research sites, show caves, or sacred places, constitute a unified underground geoheritage network that complements the mountain and volcanic landscapes described in other chapters.

3 The Myth of Amirani

Numerous myths and legends are intertwined with the caves of Georgia, the most renowned and widespread being the myth of Amirani. Amirani serves as the Georgian counterpart to the Greek Prometheus, with origins tracing back to ancient period. Unlike Prometheus, who was famously chained to the Caucasus Mountains, Amirani was confined by the gods to an inaccessible cave within the Caucasus as punishment for his defiance. The Prometheus myth, which was prevalent across the Mediterranean basin and the Caucasus, shares many similarities with the legend of Amirani, leading to their eventual synthesis (Kiknadze 2007). This convergence of narratives is why, in the early twenty-first century, the remarkable cave in Tskaltubo, Georgia, was named after both Prometheus and Amirani.

Interestingly, a parallel legend of Amirani exists in ancient Abkhazian folklore, likely serving as a prototype for the Georgian myth. According to the Abkhazian legend, the strong and smart hero Abrskili waged a valiant fight against malevolent spirits, instilling fear in the oppressors of his people. For his defiance, the gods punished Abrskili by driving him and his horse into the depths of a profound cave, securing them with iron stakes. However, the Abkhazian story ends optimistically, with the hero ultimately being freed by the people he saved. These mythological connections between the inhabitants of the southern slopes of the Caucasus and the karst caves suggest a deep historical relationship. They serve as evidence of humanity's ancient knowledge and utilization of these geological monuments.

References

Asanidze L, Lezhava Z, Tsikarishvili K, Gaprindashvili G, Chikhradze N, Polk SJ (2019) Karst map of Georgia (Caucasus region) scale: 1:1,500,000. Carbonates Evaporites 34:1205–1212

Asanidze L, Lezhava Z, Tsikarishvili K, Chikhradze N, Polk SJ (2017a) Karst Morphological Processes and Evolution of the Limestone Massif of Georgia from Depositional, Sedimentary, and Structural Investigations in Muradi Cave. In: Proceedings of 17th International Congress of Speleology, Sydney, pp 241–247

Asanidze L, Chikhradze N, Lezhava Z, Tsikarishvili K, Polk SJ, Lominadze G, Bolashvili N (2017b) Complex Speleogenetic processes and mineral deposition in the Caucasus region of Georgia. J Environ Biol 38:107–113

Kiknadze Z (2007) Georgian mythology. Bakmi, Tbilisi, pp 18–25

Kipiani S, Tintilozov Z, Okrojanashvili A, Jishkariani J (1966) The cadastre of karst caves of Georgia. Metsniereba, Tbilisi, pp 1–283

Lezhava Z, Tsikarishvili K, Asanidze L, Chikhradze N, Chartolani G, Sherozia A (2019) Karst relief development history of Zemo Imereti plateau, Georgia. Cauc Open J Geol 9(03):201–212

Lezhava Z, Asanidze L, Tsikarishvili K, Chikhradze N, Gadrani L, Chartolani G (2020) Hydrodynamic zones of fissure-karst waters of Zemo Imereti plateau in the republic of Georgia. J Environ Biol 41:337–343

Maruashvili L (1978) Tsutskhvati multi-level karst system. Exploring the caves of Colchis. Metsniereba, Tbilisi, pp 1–296

Tatashidze Z, Tsikarishvili K, Geladze G (2003) The deepest karst abysses of Georgia proceedings of Tbilisi state university. Geogr Geol 335:162–170

Tintilozov Z (1983) Akhali Atoni cave system. Mecniereba, Tbilisi, pp 1–152

Water-Shaped Geoheritage of Georgia: Geodynamically Sensitive River Gorges, Canyons, Waterfalls, and Lakes

Avtandil Okrostsvaridze, Daniel Tormey, Nana Bolashvili, and Giorgi Boichenko

Abstract

Georgia is rich in hydro-geomorphic (water-shaped) geoheritage expressed in both surface and groundwater systems. In this rugged, actively colliding orogen, rivers and water have carved abundant valleys, gorges, canyons, waterfalls, and lakes that shape landscapes emblematic of a young collision zone. These landforms are also readable indicators of past and ongoing geodynamic processes: their distribution and morphology reflect the combined influence of faulting, uplift, volcanic activity, lithologic contrasts, and erosion. Major river valleys commonly exploit regional structural trends, canyons record rapid incision responses to uplift, waterfalls often mark fracture zones or lithologic boundaries, and lakes preserve signals of tectonic, glacial, landslide, and volcanic damming processes. Beyond their scientific value, these features have strong educational and aesthetic appeal and therefore high geotourism potential. Together they form a core component of Georgia's geoheritage, with several examples likely to be of wider significance because they clearly and accessibly express active collision-zone landscape evolution.

Keywords

Water-shaped geoheritage · Georgia · River gorges · Canyons · Waterfalls · Lakes

A. Okrostsvaridze (✉) · G. Boichenko
Institute of Earth Sciences, Ilia State University, Tbilisi, Georgia
e-mail: avtandil.okrostsvaridze@iliauni.edu.ge

D. Tormey
Catalyst Environmental Solutions, Santa Monica, USA

IUCN World Commission on Protected Areas, Rue Mauverney, Switzerland

N. Bolashvili
Vakhushti Bagrationi Institute of Geography of Tbilisi State University, Tbilisi, Georgia

1 Introduction

The Caucasus is a Phanerozoic collision orogen extending for approximately 1100–1200 km between the Black and Caspian Seas. Its long-lived evolution reflects the northward displacement of the Gondwana-derived Arabian plate toward the stable Eurasian continental margin (e.g. Rolland 2017; Gamkrelidze et al. 2019). Although the intervening branches of the Tethys Ocean were progressively closed and collision became established, convergence has continued in a continental regime and today averages on the order of ~1.5–2 mm/yr. (e.g. Reilinger et al. 2006). Exhumation and development of the modern Caucasus orogen accelerated during late Cenozoic time, broadly within the last ~8–5 Ma (e.g. Wang et al. 2015). As a result of this complex history and structure, the Caucasus is among the most geodynamically active and diverse regions in Eurasia.

Georgia occupies the central part of this collision zone, between the young mountains of the Greater and Lesser Caucasus, with the Black Sea to the west and the Caspian basin to the east (approximately 41– 43° N, 44– 46° E). This setting, high relief, active neotectonics, and strong climatic gradients, supports abundant water resources. Groundwater is an important component of this hydrologic system; however, because this chapter focuses on surface-water landforms, we note groundwater only briefly. For a country of relatively small area (69,700 km^2), reported combined groundwater discharge is substantial (49.4 million m^3/day; e.g. Gugushvili 2003).

Surface waters are pervasive and provide particularly clear indicators of geomorphic change. Georgia is criss-crossed by a dense network of mountain rivers that cut spectacular valleys and gorges and generate canyons, meanders, and waterfalls. The country is also dotted with lakes of diverse origin, formed in both alpine settings and lowland basins.

The active geodynamics of the Caucasus, intense neotectonic deformation, rapid relief growth, and marked litho-

A. Okrostsvaridze and D. Tormey (eds.), *Geoheritage of Georgia*, Geoheritage, Geoparks and Geotourism,
https://doi.org/10.1007/978-3-032-27639-1_12

logic heterogeneity in Georgia together promote deeply incised river valleys, narrow canyons, and numerous waterfalls and lakes. These landforms record how surface-water systems respond to tectonic deformation and evolving topography. Accordingly, their morphology, spatial distribution, and modern dynamics provide valuable information about the country's current and recent geodynamic processes.

Beyond their scientific importance, many of these water-shaped landforms have high educational and aesthetic value, making them important components of Georgia's geoheritage. Their value is enhanced by the clear, accessible expression of active geological processes, supporting both geoscientific interpretation and geotourism.

In this chapter, we first summarize how Georgia's dense river network functions as a sensitive indicator of active uplift, fault control, and lithologic contrasts. We then present representative examples of river gorges, canyons, waterfalls, and lakes whose forms and locations reveal these controls. Finally, two integrated case studies (Javakheti volcanic plateau; Mtkvari river-gorge geometry near Tbilisi) illustrate how water-shaped landforms can be read as "living" expressions of ongoing collisional geodynamics.

2 River Gorges

On Georgia's relatively small territory (69,700 km^2), more than 26,000 rivers form a dense hydrographic network. Approximately 18,000 rivers drain westward to the Black Sea and ~8000 drain eastward to the Caspian Sea; together their combined length is about 60,000 km (Fig. 1). The total volume of surface waters in the country is reported to be up to ~100 km^3, and the average annual freshwater depth is ~760 mm. By this measure, Georgia substantially exceeds its southern neighbors, Turkey, Armenia, and Azerbaijan, where comparable values are ~215 mm, ~280 mm, and ~110 mm, respectively (Khmaladze 2009; Geladze et al. 2022).

Such abundant precipitation, coupled with high relief and strong topographic dissection, produces a network of river valleys that responds sharply to changes in uplift, deformation, and lithology. Accordingly, Georgia's river network is one of the most distinct and informative indicators of the modern geodynamic environment. Rivers respond to tectonic uplift and deformation by adjusting their longitudinal profiles and incision depths; they also respond to lithologic contrasts by developing characteristic valley shapes

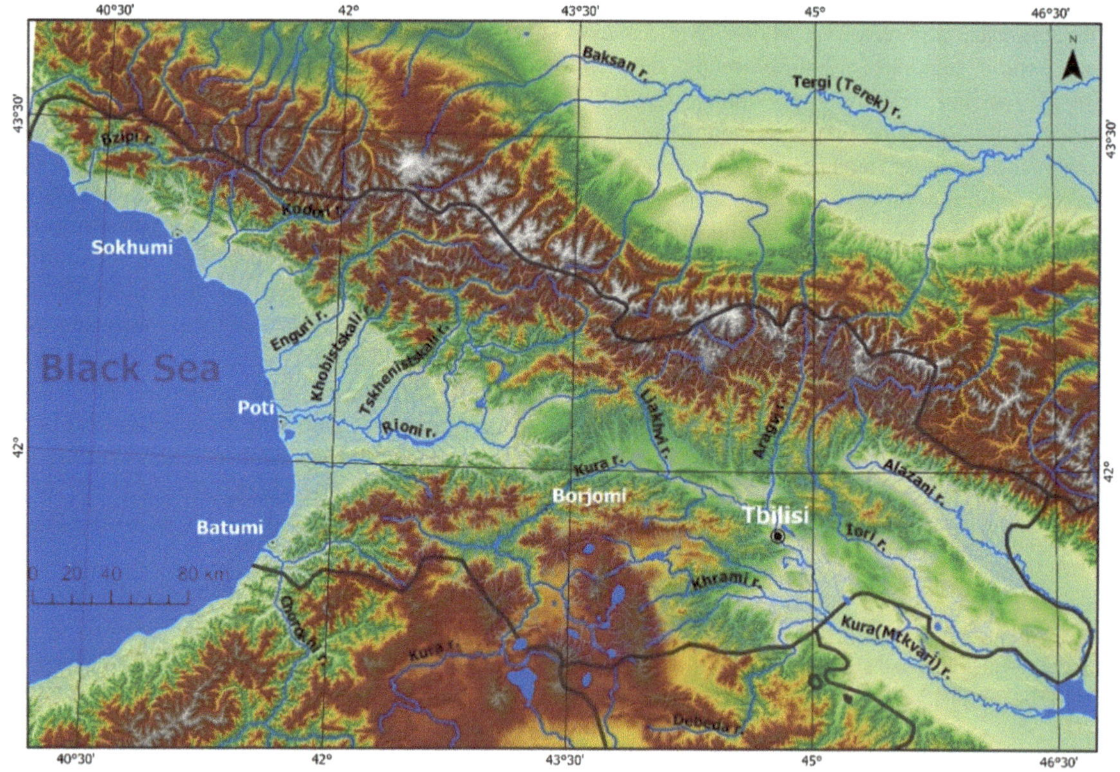

Fig. 1 The network of main rivers of Georgia

and knickpoints. In other words, these valleys function as water-cut "readouts" of active collision-zone deformation and landscape response.

Most rivers originate in the Greater and Lesser Caucasus and also in the Likhi (Likhi) Range, which links the two mountain systems. Rivers sourced in the Greater Caucasus are commonly sustained by glacial melt, groundwater, and precipitation; they typically have steep gradients and high discharge, and many carve deep valleys and local canyons. Rivers sourced in the Lesser Caucasus are fed mainly by groundwater and precipitation and generally have lower gradients; their valleys are commonly less deeply incised, with fewer canyon reaches. In all cases, river valleys evolved together with their mountain source areas and therefore preserve information about changes in relief and tectonic forcing.

As noted above, Georgia's drainage is divided between the Black Sea and Caspian Sea basins, with the Likhi Ridge forming a major watershed. Western Georgia belongs to the Black Sea basin and includes rivers such as the Chorokh, Adjara, Natanebi, and Supsa (flowing from the Lesser Caucasus); the Bzybi, Kodori, Enguri, Khobi, Cskhenist-skali, and Rioni (flowing from the southern slope of the Greater Caucasus); and the Kvirila and Dzirula, which originate on the western slopes of the Likhi Range (e.g. Gigineishvili 2000). Eastern Georgia is dominated by the Mtkvari (Kura), which receives most major tributaries of eastern Georgia and ultimately drains to the Caspian Sea. Notable tributaries include the Liakhvi, Ksani, Aragvi, Iori, and Alazani from the southern slope of the Greater Caucasus, and the Khrami from the Lesser Caucasus. In addition, the Tergi and Tusheti Alazani rivers originate on the northern slope of the Greater Caucasus, do not join the Mtkvari, and also flow to the Caspian Sea.

The river names themselves are part of Georgia's cultural and geographic identity, and we retain them here for that reason. In geoheritage terms, the discussion below emphasizes river corridors and gorge reaches that most clearly express tectonic control, uplift-driven incision, volcanic or landslide damming, and strong lithologic contrasts, features that make these landscapes both scientifically informative and visually compelling.

Georgia's rivers are integral to the country's economy, energy, ecosystems, and cultural heritage. As in many regions worldwide, settlements in Georgia commonly developed along rivers, and rivers continue to shape daily life and livelihoods. This importance is reflected in Georgian folklore and in many legends associated with rivers, waterfalls, and canyons.

Among the most impressive gorge landscapes is that of the Bzybi River in western Abkhazia, which rises near Middle Jurassic intrusive rocks on the southern slope of the Greater Caucasus and flows ~110 km to the Black Sea. Its steep gradient and discharge allow it to carve a striking canyon into massive limestone, a classic, visually accessible example of water-driven incision in a structurally complex mountain belt. Another major Abkhazian river is the Kodori. After the confluence of the Klichi, Gwandra, and Sakeni rivers, the Kodori increases markedly in discharge and, near the village of Lata, cuts a deep canyon in Cretaceous limestones. The Kodori River extends ~110 km from the Khlukhori Pass to the Black Sea.

The most water-rich river in Georgia is the Rioni, sourced on the southern slope of the Greater Caucasus in glaciers near Mount Phasis at ~2600–2700 m elevation. Its drainage basin area is ~13,400 km^2 and it reaches the Black Sea near Poti, where reported discharge is ~450 m^3/s (Zautashvili and Mkheidze 2022). Known as Phasis in Greek sources, the river is associated with the "Golden Fleece" tradition, and the pheasant's scientific name (Phasianus colchicus) reflects the historical linkage to ancient Colchis (e.g. Del Hoyo et al. 1994).

From an energy perspective, one of Georgia's most important rivers is the Enguri, which also originates on the southern slope of the Greater Caucasus at ~2600–2700 m elevation, sourced from glaciers near the Shkhara massif. The river is ~213 km long and flows through one of Georgia's most scenic gorge landscapes (Fig. 2). By contrast, the highest sediment loads in Georgia are associated with the Chorokh (Çoruh), which originates in Turkey (Eastern Pontides), flows toward the Black Sea south of Batumi, and supplies sediment that contributes to coastal delta and spit development at the Batumi Gulf. Although hydropower development in Turkey has partially reduced coarse-sediment delivery, the Chorokh still discharges large quantities of sediment (Zautashvili and Mkheidze 2022). Archaeological research and local traditions also indicate early metallurgical activity in the Chorokh valley

Fig. 2 One of the sections of the Enguri River gorge. *Photo* © Roman Tolordava

dating back to at least the second millennium BC (e.g. Okrostsvaridze 2017).

The main river of eastern Georgia is the Mtkvari (Kura), which originates in Turkey on the Ardahan volcanic plateau in eastern Anatolia. Flowing generally eastward, it crosses the Samtskhe–Javakheti volcanic plateau and the Adjara–Trialeti fold-thrust belt, enters eastern Georgia, and continues through Azerbaijan to the Caspian Sea. The Mtkvari's total length is ~1515 km, of which ~351 km lies within Georgia; nevertheless, it functions as a vital national artery, with major historic centers established along its banks, including the former capital Mtskheta and the current capital, Tbilisi.

Georgia's longest river is the Alazani (~390 km), which originates on the southern slope of the Greater Caucasus from glaciers near Mount Borjalo at ~2750 m elevation. It flows southeast along the depression between the Main Caucasus Range and the Gombori Ridge and has many water-rich tributaries that supply alluvial material to the Alazani Plain. These fertile plains form Georgia's principal wine-growing region. The Iori, also sourced near Mount Borjalo, is the second-largest river in eastern Georgia (~320 km; drainage basin ~4650 km^2; Zautashvili and Mkheidze 2022). Unlike other major rivers in Georgia, the Iori has few large tributaries and is among the most heavily regulated: reservoirs and irrigation diversions significantly reduce summer flows in its lower reaches, creating ecological pressures. Water from the Iori system is also linked to supply infrastructure for the reservoir near Tbilisi known as the "Tbilisi Sea."

The Aragvi–Tergi corridor has notable geographic, cultural-economic, and military significance because it forms a natural north–south route across the Greater Caucasus. The Tergi (Terek) rises near the Kazbegi area and cuts through the Paleozoic crystalline rocks of the Darial Gorge to enter the North Caucasus lowlands, while the Aragvi drains southward toward the Mtkvari near Mtskheta. The alignment of these valleys underpins the Georgian Military Road, historically used by both enemies and allies and today an essential transport route linking the South and North Caucasus.

Hydropower potential of rivers. Georgia's mountain rivers, commonly with elevation drops of ~2000–2500 m, have substantial hydropower potential, of which only ~22–25% is reported to be utilized. The most powerful hydropower facility is the Enguri HPP (~1300 MW), operational since 1977 and long a central element of Georgia's electricity supply. The Enguri concrete arch dam is an engineering landmark (height ~ 271.5 m; crest length ~ 728 m; Fig. 3) and represents a notable synthesis of geological setting and engineering creativity. In 2015, the Enguri arch dam was designated a cultural heritage site of national importance.

Fig. 3 General view of the Enguri hydroelectric power station dam and reservoir. *Photo* © panda travel

River valleys and biodiversity. Georgia's river valleys provide habitat for many flora and fauna species. For example, riparian forests of the Kolkheti Plain include Lafan (Pterocarya pterocarpa), Imereti oak (Quercus imeretina), river birch (Alnus barbata), white willow (Salix alba), and others. Floodplain forests along the Alazani also support diverse communities, including wild hazel groves that are regionally distinctive. Georgia's rivers host many fish (including salmon, trout, and catfish) as well as amphibians and reptiles such as the Caucasian salamander (Mertensiella caucasica), Caucasian viper (Vipera kaznakovi), and water snake (Natrix tessellata) (Red Book of Georgia 1978).

3 Canyons

In Georgia's actively uplifting collision-zone landscape, river gorges commonly include canyon reaches that are both visually striking and geodynamically informative. Canyons form where energetic mountain rivers, sustained by high discharge and steep gradients, incise into resistant bedrock. Their vertical cliffs, narrow floors, and deep channel entrenchment typically reflect a combination of strong erosive power and ongoing relief growth. Many Georgian canyons display distinctive geomorphic forms

Fig. 4 Part of the Martvili Canyon, Samegrelo region. *Photo* © Elelne Goldman

and exceptional scenic quality; several are protected as natural monuments and represent outstanding elements of the country's geoheritage. A complete inventory is beyond the scope of this chapter, so we highlight a few representative examples.

Canyons in Georgia can be grouped into two broad types. The first type is developed mainly in Paleozoic carbonate rocks along the southern slope of the western Greater Caucasus. In Imereti, Samegrelo, and parts of Abkhazia, these canyons are commonly less deeply incised than high-alpine canyons; their gorges are narrow and, set within a humid (often subtropical) landscape, they create an especially "exotic" canyon scenery. Well-known examples include the Martvili Canyon (Fig. 4), as well as Balda, Okatse (Okatse Canyon), and related sites.

The second type includes canyons formed in crystalline and other competent rocks, commonly ~100–200 m deep, that extend for long distances and reflect strong structural and tectonic controls. These canyons are best developed in the Greater Caucasus, although comparable reaches also occur within crystalline massifs of the Lesser Caucasus. Examples include canyon reaches associated with the Bzyb, Enguri, Darial, Dashbashi, and Stori systems.

A particularly instructive example is the canyon of the Khalde River (Greater Caucasus, Svaneti Region), a right-bank tributary of the Enguri River and one of the least documented canyon systems in Georgia. The Khalde canyon is exceptionally narrow, with an average depth of ~250–300 m, and is cut into Lower Jurassic clay-limestones. Notably, the canyon crosses a tectonic slip zone over a distance of nearly 1.7 km (Fig. 5). This setting provides a rare natural exposure of how river incision interacts with active structural zones, and it merits further study from geomechanical and hazard perspectives.

Fig. 5 Part of the Khalde River Canyon, Svaneti Region. *Photo* © A. Okrostsvaridze

In eastern Georgia, rivers on the southern slope of the eastern Greater Caucasus commonly have less deeply incised canyon reaches; however, an important exception is the Stori River (Fig. 6). The Stori gorge is cut into intensely weathered and silicified Lower Jurassic clay-rich strata (locally up to ~1–2 km thick) and is aligned for roughly ~25 km along a pronounced regional fault. In several segments, the river forms very narrow canyon reaches displaced by ~1–2 km, with typical depths of ~150–200 m. The Stori gorge is therefore a clear example of canyon development in an active collisional setting where incision is guided by major fault structures.

Special mention should also be made of canyon reaches on the Khrami River. The Khrami (in the apper part it is called Ktsia) originates on the southern slopes of the Adjara–Trialeti Mountains and flows southward, cutting through Quaternary continental basalts of the Javakheti volcanic plateau and into the underlying Paleozoic crys-

Fig. 6 The part of the Story River Canyon, Kakheti Region. *Photo* © A. Okrostsavridze

talline massif. Owing to strong Alpine uplift of this basement, the Khrami has carved a deeply incised valley over several tens of kilometers, including two of Georgia's most scenic canyon landscapes: Dashbashi and Samshvilde. At the confluence area where the Khrami and Chivchavi canyons meet, the medieval fortress-city of Samshvilde was established on a promontory controlling this natural corridor (Fig. 7). The site illustrates the close interaction of geology, topography, and human history, an important theme for geoheritage interpretation.

In addition to their geodynamic significance, Georgia's canyons and gorges have high geoeducational and geotourism value. Their dramatic scenery, clear exposures of geological structures, and opportunities to observe active geomorphic processes make them exceptional outdoor "classrooms." At the same time, because many canyon walls are affected by active denudation and landslide processes, careful management and protection are needed to conserve both their scientific meaning and their aesthetic qualities.

4 Waterfalls

Waterfalls are among Georgia's most prominent and dynamic geomorphological features and are closely linked to modern geodynamic processes. Their formation is commonly controlled by the interaction of tectonic structures (faults and shear zones), lithologic contrasts, and vertical river incision under conditions of active uplift, as well as by adjustments to changes in base level. For these reasons, waterfalls can serve as clear geomorphological indicators of where uplift, structure, and erosion are currently out of balance. Beyond their scientific significance, waterfalls are also outstanding geoheritage features: their high aesthetic value, spectacular appeal, and easily observed geological controls make them effective educational and geotourism sites.

Georgia has dozens of impressive waterfalls across both subtropical and high-mountain landscapes, where the surrounding vegetation, relief, and bedrock exposures together create distinctive scenery. Many sites already have established trails and viewpoints, and waterfalls contribute to local ecosystems and tourism economies. Given the scope of this chapter, we highlight only a few representative examples that illustrate different geodynamic controls.

A large proportion of Georgia's waterfalls are concentrated in the western part of the country, where vertical crustal movements are strong and atmospheric precipitation is abundant. In addition, massive Cretaceous limestone formations are widespread and intensively fractured; under humid conditions, karst processes and groundwater flow commonly generate spectacular emergent springs and falls.

A classic western example is the unique Gega Waterfall, located in Abkhazeti region, on the left tributary of the Bzyfi River, along the short (~25 km) Gega River. Fed largely by groundwater, the Gega River enters a karst fissure and, after traveling underground for approximately ~315 m, re-emerges as a waterfall from fractured limestone blocks. The fall forms multiple streams that cascade from a height of ~60–70 m, producing a striking, easily interpreted expression of karst hydrology and structural control (Fig. 8).

Fig. 7 The place where the Samshvilde and Chivchavi canyons meet and the general view of the remains of the old fortress-town of Samshvilde. *Photo* © Giorgi Boichenko

Fig. 8 Gega Waterfall, Greater Caucasus, Abkhazeti Region. *Photo* © Viva Georgia

A very different setting is represented by the Ushba (Shdugra) Waterfall in Svaneti (central segment of the southern slope of the Greater Caucasus). Fed by meltwater from the Ushba Glacier, it is among Georgia's highest waterfalls (approximately ~120 m). The waterfall's dramatic setting, steep rock walls, glaciers, and high peaks, reflects strong relief and complex geological structure. Its formation is linked to major tectonic contacts where the Paleozoic crystalline basement is thrust over Mesozoic sedimentary units, expressed in the landscape as pronounced escarpments. Local geology is further complicated by intrusion of the Ushba intermediate pluton (~179 Ma) within the Greater Caucasus tectonic shear zone (Okrostsvaridze et al. 2023). The Ushba waterfall occurs near the southern contact zone of this pluton, making it a vivid example of how structural boundaries and contrasting rock properties can localize waterfall formation (Fig. 9).

Waterfalls are relatively less common in eastern Georgia; nevertheless, their geodynamic significance can be equally strong. Here, waterfall development is commonly associated with major regional fault structures of the Greater Caucasus, consistent with rapid uplift in this segment. In

Fig. 9 Ushba Waterfall, Greater Caucasus, Upper Svaneti Region. *Photo* © A. Okrostsavridze

Fig. 10 Makhvali Waterfall, Greater Caucasus, Kakheti Region. *Photo* © A. Okrostsvaridze

such settings, waterfalls can indicate a transient imbalance between uplift and erosional adjustment, specifically, that uplift is outpacing incision at the present time. A representative example is the ~70 m high waterfall on the Makhvali River, one of the headwaters of the Alazani River, formed on the cliffs of the Speroza massif within the southern-slope zone of the Greater Caucasus in Kakheti Region (Fig. 10).

In summary, Georgia's geodynamically significant waterfalls are also visually outstanding expressions of geoheritage. However, because they commonly develop along fractured rock zones and steep escarpments, they can be vulnerable to natural hazards such as earthquakes, landslides, and rockfalls. In addition, concentrated tourism pressure at viewpoints and access routes can threaten site stability and visitor safety. These factors highlight the need for sensible, site-specific management and protection strategies for Georgia's most visited waterfall localities.

5 Lakes

Lakes, like river valleys, canyons, and waterfalls, serve as important indicators of geodynamic processes because their origin, morphology, and sedimentation reflect both the dynamics of regional tectonic settings and changes in climate. Georgia is not among the world's most lake-rich countries: most lakes are small, and total lake-water volume is estimated at ~0.72 km^3. Nevertheless, Georgia's lakes are diverse in origin, including tectonic, glacial, volcanic, karst, moraine-dammed, lagoon, and landslide lakes (Apkhazava 1975). This genetic diversity is itself a useful signal of the country's complex relief, active deformation, and varied surface processes.

More than one-third of Georgia's lakes occur on the young volcanic plateau of southern Georgia (Javakheti), which is therefore often called the "Land of Lakes." This region includes the largest lake by area, Lake Paravani (~37.5 km^2; shallow, ~3.3 m maximum depth), and very water-rich volcanic lakes such as Tabatskuri (reported water volume ~221 million m^3). These plateau lakes are closely tied to volcanism and neotectonic deformation and provide sensitive indicators of hydrologic and climatic variability.

Among the most scenic lakes is Lake Ritsa, a mountain lake on the southern slope of the Greater Caucasus in Abkhazeti region, surrounded by subtropical forest and high peaks (Fig. 11). Located at ~884 m elevation, it has an area of ~1.49 km^2, a maximum depth exceeding 100 m, and a reported water volume of ~94 million m^3. Its proximity (~40 km) to the Black Sea contributes to a distinctive local climate, and the site has long attracted visitors; for example,

Fig. 11 Lake Ritsa, Greater Caucasus, Abkhazeti Region. *Photo* © Eleonora Zaiceva

Fig. 12 Bateti Lake, Lesser Caucasus, Inner Kartli region. *Photo* © Georgian Travel Guide

Stalin's summer residence associated with the Soviet period remains on the lakeshore. In geoheritage terms, Lake Ritsa is especially notable for its genetic origin: it formed when landslides and debris flows on the northeastern slope of the Gagra Ridge dammed the Lashipse River and created a basin that gradually filled with water (Apkhazava 1975). The lake therefore records the interaction of steep relief, mass wasting, and river damming within an active mountain belt.

Another instructive example is Bateti Lake, located on the northern slope of the Trialeti Range (Lesser Caucasus) at ~1313 m elevation. Although often described as glacial in setting, Bateti is a classic landslide-dammed lake: it formed when a landslide on the left slope of the Bateti River valley (a tributary of the Zamakh River) created a natural dam and impounded water (Fig. 12). The site provides a clear, accessible illustration of how slope processes shape hydrologic systems and create lakes in tectonically active terrain.

The most water-rich lake in Georgia is commonly cited as Lake Tabatskuri, located at ~1990 m elevation, in the Javakheti volcanic plateau in the Samtckhe-Javakheti region (Fig. 13). The lake has an area of ~14.2 km^2, a maximum depth exceeding 40 m, and a reported total water volume of ~221 million m^3 (Apkhazava 1975). It is a caldera-type volcanic lake associated with andesitic volcanism and is framed by volcanic ridges, making it a strong "type example" of volcanic-lake genesis in the Javakheti highlands. Lakes of the Javakheti plateau, including Tabatskuri and Paravani, also function as important stopover and resting areas for migratory birds; accordingly, the region holds international significance for bird conservation (IBA status).

Overall, the spatial distribution and genetic diversity of Georgia's lakes reflect the complex interplay of tectonic activity, volcanism, climate, and surface processes. Many lakes respond rapidly to tectonic movements, climatic fluctuations, and mass-wasting events, making them valuable indicators of ongoing landscape change. In addition to geodynamic significance, lakes have ecological-indicator and geoeducational value. Their scenic appeal, clear genetic connections to geological processes, and sensitivity to disturbance make them important geotourism assets, while also underscoring the need for protection and sustainable management, given that lake ecosystems are vulnerable to both natural hazards and human pressures.

6 Geodynamic Activity and Landscape Response: Selected Examples

River Reorganization and Lake Formation on the Javakheti Volcanic Plateau. The Javakheti volcanic plateau is a Quaternary geodynamic system developed within a collisional orogen, where volcanism, tectonic faulting, and surface processes jointly shape the landscape. In this setting, young relief can respond rapidly to geodynamic activity, which is clearly expressed in both the geometry of the river network and the origin and arrangement of lakes.

Digital elevation model (DEM) analysis (Fig. 14) shows that an andesitic lava flow sourced from the Abul–Samsar volcanic ridge blocked the Paravani River valley and fundamentally altered valley morphology. Lava damming created upstream ponding and led to the formation of Lake Paravani, the largest lake in Georgia by area. This example illustrates how decisively lava flows can reorganize hydrologic networks and transform landscape configuration in volcanic tectonic provinces.

In the same valley system, a linear chain of lakes, Abuli, Avchala, and Sagamo, is aligned southward along the Javakheti Fault (Pasquarè et al. 2011) and is clearly delineated on the DEM (Fig. 14). Their spatial coincidence with the mapped fault trace, together with their stepwise change in elevation, indicates strong structural control. These lakes

Fig. 13 Tabatskuri lake, Javakheti volcanic plateau, Lesser Caucasus. *Photo* © Georgian Travel Guide

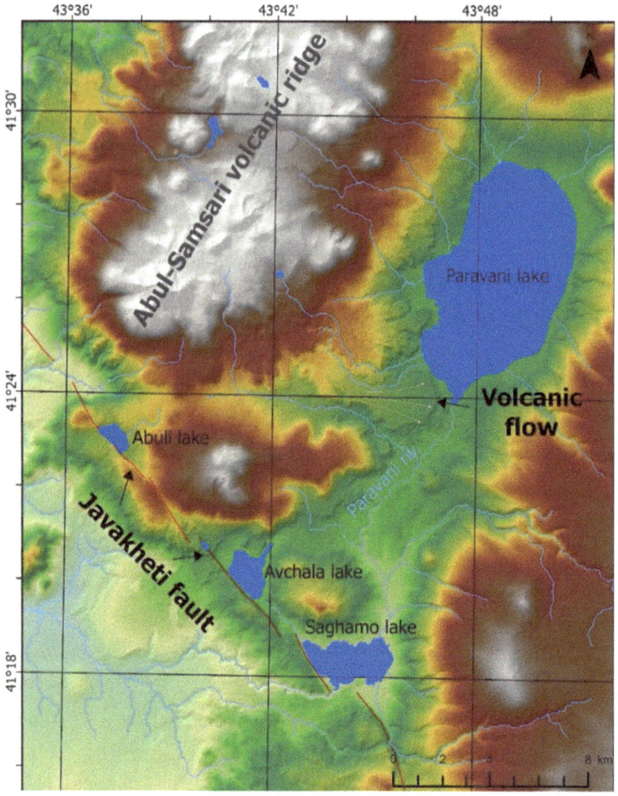

Fig. 14 Digital Elevation Model of the southeastern part of the Abul-Samsar volcanic ridge. It shows two geomorphological phenomena: 1. A river valley blocked by a young volcanic flow, which led to the formation of Lake Paravani; 2. Several lakes formed along a neotectonic fault

likely formed within local depressions created by fault-zone deformation (uplift/subsidence) interacting with active volcanism and associated surface processes.

Taken together, the Paravani lava-dam system and the fault-aligned lake chain provide a clear illustration of how volcanism and tectonics interact to drive landscape evolution on the Javakheti plateau. River damming, lake formation, and drainage reorganization represent both immediate and longer-term landscape responses to geodynamic change. These features are therefore exemplary active geoheritage, with significant scientific, educational, and geotourism value.

Regional Tectonic Faults and the Geometry of the Mtkvari River Gorge. The Mtkvari (Kura) River is one of the principal drainage systems of the South Caucasus collision orogen, and its valley geometry encodes ongoing tectonic and geodynamic processes. The modern form of the Mtkvari valley cannot be explained by fluvial erosion alone; rather, it reflects interaction among regional fault systems, orogenic structures, and neotectonic uplift and subsidence.

The Mtkvari originates in the collisional–volcanic province of eastern Anatolia and enters Georgia from the south. Within Georgia, the river transects the Samtskhe–Javakheti volcanic plateau, effectively separating its western (Samtskhe) and eastern (Javakheti) sectors. In this reach, the river valley closely follows a regional fault zone, consistent with strong structural control of channel position (Okrostsvaridze et al. 2017).

Downstream, the Mtkvari encounters the Eocene Adjara–Trialeti ridge and is unable to cross it directly. As a result, the river is deflected to follow the ridge's southern margin for a distance, receiving tributaries including the Paravani, Uravli, Kvabliani, and Fochovi, which substantially increase discharge and stream power. Near the Akhaltsikhe Depression, the river makes a pronounced bend (on the order of ~140°) and then, along the Borjomi–Kazbegi regional fault, cuts across the Adjara–Trialeti orogenic belt, effectively dividing it into its Adjara and Trialeti segments (Fig. 15).

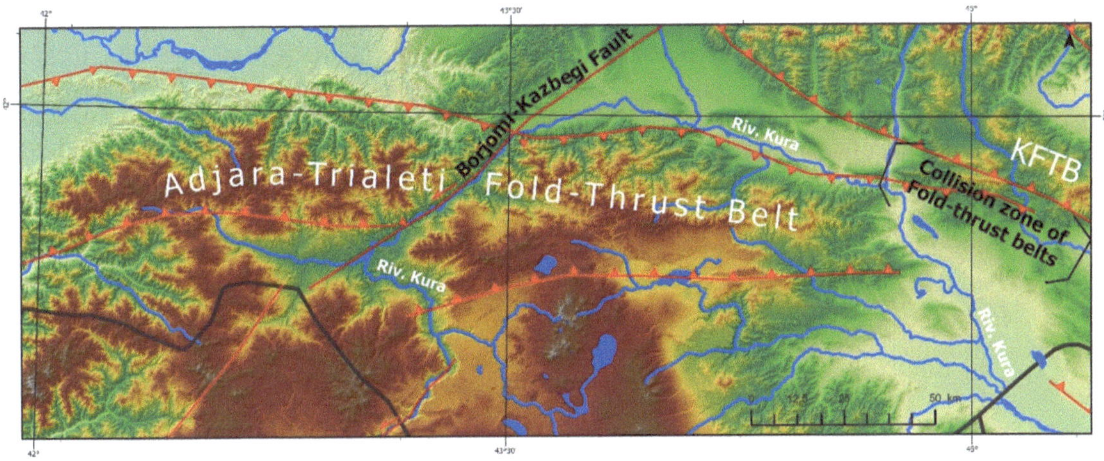

Fig. 15 Geometry of the Mtkvari River bed in Georgia. KFTB–Mtkvari Fold-Trust Belt

After emerging into the Inner Caucasus depression, the river changes direction again and flows generally eastward, following structural trends along the Trialeti range margin on the Transcaucasian microplate. Near Tbilisi, the Mtkvari again changes direction sharply and turns southward, crossing the eastern segment of the Trialeti range and continuing toward the eastern lowlands. These repeated deflections and cross-cutting relationships indicate that tectonic structure and differential uplift strongly influence the river's longitudinal course.

As illustrated in Fig. 16, the geometry of the Mtkvari channel in the Tbilisi fault zone is closely linked to interaction between two slip/fold-thrust systems with opposing vergence: the north-vergent Adjara–Trialeti fold-thrust belt (Lesser Caucasus side) and the south-vergent Mtkvari fold-thrust system on the Greater Caucasus side. Their interaction produced uplift focused in the Saguramo and Ialno ridges. As uplift intensified, the Mtkvari channel migrated from the Yalno ridge sector toward the Teleti ridge sector, shifting from east to southwest. This migration is recorded by the modern Mtkvari gorge and terrace system within and

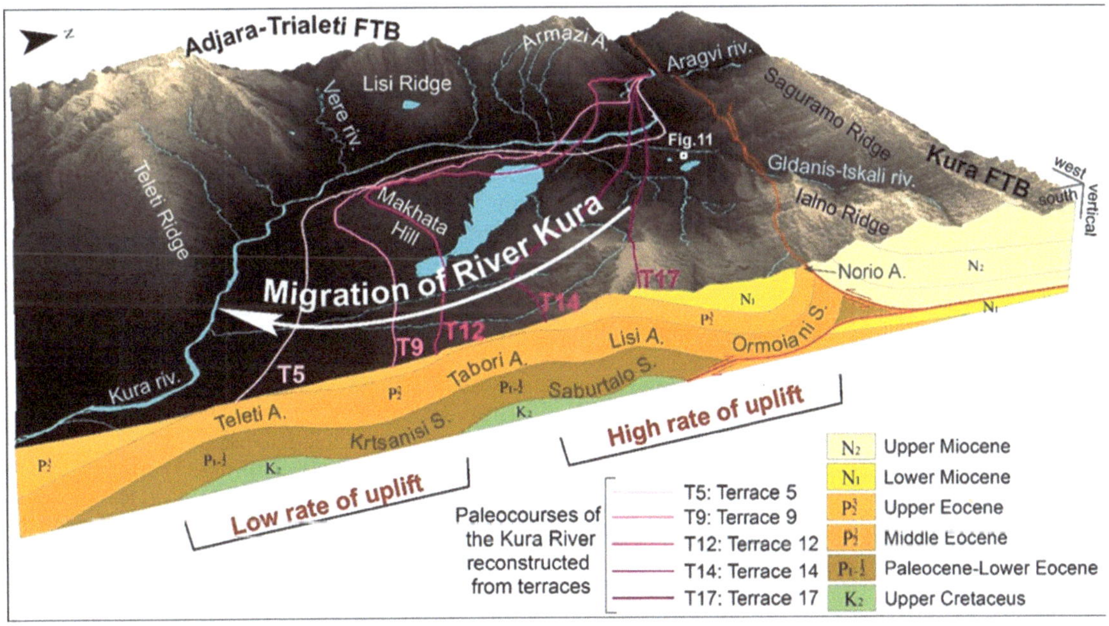

Fig. 16 Block diagram illustrating the proposed mechanism for the formation of the Kura River terraces. Terrace development is linked to a horizontal gradient in uplift rates caused by the collision between two opposing fold-thrust belts: the north-vergent frontal thrust of the Adjara–Trialeti Fold-Thrust Belt in the Lesser Caucasus and the south-vergent frontal thrust of the Kura Fold-Thrust Belt in the Greater Caucasus. Their interaction enhanced uplift in the Armazi–Norio anticline, leading to the progressive southwestward migration of the Kura River, away from the Ialno Ridge and toward the Teleti Ridge

around Tbilisi, whose morphology and structure preserve the geometry of channel movement (Boichenko et al. 2025; Fig. 16).

In summary, the Mtkvari valley's multiple changes in direction, repeated interactions with orogenic belts, and documented channel migration provide one of Georgia's clearest geodynamic indicators at the landscape scale. Especially in the Tbilisi region, the Mtkvari gorge and terrace system serves as a "living witness" to the evolving balance between active thrust systems and spatially variable neotectonic uplift within the South Caucasus collision zone.

7 Conclusion

This chapter has reviewed Georgia's river gorges, canyons, waterfalls, and lakes as water-shaped landforms that developed during the evolution of the Caucasus collision orogen. As such, they function as important geodynamic indicators that encode interactions among neotectonic movements, deformation, lithologic contrasts, and surface (exogenic) processes within an active collisional setting. Their morphology and structural context allow assessment of orogenic evolution trends and of the dynamics of modern landform development. Overall, the gorge–canyon systems, waterfalls, and genetically diverse lakes represent predominantly late Neogene–Quaternary landscapes that can respond rapidly to changes in uplift, base level, and surface processes.

At the same time, these hydrogeological and geomorphological features have long supported, and continue to support, economic, cultural, and social life. Major river valleys provide natural corridors for settlement, agriculture, and transportation, and they underpin water and energy resources. The landscapes around these settlements also contribute to local identity, traditions, folklore, and the many stories associated with rivers, canyons, waterfalls, and lakes. Georgia's mountain rivers additionally offer substantial potential for renewable hydropower, relevant to both domestic supply and regional energy exchange.

Across the book's geological "stage," the examples discussed here span the Greater Caucasus belt, the Adjara–Trialeti domain of the Lesser Caucasus, and the Javakheti volcanic highlands. Many of these landforms are also practical anchors for Georgia's candidate geoparks and emerging national georoutes (see Chaps. 2, 3 and 6).

Accordingly, Georgia's river gorges, canyons, waterfalls, and lakes should be understood not only as natural "records" of active geodynamic processes, but also as geoheritage assets of multifaceted value. Their protection and sustainable use are therefore essential both for conserving natural heritage and for supporting long-term cultural and socio-economic development, particularly given that many sites are vulnerable to hazards (e.g., landslides, rockfalls) and to localized tourism pressure.

References

Apkhazava I (1975) Georgian Lakes. Metcniereba, Tbilisi, pp 1–182

Boichenko G, Cowgill C, Godoladze T, Okrostsvaridze A, Torre R, Binnie S, Ritter B (2025) Kura (Mtkvari) river terraces record fluvial response to the collision of the greater and lesser Caucasus thrust belts, Georgia. Geomorphology 476:109656

Del Hoyo J, Elliott A, Sargatal J (1994) Handbook of the birds of the world. In: Volume 2: new world vultures to Guineafow. https://doi.org/10.5962/p.357683

Gamkrelidze I, Okrostsvaridze A, Maisadze F, Basheleisvili L, Boichenko G, Skhirtladze I (2019) Main features of geological structure and geotourism potential of Georgia, the Caucasus. Mod Environ Sci Eng 5(5):422–442. https://doi.org/10.15341/mese(2333-2581)/05.05.2019/010

Gigineishvili G (2000) Inland waters. In: The book: "geography of Georgia". Metcniereba, Tbilisi, pp 116–133

Geladze V, Bolashvili N, Karalashvili T, Machavariani N (2022) Inland waters. In: Bolashvili N, Neidze V (eds) The physical geography of Georgia. Geography of the physical environment. Springer, Charm, pp 101–125. https://doi.org/10.1007/978-3-030-90753-2_10

Gugushvili V (2003) Concept of geodynamic and geoecological lithomonitoring for mineral resources and groundwater management in Georgia. Georgian Geological Department, Tbilisi, pp 1–55

Khmaladze G (2009) Water resources of Georgia. Mecniereba, Tbilisi, pp 1–74

Okrostsvaridze A (2017) The Argonauts: a modern investigation of the mythical "Gold Sands". LAMBERT Academic Publishing, Saarbrucken, pp 1–92

Okrostsvaridze A, Chung SL, Lin YC, Skhirtladze I (2017) Geometry and zircon U-Pb geochronology of the Mtkvari ignimbrites flow, Samtske-Javakheti volcanic Highland, lesser Caucasus. Bull Georg Acad Sci 11(4):82–90

Okrostsvaridze A, Lee YH, Tormey D, Skhirtlade I (2023) U-Pb zircon chronological constraints for three stages of syn-orogenic plutonic magmatism of the greater Caucasus Svaneti segment. Georg Epis 46(2):194–210. https://doi.org/10.18814/epiiugs/2022/022023

Pasquare F, Tormay DR, Vezzoli M, Okrostsvaridze A, Tutberidze B (2011) Mitigation of the consequences of extreme events on strategic facilities: evaluation of volcanic and seismic risk affecting the Caspian oil and gas pipelines in the republic of Georgia. J Environ Manag 92(7):1774–1782

Red Book of Georgia (1978) Sabchota Sakartvelo. Tbilisi:1–256

Reilinger R, McClusky S, Vernant P, Lawrence S, Ergintav S (2006) GPS constraints on continental deformation in the Africa-Arabia Eurasia continental collision zone and implications for the dynamics of plate interactions. J Geophys Res 111:B05411. https://doi.org/10.1029/2005JB004051

Rolland Y (2017) Caucasus collisional history: review of data from geology, geophysics and thermochronology. J Asian Earth Sci 145:328–351. https://doi.org/10.1016/j.jseaes

Wang SJ, Lee YH, Chung SL, Okrostsvaridze A (2015) Cenozoic Mountain building process in the greater Caucasus, from subduction to collision. In American geophysical union, fall meeting abstracts 2014 pp T21A–4565

Zautashvili B, Mkheidze B (2022) Hydrogeology and mineral waters. In: Bolashvili N, Neidze V (eds) The physical geography of Georgia. Geography of the physical environment. Springer, Cham. https://doi.org/10.1007/978-3-030-90753-2_3

Geoheritage of the Georgian Black Sea Coast: Between the Greater and Lesser Caucasus

Avtandil Okrostsvaridze, Daniel Tormey, Giorgi Lominadze, and Zurab Janelidze

Abstract

The Black Sea coast of Georgia forms a distinctive coastal corridor situated between the collisional mountain systems of the Greater and Lesser Caucasus. Extending for approximately 310 km along the margin of the historical Colchis Plain, this coast is a geodynamically active zone where mountain uplift and exhumation, river transport, and coastal sedimentation operate as a coupled land to sea system. Over the Holocene, sediment supplied by water-rich rivers draining the Greater and Lesser Caucasus has progressively infilled parts of the coastal marine basin, producing the young alluvial and deltaic landscapes of the Colchis lowland and driving local shoreline progradation. This interaction between uplift, sediment flux, and coastal processes has generated a suite of geomorphic features that are both scientifically informative and visually accessible. In this chapter we highlight three representative examples, Cape Bichvinta (Pitsunda), Lake Paliastomi, and the magnetite rich beach sands of the Guria coast. Together, these sites illustrate the Georgian Black Sea margin as a living laboratory for collision zone to coastal system interaction, with strong educational and geotourism value and clear relevance for Georgia's emerging Key Geoheritage Area inventory and candidate UNESCO Global Geopark and georoute narratives.

A. Okrostsvaridze (✉) · Z. Janelidze
Institute of Earth Sciences, Ilia State University, Tbilisi, Georgia
e-mail: avtandil.okrostsvraidze@iliauni.edu.ge

D. Tormey
Catalyst Environmental Solutions, Santa Monica, USA

IUCN World Commission on Protected Areas, Rue Mauverney, Switzerland

G. Lominadze
Tbilisi State University, Vakhushti Bagrationi Institute of Geography, Tbilisi, Georgia

Keywords

Black Sea coast · Colchis plain · Coastal geomorphology · Sediment transport

1 Introduction

Geoheritage is often discussed through the lens of mountains, volcanism, and active tectonic zones, where geological processes are dramatic and exposures are obvious. Coastal systems, by contrast, can receive less attention in geoheritage narratives, even though the land-sea interface preserves some of the clearest and most dynamic records of landscape response to mountain building. The Georgian sector of the Black Sea coast is an excellent example. Along a relatively short corridor, the coast links the Greater Caucasus, the low relief Colchis Plain, and the Lesser Caucasus, and it records how an active collision orogen delivers sediment, reshapes drainage, and builds new coastal landforms through time.

The Black Sea coast of Georgia is a dynamic geomorphological system whose present form reflects interaction among tectonic uplift and subsidence, river sediment supply, and changing sea level. While the regional geological framework was established over Mesozoic and Cenozoic time, much of the modern coastal configuration and the adjacent Colchis lowland is Holocene in age. In this setting, rivers draining the Greater and Lesser Caucasus transport large sediment loads to the coastal margin, where deposition, shoreline migration, and lagoon development record the coupled evolution of mountains, rivers, and sea.

As a result of these interacting processes, several significant geomorphic and sedimentary features occur along Georgia's Black Sea coast. In this chapter, we focus on three representative examples that are both scientifically informative and accessible for interpretation: (1) Cape Bichvinta (Pitsunda), (2) Lake Paliastomi, and (3) the magnetite rich

A. Okrostsvaridze and D. Tormey (eds.), *Geoheritage of Georgia*, Geoheritage, Geoparks and Geotourism,
https://doi.org/10.1007/978-3-032-27639-1_13

beach sands of the Guria coast. Together these sites provide a compact set of coastal geoheritage examples with clear potential for education and geotourism and direct relevance to Georgia's emerging Key Geoheritage Area inventory and UNESCO Global Geopark and georoute concepts.

2 Brief Geological History of the Region

The Black Sea coast of Georgia is geologically complex because it lies at the intersection of the Black Sea basin system and the principal domains of the Caucasus collision orogen. Along this margin, the Greater Caucasus, the Transcaucasian microplate, and the Lesser Caucasus domains converge, and their combined tectonic evolution controls basin subsidence, sediment supply, and coastal configuration. The sedimentary cover of the Transcaucasian microplate forms the Rioni Basin, which is also known historically and geographically as the Colchis Plain (Fig. 1).

The Black Sea itself is a large basin of approximately 423,000 km^2 bordered by six countries. It includes western and eastern deep basins that reach depths of about 2–2.2 km and are underlain by oceanic or sub-oceanic crust. Sedimentary cover is very thick in both basins, and they are separated by a submerged ridge of continental affinity, the Mid Black Sea High, also covered by several kilometers of sediment (Nikishin et al. 2003). The Black Sea basin system developed mainly during Cretaceous time as a back arc basin associated with the northern active margin of the Tethyan basin. Rifting and crustal thinning intensified during the mid Cretaceous, and the region subsequently experienced episodes of compression and renewed subsidence. Modeling suggests that lithospheric bending related to compressional deformation contributed to rapid additional subsidence in parts of the basin (Nikishin et al. 2003).

Holocene sea level history of the Black Sea has also attracted wide attention, including the hypothesis of a rapid Mediterranean inflow about 8–7 ka and links made in popular accounts to the Noah's Flood narrative (e.g. Ryan et al. 2003). Subsequent work has not identified geological, paleontological, palynological, or archaeological evidence for a catastrophic flooding event of the type proposed, and the topic remains best treated cautiously and in the context of broader Black Sea sea level research (e.g. Yanko et al. 2014). For the Georgian coast, the key point is that late Quaternary and Holocene sea level fluctuations, combined with high sediment supply, have shaped shoreline position, lagoon formation, and deltaic growth.

The Caucasus collision orogen borders the Black Sea to the east and extends roughly 1200 km from the Caspian to the Black Sea. It formed during closure of Tethyan oceanic domains and collision between the northward moving Arabian Plate and Eurasia, a process that continues in a post-collisional regime (e.g. Okrostvaridze & Tormey 2011; Rolland 2017; Gamkrelidze et al. 2019). The Greater Caucasus belt to the north comprises a pre-Jurassic crystalline basement and an overlying Mesozoic to Cenozoic

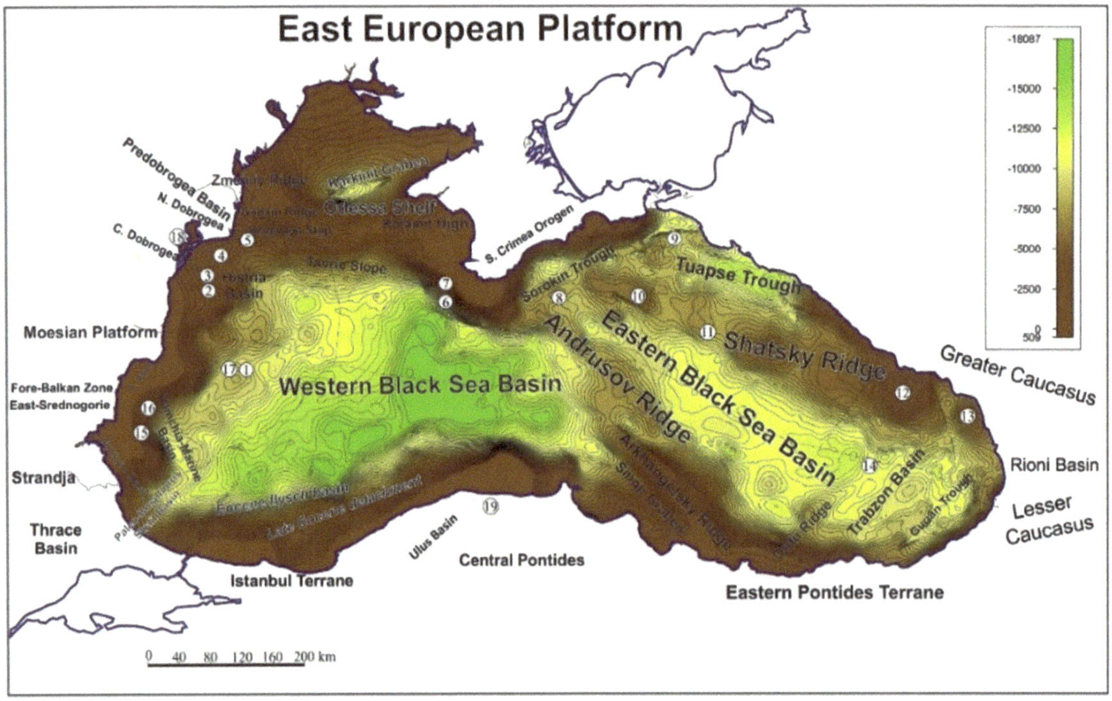

Fig. 1 Geographical map of the Black Sea and the main geological structures around it (modified after Nikishin et al. 2003)

sedimentary and magmatic cover, reflecting earlier island arc and basin evolution followed by later collision related uplift and exhumation. South of the Greater Caucasus lies the Transcaucasian microplate, whose foundation includes older metamorphic and igneous rocks overlain by thick unmetamorphosed sedimentary and volcaniclastic successions. This microplate is exposed in places within central Georgia, including the Dzirula uplift (Okrostsvaridze & Tormey 2013), and it subsides westward toward the Black Sea to form the Colchis lowland and the Rioni Basin.

To the south, the Transcaucasian microplate is bounded by major fault systems and the Adjara Trialeti belt of the Lesser Caucasus, an Eocene age volcanogenic and tectonic domain interpreted as part of a back arc to rift related system that developed on the southwestern edge of the microplate (Okrostsvaridze et al. 2021). Along the Georgian margin, this framework creates high relief contrasts between uplifted mountain belts and subsiding basins. It is this combination of ongoing uplift and exhumation, high precipitation and river discharge, and basin subsidence and sea level change that sets the stage for the young and highly dynamic coastal geoheritage features of the Georgian Black Sea margin discussed below.

3 The Geohistory of the Coastal Zone of the Black Sea of Georgia

The Black Sea coast of Georgia lies between the orogenic belts of the Greater and Lesser Caucasus and forms an approximately 310 km long corridor where sea and land interact (Fig. 2). Rapid uplift and exhumation of both Caucasus ranges intensified during the late Miocene in the post-collisional stage, approximately 7–5 million years ago (e.g. Wang et al. 2015; Rolland 2017). Along this corridor, three major geological domains meet and shape different coastal behaviors: (1) uplifted and eroding mountain belts of the Greater and Lesser Caucasus, and (2) the subsiding Transcaucasian microplate and its sedimentary cover expressed as the Colchis lowland. As a result, relationships between land, rivers, and sea vary along the coast, producing coastal landforms of different genetic types (e.g. Geladze et al. 2022).

Many coastal geoheritage features occur along the Georgian Black Sea margin. In this chapter we focus on three contrasting examples that illustrate the range of processes and heritage values: (1) Cape Bichvinta (Pitsunda), (2) Lake Paliastomi, and (3) the magnetite rich beach sands of the Guria coast. Together these sites show how uplift, sediment supply, shoreline currents, sea level change, and coastal deposition produce readable geomorphic indicators in a young and dynamic coastal system.

Bichvinta (Pitsunda) Cape. Cape Bichvinta is a coastal promontory on the Black Sea margin near the mouth of the Bzyb River. The cape covers an area on the order of 15 km^2 and forms a low relief plain that slopes gently toward the sea (Fig. 3). Much of the surface lies only a few meters above sea level and includes shallow depressions, locally marshy areas, and low mounds. The nearby Lake Inkiti is an additional coastal feature; it is commonly interpreted as a former bay or lagoon connected to the Black Sea during earlier stages of coastal evolution (Apkhazava 1975).

Cape Bichvinta is also notable for its relict pine forest along the coastal strip, a key element of the landscape identity and climate experience of the cape. A wide, light colored beach composed largely of carbonate sand encircles much of the promontory (Fig. 4). The combination of low relief, wide beaches, and sheltered waters produces a coastal environment that has long supported settlement and tourism.

From a genetic perspective, the cape is best interpreted as a Holocene age accretionary coastal feature built by the interaction of sediment supply from rivers draining the Greater Caucasus and alongshore transport and deposition. In simplified form, the controlling factors include (1) late Cenozoic uplift and intense erosion in the Greater Caucasus, (2) high precipitation and high river discharge delivering carbonate rich sediment to the coast, (3) coastal currents that redistribute sediment alongshore, and (4) local shoreline geometry and bedrock controls that promote sediment trapping and promontory growth. The cape's main growth phase likely coincided with late Holocene stabilization of sea level, when shoreline position approached its modern range and soils and vegetation developed on newly formed coastal deposits (Janelidze 2008; Janelidze 2015).

Geoheritage significance and management. Cape Bichvinta is a visually accessible example of coastal accretion and landform building in an active orogen to sea system, combined with relict ecological and cultural landscape values. It has strong educational and geotourism potential because the relationship between river sediment supply, beaches, and promontory growth can be interpreted directly in the field. The site is also sensitive to coastal erosion, storm events, and development pressure, and therefore benefits from careful management of shoreline infrastructure and visitor use. Within a national geoheritage programme, Bichvinta is a good candidate coastal geosite that could be included as a node within a future Colchis coastal Key Geoheritage Area cluster and within Black Sea georoute narratives.

Lake Paliastomi. Lake Paliastomi is located within the Colchis lowland south of the city of Poti and is separated from the Black Sea by a narrow barrier strip (Fig. 5). The lake lies very close to sea level and is shallow, with a maximum depth of only a few meters, yet it occupies a

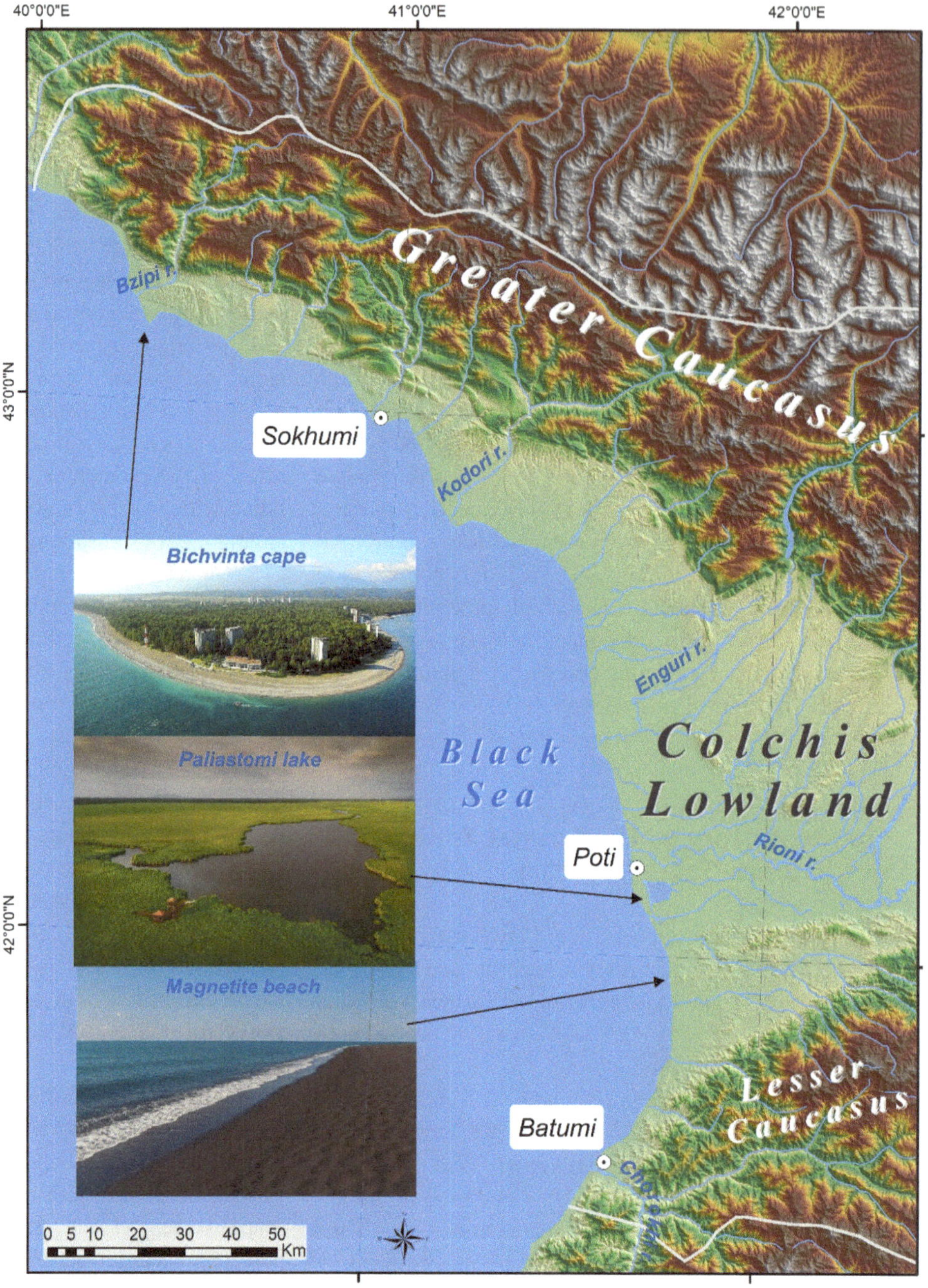

Fig. 2 Aster global digital elevation map of the Black Sea coastal area of Georgia, with geoheritage sites

significant surface area and forms a major wetland landscape. Lake Paliastomi is fed by local river inflow and is linked to the Black Sea through a controlled outlet or channel system (Janelidze 2020).

Genetically, Lake Paliastomi is best understood as a lagoon type lake formed by Holocene shoreline evolution. During late Pleistocene low sea level conditions, much of the present Colchis lowland was exposed land and the modern lagoon basin did not exist. As sea level rose in the early Holocene, coastal transgression and sediment redistribution reshaped the nearshore zone. By mid Holocene time, relative stabilization of sea level combined with shore parallel currents and barrier dune development created an isolated lagoon basin separated from the open sea. This barrier development and isolation produced the modern lake system (Janelidze 2020). Similar former lagoon systems are also recognized elsewhere on the Colchis margin, including basins that have since been filled by sedimentation (Laermanns et al. 2018).

Geoheritage significance and management. Lake Paliastomi is a classic example of lagoon formation and barrier development in a young, high sediment supply coastal plain. It is also a key ecological asset and an important component of Kolkheti National Park, with high value

Fig. 3 General view of Cape Bichvinta from the southeast. *Photo* © TutTravelRu

Fig. 4 A general view of one of the beaches on the Bichvinta coastline, which is made of fine-grained limestone sand. *Photo* © Turpomru

for migratory birds and wetland biodiversity (Fig. 6). The site therefore illustrates coupled geo ecological heritage. Its sensitivity to water level change, sedimentation, and water quality makes sustainable management essential, particularly where upstream land use, coastal infrastructure, or visitor pressure could affect hydrologic connectivity and habitat integrity. In a KGA and UNESCO Global Geopark context, Paliastomi is a natural flagship geosite within a Colchis coastal cluster that links coastal geomorphology, wetlands, and cultural landscape history.

Magnetite rich beach sands of the Guria coast. Along the Guria segment of the Georgian Black Sea coast, between the mouths of the Natanebi and Supsa rivers, sections of the beach are characterized by dark, fine grained sands enriched in heavy minerals, including magnetite and hematite. These sands represent a coastal placer system produced by erosion of mineralized source rocks inland, fluvial transport to the coast, and hydraulic sorting by waves and currents that concentrate dense minerals in beach deposits (Fig. 7).

In this sector, heavy mineral material is supplied by rivers draining inland source areas, including mineralized zones associated with Eocene magmatic and hydrothermal systems in Guria (Okrostsvaridze et al. 2022). Rivers deliver terrigenous and heavy mineral sediment to the shoreline, and alongshore currents redistribute and concentrate the heavy mineral fraction along specific stretches of beach. The result is a set of magnetite rich beach segments that vary in thickness and grade over distances of kilometers.

Fig. 5 General view of Lake Paliastomi. The Black Sea is visible above on the right. *Photo* © Georgian Travel

Fig. 6 Kaparchina River, through which Lake Paliastomi is connected to the Black Sea. *Photo* © Georgian Travel

Geoheritage significance and management. The Guria magnetite sands provide a clear and accessible example of coastal placer formation and sediment sorting processes, linking inland geology, river transport, and shoreline dynamics in a way that visitors can observe directly.

These beaches are also culturally important as recreational and wellness destinations, and local narratives commonly attribute therapeutic value to the warm dark sands. From a geoheritage perspective, interpretation should emphasize process visibility and landscape linkage rather than medical claims. Management considerations include maintaining beach integrity under coastal erosion and storm regimes, controlling unregulated extraction where it occurs, and managing visitor pressure in popular resort sectors. As part of Georgia's geoheritage programme, the Guria placer sands are excellent candidates for inclusion in a Colchis coastal Key Geoheritage Area cluster and for incorporation into a Black Sea coastal georoute.

4 Black Sea Colchis Coastal Zone Dynamics and its Influence on Cultural Heritage

Coastal dynamics along the Georgian sector of the Black Sea have strongly influenced the Colchis coastal zone from prehistory through Greek colonization and later periods. Changes in relative sea level, shoreline position, and lagoon

Fig. 7 View of the Shekvetili magnetite beach in the Guria sector of the Black Sea. *Photo* © CarTravelPaths

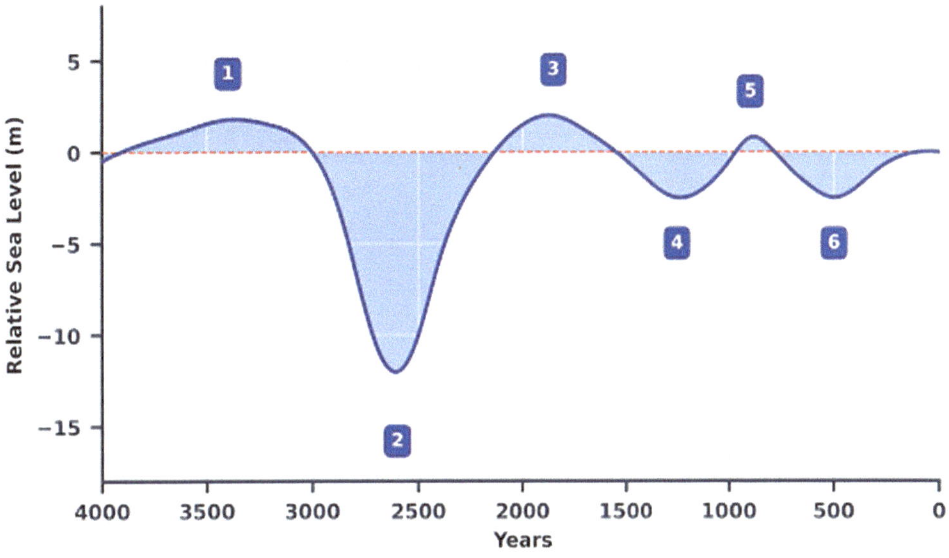

Fig. 8 Curve of Black Sea level fluctuations over the last 4000 years. Created according to Balabanov and Gaprindashvili (1987). Transgressive and regressive phases: 1- New Black Sea transgression, 2- Phanagorian regression, 3- Nymphic transgression, 4- Medieval regression, 5- Medieval transgression, 6- Late Medieval regression

and wetland development caused the coastline to advance or retreat by hundreds of meters and, in some settings, by kilometers. These shifts created recurring challenges for settlement and navigation. They also help explain why traces of cultural heritage along parts of the Colchis lowland are either destroyed by erosion and inundation or preserved beneath thick peat deposits within marshy, anoxic environments.

Black Sea water level history since the late Pleistocene has been widely studied (e.g. Janelidze 1980; Balabanov and Gaprindashvili 1987; Janelidze 2015; Dean et al. 2016). During the late Pleistocene, the Black Sea functioned as an isolated basin with low water level relative to the global ocean. Following late glacial warming and sea level rise, the Black Sea reconnected with the Mediterranean system, and by about 6 ka its level approached the modern range. Even after this time, transgressive and regressive oscillations continued and influenced lagoon formation, barrier development, and shoreline migration (Fig. 8).

For the last several thousand years, Balabanov and Gaprindashvili (1987) proposed a sequence of relative sea level fluctuations that includes a late Holocene transgressive phase reaching above modern level, followed by the Phanagoria regression, later oscillations including medieval low stands, and finally the modern transgressive tendency. Some details and magnitudes remain debated, particularly for the Phanagoria regression, but the overall pattern of repeated late Holocene shoreline shifts is consistent with broader regional sea level reconstructions, including studies from the eastern Mediterranean after the Black Sea became an open sea connected to Mediterranean sea level change. The Black Sea level fluctuation of Fig. 9 is also consistent with more recent results from Eastern Mediterranean Sea level dynamics (e.g., Dean et al. 2016).

These shoreline changes provide a plausible framework for understanding the distribution and preservation of archaeological sites on the Colchis plain. As sea level rose and wetlands expanded, parts of the lowland became increasingly marshy, and settlements were forced to adapt or relocate. In some areas, the development of peat forming wetlands created an anoxic preservation environment that can protect organic materials and wooden structures. A well documented example is the Early Bronze Age settlement of Ispani in the modern Kobuleti district. The site is covered by a peat layer on the order of 1–2 m thick and preserves wooden frame structures and associated cultural materials, including pottery fragments and bronze related artifacts (Jibladze 2007). Similar peat covered contexts are reported for other Colchis lowland sites, suggesting that peat accumulation has been a key mechanism for preserving cultural layers in areas that were repeatedly affected by wetland development and shoreline migration.

Coastal change also affected the geography of Greek colonization and later navigation. Historical accounts describe ports, river mouths, bays, and navigable river reaches whose locations are difficult to reconcile with the modern coastline if shoreline change is not considered. Recent integrated geological and archaeological work indicates that many ancient descriptions were likely accurate for their time and that the perceived inconsistencies arise from later coastal evolution rather than from unreliable reporting. In this context, the reported partial submergence of ancient coastal city remains such as Dioscurias and Pityus or Pityunt and the shifting position of harbors and river mouths can be interpreted as outcomes of changing relative sea level combined with high sedimentation and subsidence in the Colchis basin.

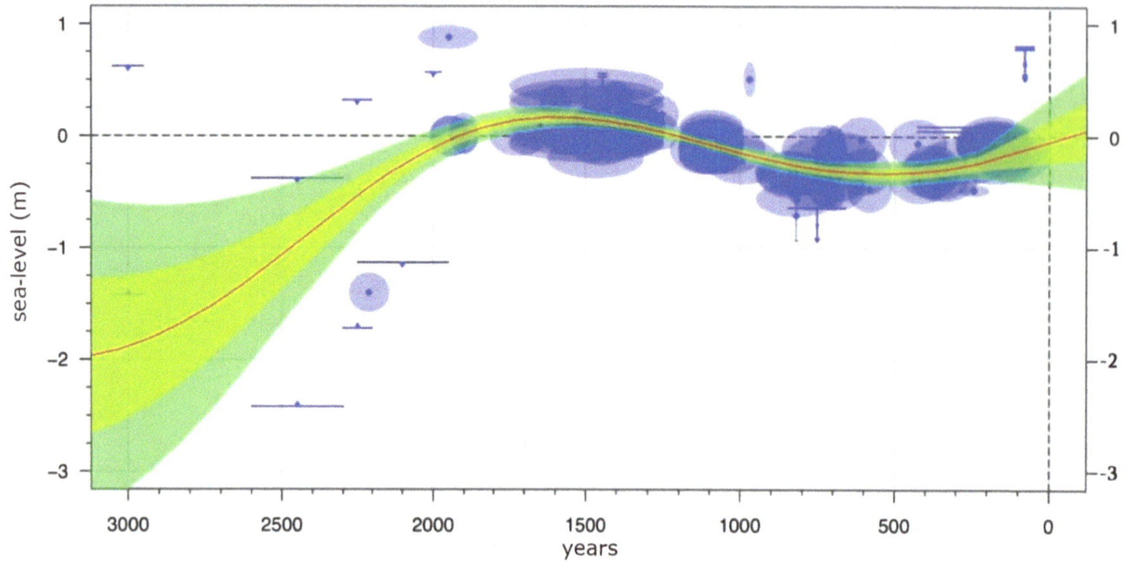

Fig. 9 An indicator of the change in sea level in the Eastern Mediterranean over the last 3000 years, based on archaeological and biological data (blue); also statistical analysis fields: yellow–68% probability, green–95% probability; the red curve corresponds to the average statistical sea level data, and zero–to the modern sea level (Dean et al. 2016)

The shoreline history also intersects with the cultural narrative of the Argonaut tradition. If relative sea level and wetland conditions differed substantially during the Bronze Age, then navigation routes, river gradients, and access to inland centers would also have differed from today. In broad terms, higher sea level and slower flowing lowland rivers could have increased upstream navigability, while later regression, coastal progradation, and basin subsidence would tend to bury or obscure earlier coastal landscapes beneath sediment and peat. Locating Bronze Age coastal zones and associated cultural layers therefore requires interdisciplinary work combining geomorphology, stratigraphy, and archaeology. Candidate Bronze Age sites reported from the Colchis lowland include Ispani, Naukvhamo, Diyaha Gudzuba, Chakvi, Anaklia, and Pitshora (Jibladze 2007), and their distribution can be reinterpreted through the lens of coastal change.

Geoheritage implications. The Georgian Black Sea margin illustrates a coupled geocultural heritage system in which active geomorphic processes directly influence cultural landscapes. For geoheritage practice, this means that coastal geosites and associated archaeological resources should be interpreted together, and that management should recognize both vulnerability and preservation potential. Erosion, inundation, and coastal development can destroy heritage, while wetlands and peat can preserve it. This dynamic context strengthens the case for a Colchis coastal Key Geoheritage Area cluster and for a thematic Black Sea coastal georoute that integrates coastal geomorphology, wetlands, and the cultural history of Colchis within a single interpretive framework.

5 Discussion

Geoheritage along the Georgian Black Sea margin is best understood not as isolated objects, but as an interacting system that records long term coupling between a collisional orogen and a marine basin. The system includes uplift and exhumation of the Greater and Lesser Caucasus, subsidence of the Transcaucasian microplate and the Colchis basin, intense erosion and river transport, Holocene sediment accumulation, and active coastal morphodynamics. Together these linked processes create an excellent natural laboratory for scientific observation and for public education about how mountains, rivers, and coasts evolve as a connected land to sea continuum.

The three examples highlighted in this chapter represent contrasting but complementary process types. Cape Bichvinta records coastal accretion and shoreline growth in a high sediment supply setting. Lake Paliastomi records lagoon formation and barrier development linked to relative sea level change and coastal sediment redistribution. The Guria magnetite sands record coastal placer formation and the selective concentration of heavy minerals driven by fluvial delivery and wave and current sorting. Considered together, these sites provide a compact and readable set of geoheritage features that can be interpreted in the field and linked directly to the broader tectonic and geomorphic evolution of the region.

A common question raised by the Georgian coast is why large natural bays suitable for sheltered harbors are comparatively limited. In large part, this reflects the alluvial and accretionary character of the Colchis coastal plain. Rapid late Cenozoic uplift of the Greater and Lesser Caucasus increased erosion rates, and high river discharge deliver large sediment loads to the coastline. This sediment tends to infill shallow depressions and smooth coastal irregularities, while alongshore currents redistribute material and further reduce the persistence of embayments. In addition, along sectors where bedrock is present, the resistance contrast of tectonized crystalline rocks can shape smaller coves, but the overall coastline is dominated by sedimentary construction and smoothing rather than by stable rocky bays. For geoheritage interpretation, this is not a deficiency but a defining process signature of a young, active, sediment rich collision zone coast.

6 Conclusion

This chapter shows that the Georgian Black Sea coast represents a distinctive land to sea geodynamic and geomorphic system linked to the Caucasus collision orogen. Uplift and exhumation of the Greater and Lesser Caucasus, combined with high precipitation and high river discharge, drive intense sediment delivery to the coastal margin. In turn, basin subsidence, sea level fluctuations, and coastal currents organize deposition and shoreline migration, producing the young alluvial and deltaic landscapes of the Colchis lowland. Within this setting, major coastal geoheritage features include the accretionary Cape Bichvinta promontory, the lagoon origin Lake Paliastomi system, and the Guria magnetite rich beach sands. These sites have strong educational and geotourism value because the controlling processes can be read directly in the landscape.

From a national geoheritage programme perspective, the Georgian Black Sea margin is well suited to treatment as a coherent coastal geoheritage cluster. It offers a practical basis for defining a Colchis coastal Key Geoheritage Area that integrates active geomorphology, wetlands, and the coupled geo cultural history of coastal change. It also provides excellent content for a thematic Black Sea coastal georoute and for inclusion as a coastal module within broader candidate UNESCO Global Geopark narratives that connect inland mountains, rivers, karst and wetlands to the dynamic coast. Because these coastal landforms and associated cultural resources are sensitive to erosion, storms, development pressure, and climate change, their protection and sustainable management should be treated as a priority for long term conservation and sustainable tourism.

References

Apkhazava I (1975) Georgian Lakes. Metcniereba, Tbilisi, pp 1–182

Balabanov IP, Gaprindashvili MV (1987). On the location of the ancient Greek cities of Pitiunt and Dioscuria. Magazine "Matsne", series "history, archeology, ethnography and history of art", Tbilisi, pp 151–160. (In Russian)

Dean S, Sivan D, Evelpidou E (2016) Re-assessing East Mediterranean Sea-level trends: 3000 years of archaeological indicators in Greece and Israel. In: Proceedings of the fourth plenary conference of the IGCP 610 project, Tbilisi, pp 48–51

Gamkrelidze I, Okrostsvaridze A, Maisadze F, Basheleisvili L, Boichenko G, Skhirtladze I (2019) Main features of geological structure and geotourism potential of Georgia, the Caucasus. Mod Environ Sci Eng 5(5):422–442. https://doi.org/10.15341/mese(2333-2581)/05.05.2019/010

Geladze V, Bolashvili N, Karalashvili T, Machavariani N (2022) Inland waters. In: Bolashvili N, Neidze V (eds) The physical geography of Georgia. Geography of the physical environment. Springer, Cham, pp 101–125. https://doi.org/10.1007/978-3-030-90753-2_10

Janelidze CP (1980) Paleogeography of Georgia in the Holocene. Metsniereba, Tbilisi, pp 1–178

Janelidze Z (2008) The age of the land part of the Black Sea coastal zone of Georgia. Proceedings of the Vakhushti Bagrationi Institute of Geography 2(81). Tbilisi, pp 176–179

Janelidze Z (2015) The Black Sea coastal zone of Georgia: retrospective-prognostic analysis. Tbilisi, Ilia State University Press, pp 1–135 (In Georgian)

Janelidze Z (2020) The history of Lake Paliastomi. J Amirani 23:355–361

Jibladze L (2007) Settlements of the Colchis plain of the III-I millennia BC. Doctoral dissertation. Tbilisi, pp 1–247. (In Georgian)

Laermanns H, Kelterbaum D, May et al (2018) Mid-to late Holocene landscape changes in the Rioni delta area (Kolkheti lowlands, W Georgia). Quat Int 45:85–98

Nikishin AN, Korotaev MV, Ershov AV, Brunet MF (2003) The Black Sea basin: tectonic history and Neogene–quaternary rapid subsidence modeling. Sediment Geol 156(1–4):149–168

Okrostsvaridze A, Tormey D (2011) Evolution of the Variscan orogenic plutonic magmatism of the greater Caucasus. Journal of Nepal Geological Society, Special Issue 403:45–52

Okrostsvaridze A, Tormey D (2013) Phanerozoic continental crust evolution of the inner Caucasian microplate: the Dzirula uplift. Episodes 36(1):31–39

Okrostsvaridze A, Chang YH, Chung SL, Rabairol F, Gogolaze S (2021) Paleozoic xenoliths in Eocene plutons: evidence for the destruction of pre-Jurassic crystalline basement beneath the Adjara-Trialeti belt. Lesser Caucasus Geologica Carpathica 72(4):299–315

Okrostsvaridze A, Akimidze K, Bluashvili D, Boychenko G, Gabrielashvili R (2022) Thorium-an energy of the future and its ore occurrences in Georgia. Universal, Tbilisi, pp 1–222

Rolland Y (2017) Caucasus collisional history: review of data from geology, geophysics and thermochronology. J Asian Earth Sci 145:328–351. https://doi.org/10.1016/j.jseaes

Ryan WBF, Major CO, Goldstein SL (2003) Catastrophic flooding of the Black Sea. Annu Rev Earth Planet Sci 31:525–554

Yanko V, Mudle PJ, Kadirin LE (2014) Holocene marine transgression in the Black Sea: new evidence from the northwestern Black Sea shelf. Quat Int 345(6):100–118. https://doi.org/10.1016/j.quaint.2013.07.027

Wang SJ, Lee YH, Chung SL, Okrostsvaridze A (2015) Cenozoic Mountain building process in the greater Caucasus, from subduction to collision. American Geophysical Union, fall meeting abstracts 2014, pp T21A–4565

Stunning Geoheritage of Georgia: Natural "Freezers"

Nana Bolashvili, Kukuri Tsikarishvili, and Tamazi Karalashvili

Abstract

The issues of origin, spreading, and geomorphological features of natural freezers and unique geoheritage objects have been analyzed in the work on the example of Georgia. This natural phenomenon is distinguished by its paradoxical thermal regime, ice formation in summer. The geological and climatic prerequisites for the formation of natural freezers are discussed, which are related to the relief, rock composition, and microclimatic factor interactions. Two main types of freezers are described in the example of Georgia: cleft systems in the fallen rocks of Eastern Georgia and karst formations of Western Georgia. The work is based on historical sources, including the description by Vakhushti Bagrationi, which confirms the existence of natural freezers in the territory of Georgia and presents their socio-economic purpose. Within Georgia's emerging Key Geoheritage Areas framework, these natural freezers represent an important geoheritage resource with high potential for geotourism, education and inclusion in future UNESCO Global Geopark proposals and national georoutes.

Keywords

Georgia · Natural "freezer" · Ice cave · Geoheritage

N. Bolashvili (✉) · K. Tsikarishvili · T. Karalashvili
Tbilisi State University, Vakhushti Bagrationi Institute of Geography, Tbilisi, Georgia
e-mail: nana.bolashvili@tsu.ge

1 Introduction

The mention of ice is directly associated with cold and winter. However, in many countries of the world there are also natural phenomena ("freezers") where ice forms in summer. Its thickness is greater, the higher is air temperature. On the contrary in winter, the colder it is outside, the warmer it is in the "freezer". The population often calls it "mysterious" caves.[1]

Indeed, there is something mysterious in them, as if an invisible force that disrupts the normal order of nature. Freezers are mainly formed in temperate climate by combination of local relief and climate in volcanic rocks and Upper Pliocene, dolerite lavas, large-volume giant lava rocks formed as a result of tectonic or gravitational processes. Similar freezers are observed in different regions of Georgia, and consequently their morphological features and formation conditions are different, and the main signs are that ice formation occurs in all of them in summer.

2 Natural Freezers (Ice Caves)

In eastern Georgia, ice caves natural (ice caves) freezers are found mainly in the fallen rocks of the steep slopes, where there are small impassable depths, sometimes they form labyrinths, on the "walls and floors" of which ice has been forming. In western Georgia, the shape of natural freezers is also different, as are the processes of ice formation: they are associated with karst wells and caves formed in the limestone belt.

Such natural freezers (ice caves) were described in Georgia as early as in the eighteenth century. Vakhushti's work "Description of the Kingdom of Georgia" (1745). It is remarkable that Vakhushti himself was amazed by this

[1] (https://www.amusingplanet.com/2016/09/a-mysterious-ice-cave-that-produces-ice.html).

A. Okrostsvaridze and D. Tormey (eds.), *Geoheritage of Georgia*, Geoheritage, Geoparks and Geotourism,
https://doi.org/10.1007/978-3-032-27639-1_14

integral components of these KGAs provides a scientific and management basis for including them in future UNESCO Global Geopark proposals and national georoutes, and for developing targeted interpretive material that links their microclimates to wider geological and climatic processes in Georgia.

4 Conclusion

A large part of the ice caves of the karst zone of Georgia have been described, including those on Khvamli (Boga), on the Racha ridge (Skhvava), on the southern slope of Satsaliki, near the village of Nikortsminda, in Samegrelo and Abkhazeti (freezers of the Askhi and Arabica massifs). Some of the freezers of Georgia are protected as unique geoheritage sites and are included in the "Red Book of Georgia" (1982) ("Skhvavasa" and "Khorkhebi" freezers and "Snowy Gap". It is likely that their number will increase in future), each of them requires universal care and protection. They are special phenomena of inorganic nature, and these rare formations, created by nature for millennia, are of great scientific value and aesthetic significance.

Within the national geoheritage framework outlined in Chap. 2, ice caves and natural freezers form a distinctive thematic layer across several Key Geoheritage Areas. Their combination of scientific interest, aesthetic appeal and deep cultural associations—from royal ice supply and food storage to wine cellars and local beliefs about healing powers—makes them priority sites for targeted conservation, monitoring and interpretation. As climate warms and visitation increases, integrating natural freezers into geopark proposals and geotourism routes will be essential to ensure that this fragile "natural refrigeration" heritage is both protected and used to communicate wider messages about geology, climate change and human adaptation in Georgia.

References

Vakhushti (1745) Description of the kingdom of Georgia, 1941, TSU Publishing, Tbilisi, pp 1–356. (in Georgian)

Kiknadze TZ (1963) Formation of cave ice in the karst abyss of Vakhushti Bagrationi. Proc Acad Scien GSSR 31(2):325–331. (in Russian)

Kipiani S (1976) Patterns of distribution of ice and snow caves in Georgia. Caves of Georgia:91–99. (in Georgian)

Pirpilashvili P (1960) The natural ice chamber of Skhvara Village. Buin Georg Acad Sci 25(4):790–798. (in Georgian)

Rakviashvili K (1965) Issues in the study of Georgia's ice caves. Caves of Georgia 3:48–51. (in Russian)

Red Book of Georgia (1982) Tbilisi, pp 1–255. (In Georgian)

Tavartkiladze K, Kuznetsov A (1987) Phenomenological model of ice formation in Caucasus valleys during summer. Bul Geor Acad Scien 130(3):545–548. (in Russian)

Tsikarishvili K, Bolashvili N, Janelidze Z, Karalashvili T, Kutateladze (2010) Ice and snow caves and natural ice chambers of Georgia. Sci Technol 7(9):28–34. (in Georgian)

Seismicity of Georgia and Historical Earthquakes

Zurab Javakhishvili

Abstract

Georgia lies within one of the most actively deforming continental collision zones in Eurasia, between the converging Arabian and Eurasian plates. The resulting tectonic interactions produce frequent low- to moderate-magnitude earthquakes and occasional significant, damaging events. Recent advances in regional earthquake relocation, catalog compilation, paleoseismological investigations, and geodetic measurements have significantly improved the understanding of seismicity and active tectonics in Georgia and the broader Caucasus region. This chapter synthesizes findings from recent key studies to describe the tectonic setting, historical seismicity, and modern earthquake patterns in Georgia. It also highlights how these seismic patterns intersect with Georgia's emerging geoheritage framework and risk-aware interpretation in Key Geoheritage Areas and potential UNESCO Global Geopark territories.

Keywords

Active faults · Seismicity · Earthquake · Historical and instrumental periods

1 Tectonic Setting of the Region

The geology and tectonics of Georgia and the Caucasus region is described in Chap. 5; here we focus on seismicity resulting from this setting. The tectonic boundary between the Arabian and Eurasian plates is a complex zone of compressional tectonics that gives rise to dominantly thrust and reverse faulting within the Greater Caucasus and Lesser Caucasus, and normal faulting along with other types of faulting in the foreland basins, which are relatively less folded. Notably, a larger proportion of this convergence is generally accommodated by shortening in the Greater Caucasus than in the Lesser Caucasus (Forte et al., 2014; Onur et al., 2020). However, in the Tbilisi region, deformation appears concentrated within the Lesser Caucasus (Sokhadze et al., 2018). In addition, to the east of the country, evidence indicates the possibility of a remnant subducting slab that causes earthquakes deeper than 40 km (Mellors et al. 2012; Cowgill et al. 2016; Mumladze 2015).

Several active faults are capable of generating large-magnitude earthquakes in the region, though subsurface geometry and reliable slip-rate data are generally unavailable. The tectonic setting and main active faults of the region are given in the Fig. 1.

The resulting seismicity of Georgia is primarily associated with known faults and structures and is consistent with regional tectonics. The Greater Caucasus fold-and-thrust belt extends from the Black Sea to the Caspian Sea, generating the region's largest earthquakes. The frontal branch of the Greater Caucasus thrust, known as the Frontal thrust of Caucasus, constitutes a complex system of faults and folds (Godoladze et al. 2024). The geometry of this structure varies from west to east, exhibiting different patterns of seismicity. This difference could potentially be attributed to the proposed Borjomi-Kazbegi strike-slip fault zone, which crosses the Adjara-Trialeti fault thrust belt, Dzirula massive, and Greater Caucasus thrust (Godoladze et al. 2024; Philip et al. 1989).

2 Seismicity

Earthquake catalogs typically consist of two parts: historical (pre-instrumental) and instrumental. The historical catalog dates back to ancient times, and the intensity data of historical earthquakes is based on contemporary documen-

Z. Javakhishvili (✉)
Ilia State University, Tbilisi, Georgia
e-mail: zurab_javakhishvili@iliauni.edu.ge

A. Okrostsvaridze and D. Tormey (eds.), *Geoheritage of Georgia*, Geoheritage, Geoparks and Geotourism,
https://doi.org/10.1007/978-3-032-27639-1_15

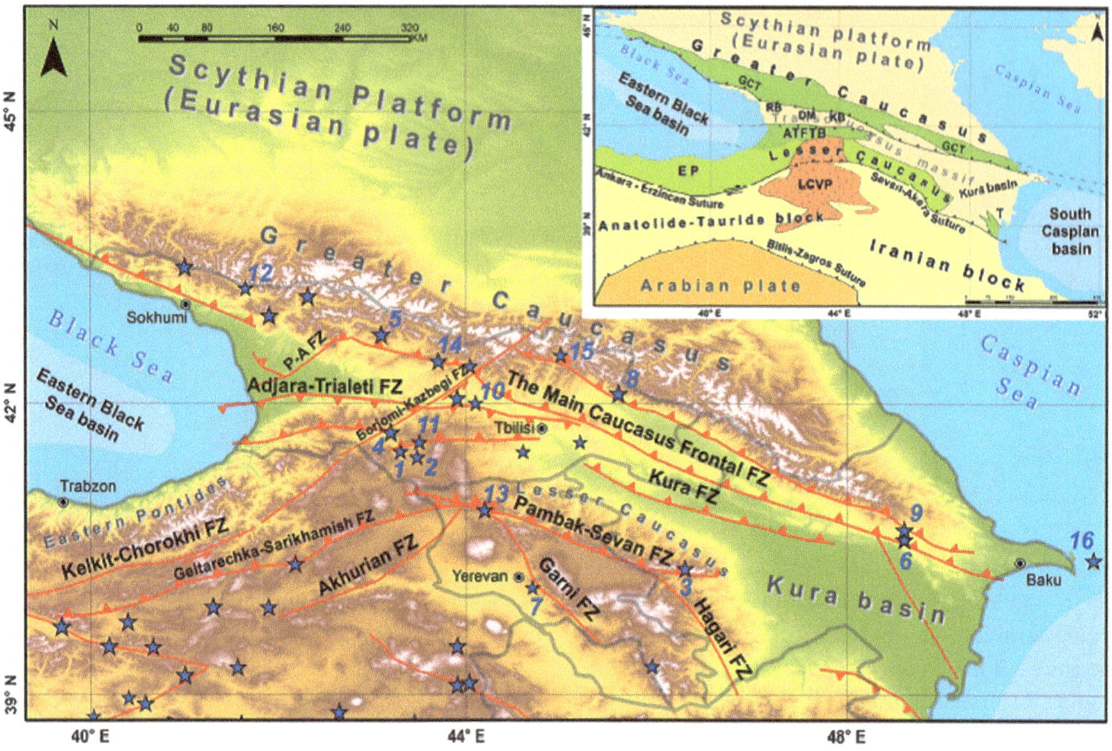

Fig. 1 Tectonic setting of the Caucasus region (Godoladze et al. 2024). Earthquake epicenters are shown with stars

tary descriptions of the damage caused by earthquakes. However, the accuracy of the older events may be affected by substantial errors in location and date. The accuracy of the nineteenth-century and early instrumental period parameters is much higher. The instrumental period in the Caucasus started in 1899, with only a few seismic stations installed in Georgia in the early twentieth century, equipped with low-sensitive mechanical apparatus.

2.1 Pre-Instrumental Period

The early instrumental period (until 1930) has nearly the same quality as in the nineteenth century, and the catalog of this period is considered historical. The instrumental period can be further divided into the early instrumental period (1930–1961), when several stations with different seismographs were installed, and the modern analog instrumental period, which began in the Caucasus in 1962. Since then, highly sensitive standard seismic equipment has been used, leading to improved accuracy of earthquake location parameters and unified magnitude estimation for the former SU and neighboring Eastern European countries. Until 90-ies the number of stations increased to 40. After the dissolution of the Soviet Union, the seismic stations gradually failed due to a lack of supplies and resources to maintain them. By 2002, there were only two operational stations, but a new era of digital seismology started in Georgia following the 2002 Mw 4.5 Tbilisi earthquake. Because of limited resources, the growth of the Georgian seismic network was sporadic between 2003 and 2015. By 2015 the Georgian network operated 25 permanent seismic stations (Godoladze et al. 2024) (see Fig. 2).

Historical records reveal multiple destructive earthquakes across Georgia. Here is a brief description of the strongest events (with Magnitude $M > 5.5$). For the oldest events (50, 400, 742), we have very limited information; accuracy is very low (errors in epicenter estimation are about 200 km, and for magnitude and intensity more than 1 unit (Kondorskaya, Shebalin 1982). The strong events of 1350 and 1742 occurred in the Greater Caucasus Range. The magnitude of these earthquakes is estimated as 7 ± 0.5, and the intensity at the epicenter is 9 ± 1 on the MSK scale. However, as noted above, the accuracy of these estimates is too low (errors in epicentre estimation may exceed 50 km for magnitude 0.5 and intensity 1 unit). The 1530 event occurred close to the 1742 earthquake, but was smaller ($M = 5.7$) and hence it had less effect.

The earthquakes of 1088 ($M = 6.5$, $I_0 = 9.0$ on the MSK scale) and 1283 ($M = 7.0$, $I_0 = 9–10$ on the MSK scale) are better defined. The 1614 event has very sparse data. It affected the Tsaishi church, and hence the estimated epicentre is located close to this area.

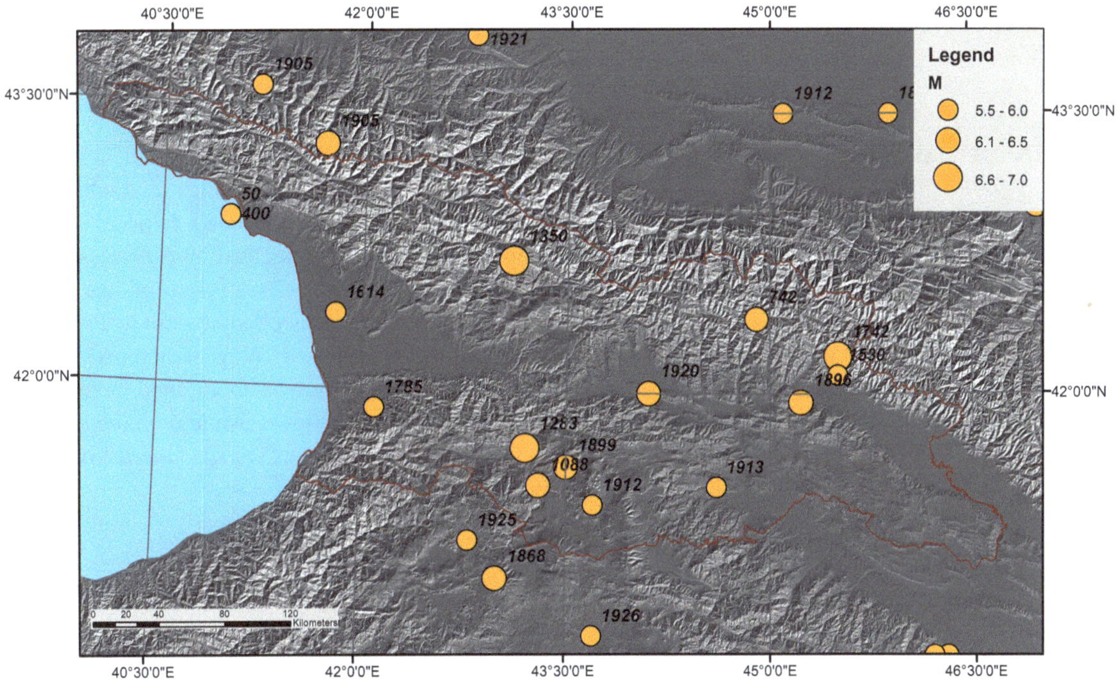

Fig. 2 Epicenters of strong earthquakes of the historical and early instrumental period (till 1930) with M ≥ 5.5 on the territory of Georgia. 2.1 Major Historical (pre-instrumental) Events. Strong earthquakes are shown in yellow

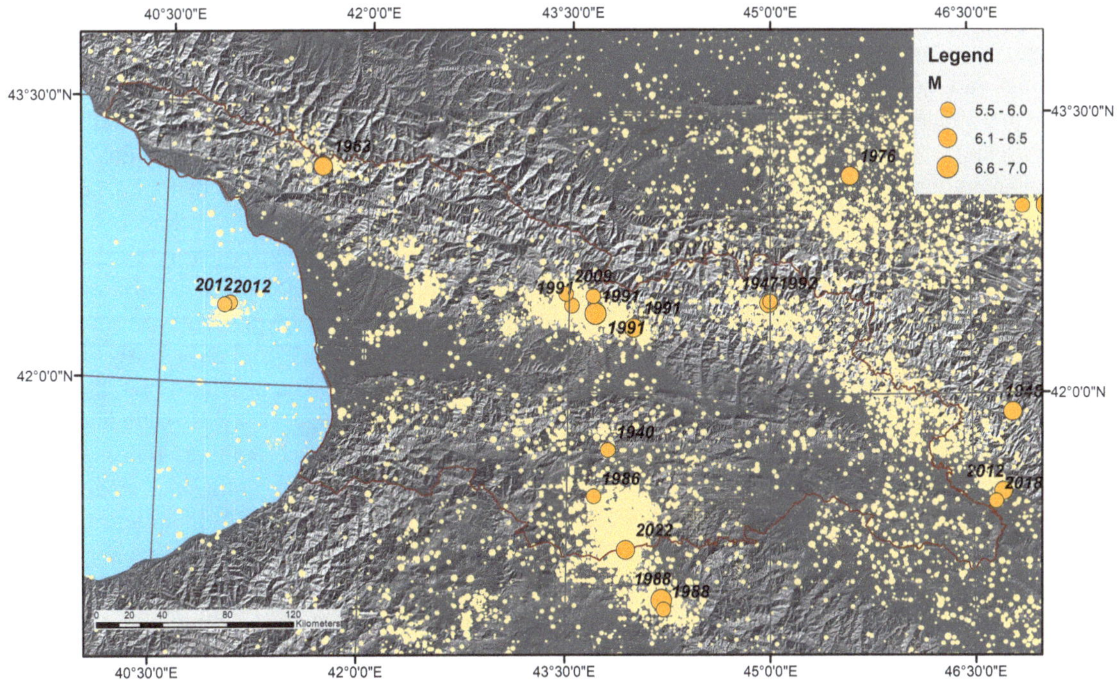

Fig. 3 Epicentres of earthquakes with M ≥ 3.0 during the instrumental period (1950–2024) on the territory of Georgia. Strong earthquakes are shown in yellow

Some late-nineteenth-century historical events and early instrumental earthquakes (e.g., 1896 M = 6.2, I_0 = 7–8; 1899 Akhalkalaki earthquake, M = 6.2, I_0 = 9; and 1920 Gori earthquake, M = 6.2, I_0 = 8–9) were more accurately characterized. The earthquakes of 1912 (M = 5.6, I_0 = 6–7) and 1913 (M = 5.6, I_0 = 6) were defined more poorly due to the lack of data. The data on the 1905 event are even more discrepant than the medieval data (see Fig. 3).

2.2 The Instrumental Period

The region was seismically even more active during the instrumental period. The **1991 Racha earthquake (Mw 7.0)**, which occurred on 29 April 1991 on the southern flank of the Greater Caucasus, remains the largest instrumentally recorded event in the country. The earthquake killed more than 200 people, left approximately 60,000 homeless, and caused damage over thousands of square kilometers. A maximum intensity of 9 on the MSK scale was observed. The mainshock was followed by aftershocks that extended over several months. Among them, there were three strong aftershocks with magnitudes greater than Ms5.5: April 29, at 18:30, Ms. = 6.1, May 3 at 20:19, Ms. = 5.5, and June 15 at 00:59, $M_s = 6.2$. These events caused further damage and casualties. The loss was 10 billion Soviet Rubles. This area has been seismically very active over the last few decades. On September 7, 2009, a strong earthquake M = 6.0 occurred here. Hundreds of aftershocks followed the earthquake. This earthquake and its aftershocks are a key case study for later chapters on hazards, landslides and the "Dynamic Georgia" overlay.

Another strong earthquake occurred on the Adjara-Trialeti belt in 1940 (Tabatskuri earthquake, M = 6.0, I0 = 8 on the MSK scale).

The strong Chkhalta earthquake of 1963 ($M_s = 6.4$, $I_0 = 9$), occurred in the western part of the Caucasus Main Ridge. Another strong earthquake occurred in the central Caucasus in 1992, the Barisakho earthquake, M = 6.5, $I_0 =$ 7–8 MSK scale.

Seismically active Javakheti plateau is situated in South Georgia. This area is characterized by frequent small earthquakes (2011; Tseng et al. 2016) and by significant volcanic activity from the Neogene to the Quaternary (Rebai et al. 1993). The present seismicity seems more active in the vicinity of the Javakheti range near longitude 44°E (Tseng et al. 2016). The most significant event (M = 5.6, I0 = 7–8 on the MSK scale) in the highland region occurred in 1986 (the Paravani earthquake). This earthquake is followed by the 1990 M = 5.5 earthquake near Paravani Lake again (See Fig. 3). The Javakheti highland is considered one of the major seismic hazard zones, other than the Greater Caucasus, in Georgia. The nature of earthquake swarms and their relation to large earthquakes has not been well studied.

There is another cluster of strong earthquakes in Black Sea in 2012 (M = 6.0, M = 5.6) at a distance of about 30 km from the Georgian seashore. Seismicity and source zones in the eastern Black Sea region are poorly studied.

3 Modern Seismicity Patterns, Relocation Studies, Focal mechanisms and stress regime

Bondár et al. (2023) and Godoladze et al. (2024) provide high-resolution relocated catalogs that clearly illuminate active structures. The improved hypocenter distributions identify several key relocated hypocenters that typically fall within the upper 20–25 km of the crust, consistent with a continental collision setting. Precise alignment of earthquakes along dipping structures in the Greater Caucasus indicates active thrust and reverse faulting at depth. Distributed shallow seismicity in the Lesser Caucasus points to more complex block interactions and crustal heterogeneity.

The relocated seismicity clearly demonstrates a change in the depth distribution of earthquakes as we move from western Georgia to the Kura basin. In western Georgia and Javakheti, earthquakes are primarily located in the upper crust. However, in the Kura depression, starting from eastern Georgia (Kakheti), earthquake depths increase to 30–40 km. In the northeastern area, the earthquake depth distribution reaches 150 km, indicating active subduction processes. Notably, we have observed deeper earthquakes (30–40 km) near the Tbilisi area, which may signify the presence of more complex and unexplored features associated with active tectonics. (See Fig. 4). Focal mechanism solutions derived from regional and global datasets indicate that the present-day stress field in Georgia is dominated by compressional tectonics, consistent with Arabia–Eurasia convergence. Earthquakes in the Greater Caucasus are characterized primarily by thrust and reverse faulting mechanisms, with nodal planes striking predominantly west–east to northwest–southeast, in agreement with the orientation of major fold-and-thrust structures. In the Lesser Caucasus and Adjara–Trialeti belt, focal mechanisms display a broader range of styles, including thrust, strike-slip, and locally normal faulting, reflecting strain partitioning within a complex collisional environment (Tseng et al., 2016; Onur et al., 2020). In eastern Georgia and the Kura Basin, focal mechanisms commonly show oblique-slip solutions, indicating the coexistence of compressional and strike-slip deformation within the foreland domain. The along-strike variation in focal mechanisms mirrors the observed changes in seismogenic depth and supports interpretations involving inherited lithospheric structure and possible remnant subduction beneath the region.

4 Geoheritage and Hazard Implications

For Georgia, seismicity is not only a topic for scientific and engineering concern; it is also an integral part

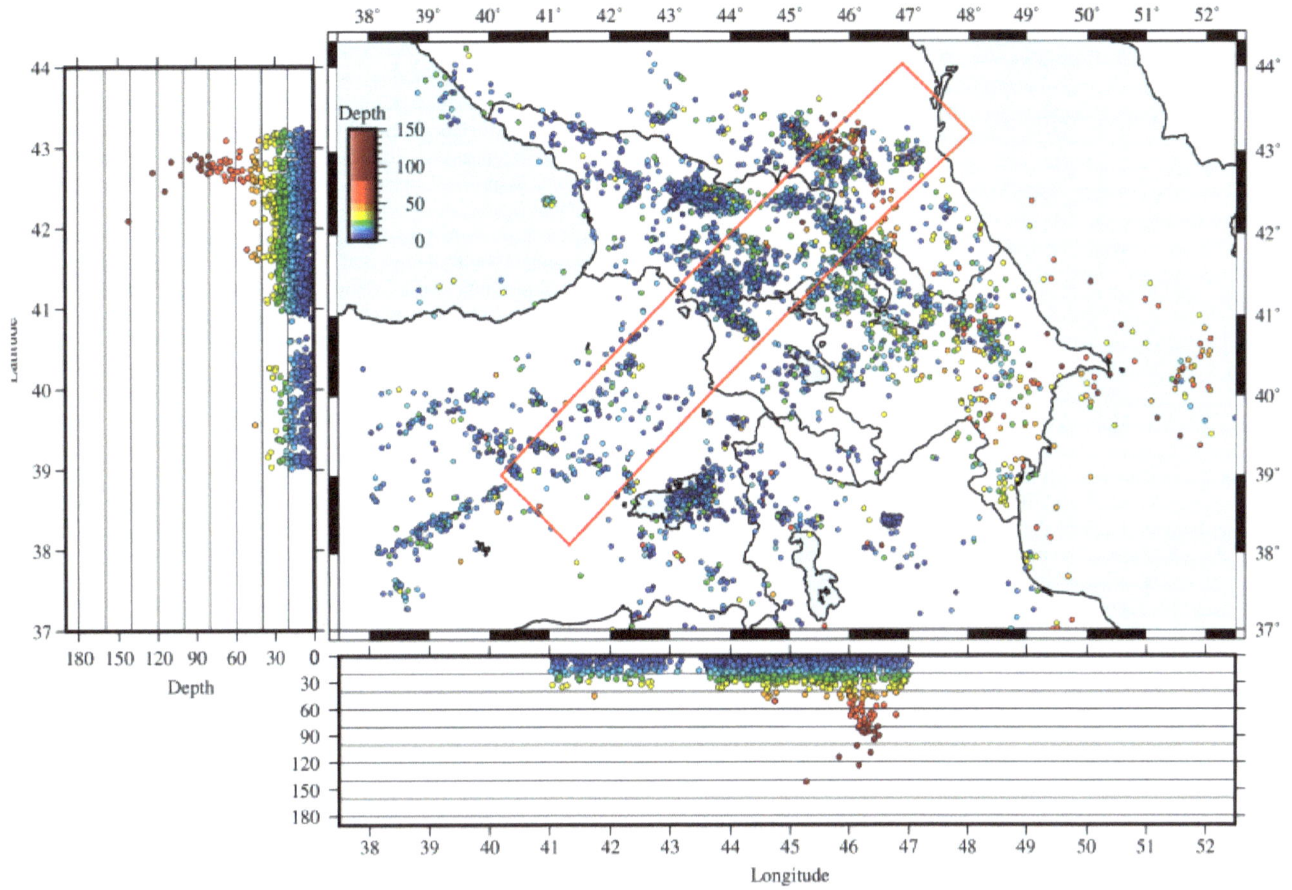

Fig. 4 Map of earthquakes with azimuth gap < 180° and distance to the nearest station less than 30 km, color coded by depth. The red box marks the cross-sectional area. The cross-section panels to the left and below the map display earthquake depth distribution along the path (from Godoladze et al. 2024)

of the country's geoheritage story. The same tectonic processes that have built the Greater and Lesser Caucasus and the Inner Caucasian (Transcaucasian) Microplate KGAs continue to generate earthquakes, uplift, landslides and incision that keep many geoheritage sites dynamic and visually striking. In this sense, seismic hazard and geoheritage are two expressions of the same Earth-system energy.

The main seismic source zones identified in this chapter correspond closely to the geological belts outlined in the geology chapter. Along the Greater Caucasus high-peak KGA, seismicity is concentrated beneath the southern slope and frontal regions, where crustal shortening and thickening are most active. In the Lesser Caucasus volcanic highlands KGA, seismicity is generally shallower and more distributed, but still significant, particularly around Javakheti and Samtskhe. Beneath the Inner Caucasian Microplate KGA, hypocentres deepen to 30–40 km in the Kura Depression and reach ~150 km in the north-east, indicating ongoing subduction-related processes. These patterns mean that many of Georgia's key geoheritage and cultural landscapes—mountain valleys, volcanic plateaus, intermontane basins and spa regions—lie in zones of non-trivial seismic risk.

For a national geoheritage programme and future UNESCO Global Geopark initiatives, this has several implications. First, seismic hazard maps and recurrence information should be integrated into the planning and management of Key Geoheritage Areas and candidate geopark territories. Second, interpretation in mountains, basins and spa towns can explicitly connect visible landforms to active deformation and earthquake processes, helping visitors understand that these landscapes are still being made. Third, education programmes can use recent events, such as the 1991 Racha earthquake as case studies in how communities experience and respond to earthquakes, linking geoheritage appreciation with risk awareness and preparedness.

Finally, this chapter reinforces the message developed in the geology Chap. 5's "Dynamic Georgia" section and in the geopark framework chapters: active processes are part of the heritage, not just background hazards. Uplift, earthquakes and associated exogenous processes refresh exposures, sustain dramatic relief and occasionally damage what people build. Embracing this reality in geoheritage narratives—and designing geosites, geoparks and visitor experiences with it in mind—will make Georgia's geoheritage programme both more honest and more resilient.

References

Bondár I, Godoladze T, Cowgill E, Yetirmishli G, Myers SC, Gunia T, Buzaladze A, Czecze B, Onur T, Gök R (2023) Relocation of the seismicity of the Caucasus region. Bull Seismol Soc Am. https://doi.org/10.1785/0120230155

Cowgill E, Forte N, Niemi B, Avdeev A, Tye C, Javakhishvili Z, Elashvili M, Godoladze T (2016) Relict basin closure and crustal shortening budgets during continental collision: an example from Caucasus sediment provenance. Tectonics 35(12):2918–2947

Forte AM, Sumner E, Cowgill E, Kangarli M, Elashvili M, Godoladze T, Javakhishvili Z (2014) Late Miocene to Pliocene stratigraphy of the Kura Basin, a subbasin of the South Caspian Basin: implications for the diachroneity of stage boundaries. Basin Res. https://doi.org/10.1111/bre.12069

Godoladze T, Gök R, Onur T, Gunia I, Dzmanashvili M, Boichenko G, Bondár I, Ratiani L, Rostomashvili T et al (2024) Compilation of a comprehensive earthquake catalog and relocations in the Caucasus region. Seismol Res Lett 95:1066–1081. https://doi.org/10.1785/0220230206

Kondorskaya NV, Shebalin NV (1982) New catalogue of the strong earthquakes of the USSR from ancient times through 1977. World Data Center A, Colorado, USA

Mellors RJJ, Jackson J, Myers S, Gok R, Priestley K, Yetirmishli G, Turkelli N, Godoladze T (2012) Deep earthquakes beneath the northern Caucasus: evidence of active or recent Subduction in Western Asia. Bull Seismol Soc Am 102(2):862–866. https://doi.org/10.1785/0120110184

Mumladze T, Forte AM, Cowgill ES, Trexler CC, Nathan A, Niemi NA, Yıkılmaz MB, Kellogg LK (2015) Subducted, detached, and torn slabs beneath the greater Caucasus. Geo Res J 5:36–46. http://creativecommons.org/licenses/by-nc-nd/3.0

Philip H, Cisternus A, Gvishiani A, Gorshkov A (1989) The Caucasus: an actual example of the initial stages of a continental collision. Tectonophysics 161:1–21

Rebai S, Philip H, Dorbath L, Borissoff B, Haessler H, Cisternas A (1993) Active tectonics in the lesser Caucasus: coexistence of compressive and extensional structures. Tectonics 12(5):1089–1114

Sokhadze G, Floyd M, Godoladze T, King R, Cowgill ES, Javakhishvili Z, Hahubia G, Reilinger (2018) Active convergence between the lesser and greater Caucasus in Georgia: constraints on the tectonic evolution of the lesser–greater Caucasus continental collision. Earth Planet Sci Lett 48:154–161. https://doi.org/10.1016/j.epsl.2017.10

Tseng TL, Hsu HC, Jian PR, Huang BS, Hu JC, Chung SL (2016) Focal mechanisms and stress variations in the Caucasus and Northeast Turkey from constraints of regional waveforms, Tectonophysics 691, Part B:362–374. https://doi.org/10.1016/j.tecto.2016.10.028

Mudslides and Landslides in Georgia: Historical Hazards

Lasha Sukhishvili

Abstract

Georgia sits at the junction of the Greater and Lesser Caucasus and the Black Sea, where active tectonics, steep relief, rapidly changing climate, and intensive land use combine to produce one of Eurasia's densest mosaics of landslide and debris flow hazards. This chapter synthesizes the main geological, geomorphological, climatic, and anthropogenic drivers shaping mudslide, landslide, debris-flow, and rock–ice avalanche processes across the country, supported by documented case studies. Steep, fractured slopes in actively uplifting mountain belts favour both shallow failures and deep-seated slope instabilities, while active seismicity periodically resets slope stability, as demonstrated during the 1991 Mw≈7 Racha earthquake. Intensifying heavy-rainfall episodes, rapid glacier retreat, and permafrost degradation are increasing the likelihood of high-mobility mass movements and outburst floods in high mountain catchments. Historic and contemporary examples—including Devdoraki (2014), Vere (2015), Vashlijvari (2021), and the 2023 Shovi disaster, illustrate how these processes intersect with settlements, infrastructure corridors, and tourism resorts. From a geoheritage perspective, such "hazard landscapes" form part of Georgia's living geoheritage of active processes and are candidate components of national Key Geoheritage Areas and future geopark/georoute networks, provided that risk is carefully managed. The chapter concludes by identifying persistent challenges related to monitoring capacity, data gaps, land-use regulation on unstable terrain, and the need for decision-ready risk mitigation tools that can support safe, sustainable development and geotourism.

L. Sukhishvili (✉)
Institute of Earth Sciences and National Seismic Monitoring Center, Ilia State University, Tbilisi, Georgia
e-mail: lasha.sukhishvili@iliauni.edu.ge

Keywords

Landslide · Debris flow · Rock-ice avalanche · Geoheritage · The caucasus · Georgia

1 Introduction

The geology of Georgia and the Caucusus is described in Chap. 5. Active tectonics, steep relief, and sharp climatic gradients converge to produce one of Eurasia's densest mosaics of landslide processes. National-scale assessments consistently identify slope instability as a leading natural hazard, while also noting persistent data gaps and the need for decision-ready risk mitigation products. Intense convective bursts, persistent spring and autumn rains, rapid snowmelt, and rain-on-snow events all contribute to transient pore pressure rise and debris and mud flow initiation in steep headwaters. On the climatic horizon, multi-model projections point to increases in the frequency and intensity of extreme precipitation, sharpening the hydrologic triggers that already dominate landslide occurrence in many Georgian basins. On the seismotectonic side, earthquakes periodically reset slope stability, most dramatically in the 29 April 1991 Racha Mw≈7 event, during which landslides (including delayed failures) caused a large number of casualties.

Georgia's Greater Caucasus glaciers are retreating rapidly, driving the growth of proglacial and supraglacial lakes and the spread of debris-covered ice. These changes destabilize glacier termini and steep headwalls. With fast glacier retreat, permafrost degradation, and steep relief, Georgia's high mountain belts are primed for ice–rock avalanches, debris flow transformations, glacial lake outburst floods (GLOFs), and thaw-related slope failures.

In addition to natural factors, rapid urbanization, unregulated infrastructure expansion, and other human activities have exacerbated slope instability, serving as trig-

A. Okrostsvaridze and D. Tormey (eds.), *Geoheritage of Georgia*, Geoheritage, Geoparks and Geotourism,
https://doi.org/10.1007/978-3-032-27639-1_16

Fig. 1 Itkhvisi (Central North Georgia) landslide, April 2022

gers for multiple landslides observed in recent years (e.g. Fig. 1). Construction on marginally stable slopes, inadequate drainage, and encroachment into narrow valleys amplify the consequences of otherwise moderate events, especially in and around major cities such as Tbilisi.

These unstable slopes and valleys are not only sources of risk, but also part of Georgia's geoheritage of active geomorphological processes. Certain corridors and basins, such as Devdoraki, Telavi, Duruji, the Vere gorge in Tbilisi, and the Shovi resort area, record repeated, often catastrophic mass-movement events and the ways communities, infrastructure, and policy have responded over time. From the perspective of the emerging national Key Geoheritage Area inventory (see Chap. 2), such "hazard landscapes" are candidate process-geoheritage sites and, where safety allows, can form components of future geoparks and geotourism routes.

This chapter synthesizes the main geological, geomorphological, climatic, and anthropogenic drivers of mudslides, landslides, debris flows, and rock–ice avalanches in Georgia, and illustrates them through selected historical and recent case studies. It then discusses implications for risk governance, monitoring, land-use planning, and highlights how strengthened hazard knowledge is essential both for protecting people and for safely conserving and interpreting Georgia's outstanding geoheritage.

2 Devdoraki Debris Flows

The Kazbegi–Jimara massif, second only to Mount Elbrus in glacier extent, forms the main glaciation centre of the upper Tergi River basin in the Greater Caucasus (Tielidze, Wheate 2018). Mount Kazbegi (Mkinvartsveri; 5054 m asl) is a dormant stratovolcano whose last major eruptions occurred about 185 ± 30 ka, with a younger parasitic eruption at around 6 ka BP (Chernyshev et al. 2002). Geochemical and geothermal evidence indicates ongoing magmatic degassing beneath the massif (Evans et al. 2009), which some authors have linked to destabilisation processes in nearby glacierized headwalls, including the Kolka rock–ice slide of 20 September 2002 on the northern flank of Mkinvartsveri (Haeberli et al. 2004).

Above about 3,400 m asl, the climate is moderately humid, with 1,000–1,200 mm of annual precipitation (mostly snow), mean winter temperatures around –10 °C, and summer temperatures near 2 °C (Abdaladze et al. 2015). These conditions favour the development of thick snowpacks and glacier ice on steep rock walls and narrow cirques, preconditioning the terrain for rock–ice avalanches, icefalls, and debris flows.

The surging-type Devdoraki Glacier, situated on the northern slope of the Kazbegi massif, is historically one of the most active and well-documented glaciers of the Greater Caucasus. It is renowned for the recurrent "Kazbegi blockages" of the 18th–19th centuries, when massive ice–rock avalanches periodically dammed the Tergi River, forming temporary lakes and destructive outburst floods. The largest known event occurred in 1832, when an estimated 20–22 × 10^6 m^3 of ice and debris blocked the valley for several hours, severely impacting traffic along the Georgian Military Road (Tielidze et al. 2019).

Throughout the nineteenth and twentieth centuries, Devdoraki Glacier underwent multiple advance–retreat cycles, including partial surges and temporary stagnation phases. While most Caucasus glaciers were retreating, Devdoraki remained comparatively stable or advanced during periods of increased accumulation (1950s–1970s). Its rapid flow, confinement within a narrow gorge, and tendency to form ice–debris dams make it one of the key glaciers associated with recurrent mass-movement hazards in the Kazbegi region (Tielidze et al. 2019).

On 17 May 2014, a large debris flow composed of rock and ice descended from the headwall of Mt. Mkinvartsveri following the collapse of part of the Devdoraki Glacier. The debris travelled along the glacier tongue and into the Dariali Gorge, destroying sections of the Georgian Military Road, a gas pipeline, and border infrastructure near the Russian frontier. A second major event in August 2014 further impacted border facilities and damaged a small hydropower plant (Tobler et al. 2016).

The May 2014 collapse was triggered by a combination of gravitational failure along pre-existing fractures, permafrost degradation, glacier melting, and possibly volcanic-related thermal and seismic activity. The Institute of Earth Sciences and the National Seismic Monitoring Center of Ilia State University recorded a minor tremor shortly before the event. Rising mean annual air temperatures have likely accelerated permafrost degradation, increasing slope instability (Noetzli and Gruber 2009). Observations of fractured and deformed glacier tongues on Mt. Mkinvartsveri in 2015 indicate ongoing structural instability within the upper Devdoraki glacier system (Tobler et al. 2016).

Since these processes, an early warning system has been installed comprising a glacier monitoring station and a debris flow alarm system designed to detect potential hazards in the Devdoraki area. The glacier monitoring component includes cameras that regularly capture images of critical glacier zones to track surface movement and identify early signs of instability. The debris flow monitoring system, located downstream in the valley, integrates multicomponent instruments to detect flow activity and automatically alert authorities. Key challenges for the system include the large size of the channel and harsh environmental conditions, which may affect the reliability of measurements and equipment performance during winter (Tobler et al. 2016).

From a geoheritage perspective, Devdoraki and the Dariali Gorge represent a high-mountain "process geosite" where the interaction of active volcanism, rapidly changing glaciers, permafrost degradation, and extreme mass movements can be directly observed. As discussed in Chap. 2, such sites are strong candidates for inclusion in Georgia's emerging Key Geoheritage Area inventory as active-process KGAs, and, where safety permits, can support carefully managed geotourism and educational visits along the Georgian Military Road corridor (Figs. 2 and 3).

Fig. 2 Millions of cubic meter debris accumulated in the Dariali Canyon (June 2014)

Fig. 3 Loose material and debris still located at the Devdoraki proglacial area (June 2014)

3 Telavi Debris Flows

The Telaviskhevi River originates on the northeastern slope of the western Kura Fold–Thrust Belt (Gombori Range) and flows northward through the regional capital city of Telavi. It is fed by both snowmelt and rainfall. Floods typically occur in spring, while flash floods are common in summer and autumn. The river is about 15 km long, and its catchment area (including tributaries) covers roughly 80 km^2.

During intense rainfall, powerful flows form from the very headwaters within the degraded basin, capable of transporting large volumes of solid material. The debris flow, composed mainly of clay, Plio–Pleistocene conglomerates, and biomass, is joined downstream by torrential tributaries and enters Telavi as a mud and debris flow. The destructive power of such events increases markedly when the flow is constricted within narrow valley sections or when lateral landslides collapse into the channel. One of the most destructive events occurred on the night of 14 June 1977, when a debris flow from the Telaviskhevi River devastated parts of the city of Telavi (CENN 2020). In 1978, three debris flow mitigation structures (known as the Kherkheulidze Dams) were built in the middle section of the Telaviskhevi Gorge in an attempt to trap coarse material and reduce downstream impacts.

From a geoheritage perspective, the Telaviskhevi valley exemplifies an urban "hazard landscape" where repeated debris flows, slope instability, and engineered responses are closely intertwined. The 1977 event, the subsequent construction of the Kherkheulidze Dams, and ongoing risk to the city together form part of Georgia's living process geoheritage. Within the emerging national Key Geoheritage Area inventory (see Chap. 2), Telaviskhevi is a candidate low-altitude active-process site whose story can be inter-

Fig. 4 Lower 'Kherkheulidze Dam' located in the Telaviskhevi Gorge (July 2025)

Fig. 5 Lower part of the river Duruji, and the protective wall (to the right)

preted alongside the broader wine-region landscape and future geotourism routes, provided that risk is carefully managed (Fig. 4).

4 Duruji Mud Flows

The Duruji River is among the most active debris and mudflow prone rivers in the mountainous regions of the Southern Caucasus. Its headwaters drain steep, highly dissected slopes composed of weak, easily erodible rocks, and the river has built a large alluvial fan on which the town of Kvareli is located. During intense rainfall, particularly when preceded by prolonged wet periods or rapid snowmelt, highly concentrated mud and debris flows are generated in the upper catchment and propagate rapidly downstream.

Historical records and engineering reports document repeated destructive events: a major flood/mudflow in 1832 devastated Kvareli, followed by further catastrophic events in 1889, 1904, 1949, 1990, and 2002. The scale of these events is reflected in the volume of transported material, the height of deposited levees, and the extent of damage to agricultural land, buildings, and infrastructure. The Duruji system has therefore become a classic example in regional literature of a debris and mudflow river with a long record of impacts on a settled fan (Gavardashvili and Ayyub 2011).

In response to these recurrent disasters, various protective works have been constructed along the Duruji River. In 1906, under the leadership of public figure Ilia Chavchavadze, a series of protective dams were built along the channel to reduce the frequency and severity of damaging flows. Later, a large trapezoidal protective wall was constructed along the lower Duruji, measuring approximately 5.5 km in length, 7 m in height, and at least 2 m in width at the base. These structures are designed to confine the main flow and prevent lateral avulsions across the fan surface, but they require constant maintenance and regular clearing of accumulated sediment to remain effective.

From a geoheritage perspective, the Duruji valley–fan system illustrates a low-altitude "hazard landscape" in which active debris and mudflow processes, a densely settled fan, and a century of engineering responses are closely intertwined. It is an outstanding process geosite for understanding how geomorphic activity shapes settlement patterns and risk governance in eastern Georgia. Within the emerging national Key Geoheritage Area inventory (see Chap. 2), Duruji and the Kvareli fan are strong candidates for inclusion as an active-process KGA linked to the wine-region landscape and potential geotourism routes, provided that risk is carefully managed (Fig. 5).

5 Tbilisi Disaster 2015

On the night of 13–14 June 2015, the capital city of Tbilisi experienced one of the most destructive floods in its modern history. Several hours of intense rainfall in the Vere River basin triggered a large landslide above the village of Akhaldaba, west of the city. Approximately 1 million m^3 of soil, rock, and trees mobilised into the Vere gorge, where the mass movement transitioned into a debris flow and flood that propagated downstream toward the city along the narrow gorge.

Before entering the city centre, the flow was temporarily obstructed at several constrictions, including bridges and the corrugated-steel tunnel sections of the Vere channel. These blockages caused rapid ponding and an abrupt rise in water and debris level within the confined channel. When the temporary dams failed, a highly destructive debris-rich flood wave surged through the Vere valley corridor into the urban area, overwhelming the channel and adjacent infrastructure.

Fig. 6 Landslide source zone, view to the North—river Vere gorge (June 2015)

The Tbilisi Zoo, Mziuri Park, sections of Heroes' Square, roads, and residential buildings along the flood path were severely damaged or destroyed, and numerous vehicles and structures were swept away.

The peak discharge at the city centre, estimated at about 500 m^3/s, was several times higher than typical flood flows for the Vere River and far exceeded the design capacity of the existing flood-protection works. The disaster caused significant loss of life, left many families homeless, and inflicted substantial economic damage. It also exposed critical weaknesses in land-use planning, slope and drainage management in the upper catchment, and the reliance on heavily engineered, partly enclosed channels in the lower Vere valley.

From a geoheritage perspective, the 2015 Vere River disaster revealed an urban "hazard landscape" where active geomorphic processes, historical flood records, and modern infrastructure are tightly coupled. The Vere valley in Tbilisi is a key process geosite for understanding how flash floods, debris flows, and landslides can interact with a rapidly growing city. Within Georgia's emerging Key Geoheritage Area inventory (see Chap. 2), the Vere corridor is a candidate urban active-process site whose story can inform safer river-corridor planning and, in time, carefully designed educational trails and interpretive points that commemorate the event while prioritising risk reduction (Fig. 6).

6 Shovi Disaster

In August 2023, a massive rock–ice avalanche, with an estimated volume of about 1×10^6 m^3, detached from a steep permafrost-affected slope near the Tbilisa Glacier at around 3,850 m asl in the upper Bubistskali River catchment, at the southern foothills of the Greater Caucasus. The initial failure occurred in crystalline basement rocks forming a nearly vertical headwall, where long-term glacier downwasting and permafrost degradation had reduced slope stability. The collapsing rock and ice mass impacted the glacier surface, entrained additional debris and meltwater, and transformed into a high-mobility debris flow that followed the Bubistskali gorge toward the Shovi resort area (Sukhishvili et al. 2024).

The resulting debris flow reached the Shovi resort in the Oni Municipality within minutes, destroying cottages, roads, bridges, and other infrastructure, and burying large parts of the settlement under several metres of mud, boulders, and woody debris. Cobbles and boulders exceeding tens of tonnes were transported and deposited in and around the resort area, with individual blocks found up to 10 km downstream from the source zone, attesting to the extreme energy of the flow. The disaster caused more than 30 fatalities, left one person missing, and required the evacuation of over 200 people from the surrounding area.

Meteorological records show no exceptional rainfall immediately preceding the failure, indicating that glaciological and permafrost-related factors were the primary triggers. Satellite imagery and field observations reveal progressive fracturing, meltwater outflow, and surface deformation in the weeks before the event, suggesting that the slope and glacier system had entered a state of advanced instability.

From a geoheritage perspective, Shovi exemplifies a high-mountain "hazard landscape" where active cryosphere-related processes, steep crystalline bedrock, and human use of a mountain resort converge. The site is a key process geosite for understanding how permafrost degradation, glacier retreat, and rock–ice avalanches can rapidly cascade into catastrophic debris flows. Within Georgia's emerging Key Geoheritage Area inventory (see Chap. 2), the upper Bubistskali–Shovi system is a strong candidate for inclusion as an active-process KGA, provided that any future use of the area prioritises rigorous hazard assessment, real-time monitoring, and risk-informed planning. Carefully designed educational materials and remote or controlled access viewpoints could, in time, allow the Shovi story to be interpreted as part of Georgia's living geoheritage of climate-sensitive mountain hazards (Figs. 7 and 8).

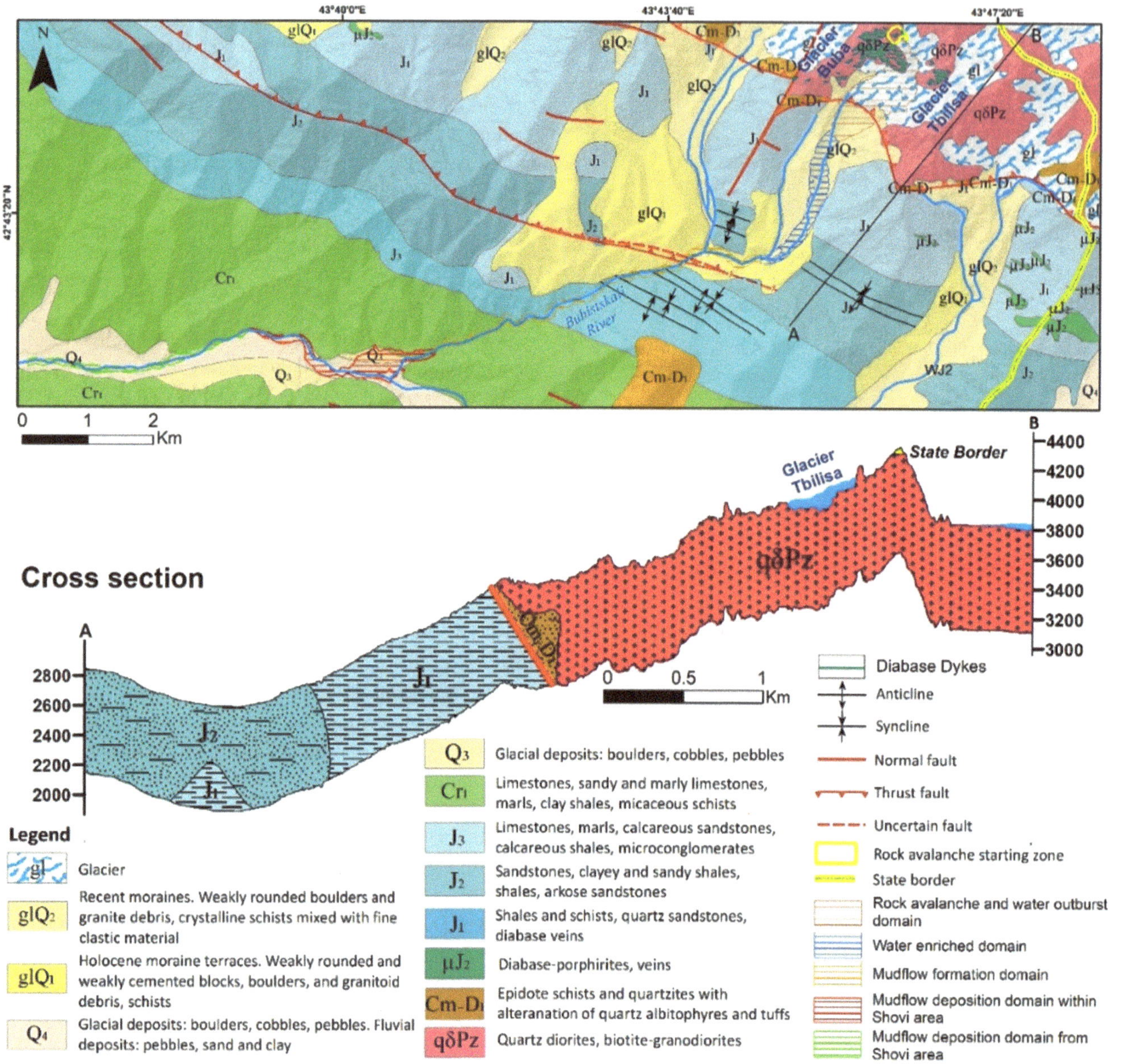

Fig. 7 Simplified geology map of the Shovi area, modified after L. Sukhishvili et al. (2024)

7 Discussion

Georgia's landslide and debris flow regime reflects the convergence of tectonic uplift, steep relief, and rapidly changing hydro-cryospheric conditions. The case studies presented here show that while landslide activity in the Greater and Lesser Caucasus has long been driven by climatic seasonality and seismic loading, recent events reveal accelerating cryosphere-related instability and growing anthropogenic influence on slope failure. Retreating glaciers, permafrost thaw, and progressive fracturing of high-altitude headwalls are increasingly recognised as precursors to rock–ice avalanches and rapid mass-flow transformations, as evidenced at Devdoraki and Shovi. These cryospheric processes act in parallel with more traditional hydrometeorological triggers, intense convective storms, prolonged rainfall, and rain-on-snow events, highlighting an evolving multi-hazard environment in Georgian basins.

Urban expansion on marginally stable slopes, insufficient drainage planning, and unregulated infrastructure development substantially elevate vulnerability, as demonstrated by the Tbilisi and Vashlijvari failures and by repeated damage in Telavi and Kvareli. Although monitoring and early warning systems have begun to emerge, current capacities remain localised and constrained

Fig. 8 Large boulder deposited at Shovi resort by a debris flow originating 10 km from the source zone

by rugged terrain, instrument reliability, limited maintenance, and incomplete hazard mapping. Strengthening hazard observation networks, integrating multi-sensor satellite products with ground data, and embedding hazard information into land-use regulation in unstable terrain are pressing requirements. A national framework that more tightly connects scientific monitoring, municipal planning, and emergency response is essential to mitigate cascading risks that increasingly evolve on short timescales.

At the same time, these unstable slopes and valleys form part of Georgia's geoheritage of active geomorphological processes. High-mountain sites such as Devdoraki and Shovi, urban corridors like the Vere gorge in Tbilisi, and debris-flow fans such as Telaviskhevi and Duruji record not only repeated mass-movement events but also the historical and contemporary ways communities and institutions have adapted to them. From the perspective of the emerging national Key Geoheritage Area inventory (see Chap. 2), such "hazard landscapes" are strong candidates for classification as active-process KGAs. Many lie along major tourism corridors, including the Georgian Military Road, the Kakheti wine region, and mountain resort areas, where future geopark initiatives and geotourism routes will need to integrate rigorous hazard assessment, real-time monitoring, and risk-informed design. Interpreting these landscapes as living process geoheritage, while prioritising safety, offers an opportunity to link disaster-risk reduction with education and sustainable tourism development.

8 Conclusion

Georgia's position at the junction of the Greater and Lesser Caucasus, together with steep relief, active tectonics and rapidly changing climate, creates a landscape highly prone to landslides, debris and mud flows, and rock–ice avalanches. The Devdoraki, Telaviskhevi, Duruji, Tbilisi (Vere valley), and Shovi case studies show how these processes repeatedly intersect with settlements, infrastructure corridors and tourism areas, and how climatic, cryospheric, seismic and anthropogenic drivers combine to generate complex cascading hazards. Recent events underscore the growing importance of cryosphere-related processes, glacier retreat, permafrost degradation and headwall destabilization, which increasingly act alongside intense rainfall and seismic shaking to trigger catastrophic mass movements.

Despite improvements in scientific understanding and the first examples of early-warning systems, significant challenges remain. Hazard monitoring is spatially limited, maintenance of instruments and protective structures is difficult in rugged terrain, and hazard and risk information are not yet systematically embedded in land-use planning and infrastructure design. Addressing these gaps will require sustained investment in observation networks, data integration, and decision-ready tools that can support authorities at local and national scales in regulating development on unstable terrain and in preparing for rapid-onset events.

At the same time, the unstable valleys, fans and gorges described in this chapter form part of Georgia's living geoheritage of active geomorphological processes. Sites such as Devdoraki and Shovi, the Vere corridor in Tbilisi, and the Telaviskhevi and Duruji systems are exemplary "hazard landscapes" where dynamic processes, cultural memory and engineering responses are closely intertwined. Within the emerging national Key Geoheritage Area inventory (see Chap. 2), these areas are strong candidates for inclusion as active-process KGAs, many of them lying along key tourism routes and within potential geopark clusters. Recognising and interpreting them as process geoheritage, while prioritising safety, rigorous hazard assessment and risk-informed planning, offers a way to connect disaster-risk reduction, education and sustainable geotourism within Georgia's broader geoheritage framework.

References

Abdaladze O, Nakhutsrishvili G, Batsatsashvili K (2015) Sensitive alpine plant communities to the global environmental changes (Kazbegi Region, the Central Great Caucasus). Am J Environ Prot 4(3):1–93. https://doi.org/10.11648/j.ajep.s.2015040301.25

Chernyshev VA, Lebedev SN, Bubnov M (2002) Isotopic geochronology of quaternary volcanic eruptions in the greater caucasus. Geochem Int 40(11):1042–1055

Evans SG, Tutubalina V, Drobyshev V et al (2009) Catastrophic detachment and high-velocity long-runout flow of kolka glacier, caucasus mountains, Russia in 2002. Geomorphology 105(3–4):314–321. https://doi.org/10.1016/j.geomorph.2008.10.008

Gavardashvili G, Ayyub B, M (2011) The field investigation of erosion and debris flow processes in catchment basin of the duruji river. In: International conference on debris-flow hazards mitigation: mechanics, prediction, and assessment, proceedings, pp 63–70 https://doi.org/10.4408/IJEGE.2011-03.B-008

Haeberli W, Huggel C, Kääb A, Polkvoj A et al (2004) The Kolka-Karmadon Rock/Ice Slide of 20 September 2002: an extraordinary event of historical dimensions in north ossetia, russian caucasus. J Glaciol 50(171):533–546. https://doi.org/10.3189/172756504781829710

Kereselidze D, Alaverdashvili M, Trapaidze V (2015) What happened on June 13, 2015, in the Vere River Gorge in Tbilisi (in Georgian) Tbilisi. Publishing House of the Institute of Hydrometeorology, Tbilisi, 2015. ISBN: 978-9941-0-8239-9

Noetzli JS, Gruber G (2009) Transient thermal effects in alpine permafrost. Cryosphere 3(1):85–99. https://doi.org/10.5194/tc-3-85-2009

Tielidze L, Roman G, M. Kumladze M (2019) The devdoraki glacier catastrophes, georgian caucasus. Hung Geogr Bull 68(1):21–36 https://doi.org/10.15201/hungeobull.68.1.2

Tielidze L, Roger D (2018) The greater caucasus glacier inventory (Russia, Georgia and Azerbaijan). Cryosphere 12(1):81–94. https://doi.org/10.5194/tc-12-81-2018

Tobler D, Haemmig C, Lorenz M (2016) Early warning systems for glacier lake outburst floods and debris flows case studies from China and Georgia. In: UNESCO chair in technologies for development: from innovation to social impact 2–4 May 2016, SwissTech Convention Centre, EPFL, Lausanne, Switzerland 1 pp 1–12

Desert Geoheritage of Georgia: Colorful Hills, Mud Volcanoes and Rock-Cut Monastery Complex

Avtandil Okrostsvaridze, Giorgi Boichenko, and Salome Gogoladze

Abstract

The Gareja Desert in eastern Georgia is a compact but remarkably diverse geoheritage area where semi-arid landscapes, vivid "colorful hills," active mud volcanoes, and a key Upper Miocene volcanic ash layer are tightly clustered within a few hundred square kilometers. These features record the evolution of the Iori Plateau and adjacent Kura Basin, from marine and fluvio-lacustrine sedimentation to syn- and post-collisional deformation and Neogene volcanism. The desert also preserves rich Late Miocene terrestrial vertebrate faunas that document faunal turnover in the Eastern Paratethys region. Cut into the folded Miocene sandstones on the plateau's southern margin, the David Gareja rock-cut monastery complex adds an exceptional cultural dimension, illustrating how geology, landscape, and spiritual life are intertwined along Georgia's eastern frontier. Taken together, the Gareja Desert and David Gareja complex form a distinctive geocultural ensemble where geological, paleontological, and cultural heritage are closely integrated. This chapter outlines the main geoheritage elements of the Gareja Desert and highlights its potential as a future UNESCO Global Geopark, complementing mountain, volcanic, karst, and wine-landscape geoheritage elsewhere in Georgia.

Keywords

Gareja Desert · Geoheritage · Colored hills · Mud volcanoes · Volcanic ash layer

1 Introduction

In Georgia, a country framed by the Greater and Lesser Caucasus and covering only 69,700 km^2, it is striking that a true semi-desert exists at all. The Gareja Desert, on the Iori Plateau along the border with Azerbaijan, forms a compact enclave of badlands, colorful hills, and sparse vegetation that contrasts sharply with the country's humid lowlands and forested mountains. This small territory condenses several major geological stories: the filling and deformation of the Kura intermontane basin, late Neogene uplift and incision, active mud volcanism, and the imprint of distant volcanic eruptions in the form of ash layers.

Within this landscape, a series of vividly banded "colorful hills" developed by the weathering of Miocene–Pliocene clays and sandstones provide a textbook example of how mineralogy and climate interact to shape terrain and color. Nearby, mud volcano fields discharge fluids and gases from depth, recording ongoing basin dynamics and offering resources for therapeutic use. An Upper Miocene volcanic ash layer, traceable across the Eastern Caucasus, crosses the desert and provides a precise time marker that links the local succession to large-scale volcanic events in the Lesser Caucasus. The same semi-arid environment preserves rich terrestrial vertebrate faunas that document Late Miocene ecosystems and faunal change in the Eastern Paratethys region.

Cut into folded Miocene sandstones on the southern margin of the desert, the David Gareja rock-cut monastery complex adds a major cultural dimension. Its churches, cells, refectories, and frescoes testify to centuries of monastic life lived in close dialogue with the surrounding hills and cliffs. Together, the Gareja Desert and David Gareja complex form a distinctive geocultural ensemble on Georgia's eastern frontier, where geological, paleontological, and cultural heritage are tightly interwoven.

A. Okrostsvaridze (✉) · G. Boichenko · S. Gogoladze
Institute of Earth Sciences, Ilia State University, Tbilisi, Georgia
e-mail: avtandil.okrostsvaridze@iliauni.edu.ge

A. Okrostsvaridze and D. Tormey (eds.), *Geoheritage of Georgia*, Geoheritage, Geoparks and Geotourism,
https://doi.org/10.1007/978-3-032-27639-1_17

This chapter introduces the main geoheritage elements of the area and considers their significance within Georgia's emerging network of geopark and geoheritage sites.

2 A Brief Geological Description of the Gareja Desert

The Gareja Desert occupies mainly the eastern part of the Iori Plateau. It is bounded to the northeast by the right bank of the Iori River, to the west by a meridian passing through the town of Gardabani, and to the south by the political border with Azerbaijan. Today, the desertified area covers approximately 205,000 hectares, with elevations ranging between 400 and 800 m above sea level. The sedimentary cover is gently tilted and displaced toward the southwest. Because the area is almost entirely treeless, rock outcrops are exceptionally well exposed, providing favourable conditions for field geological investigations (Fig. 1).

Occurrences of oil and gas accumulations have been documented within the Gareja Desert and adjacent areas, giving the region a long history of geological exploration and adding a further dimension to its geoheritage. From a geotectonic perspective, the Gareja Desert forms part of the Kura intermontane molasse basin and lies within the eastern subsidence zone of the Adjara–Trialeti paleorift structure. Geophysical data and deep borehole records show that the sedimentary succession here exceeds 10 km in thickness (e.g. Buleshvili 1960).

The oldest rocks exposed at the surface are Upper Eocene deposits, consisting of arkosic and greywacke sandstones, argillites, and clays that crop out in small areas in the extreme northwestern part of the desert. More extensive outcrops belong to Lower and Middle Miocene strata forming the anticlinal core of the NW–SE-trending Gareja Ridge; these units comprise greywackes, arkosic sandstones, conglomerates, marls, limestones, and clays. However, most of the desert is underlain by Upper Miocene and Pliocene sandstones, conglomerates, clays, and volcanic ash layers. Along the eastern and western flanks of the ridge, these rocks are overlain by Quaternary alluvial, deluvial, colluvial, and proluvial deposits derived from the Iori and Kura basins, respectively.

Overall, the Gareja Desert lies within a zone of intense tectonic deformation, which has produced folding, displacement, and fracturing of the sedimentary succession. At the surface, arid-weathering processes act on these deformed strata, together generating the sharp cuestas, ravines, and other strikingly exotic landscapes that characterise the region and set the stage for its geoheritage features.

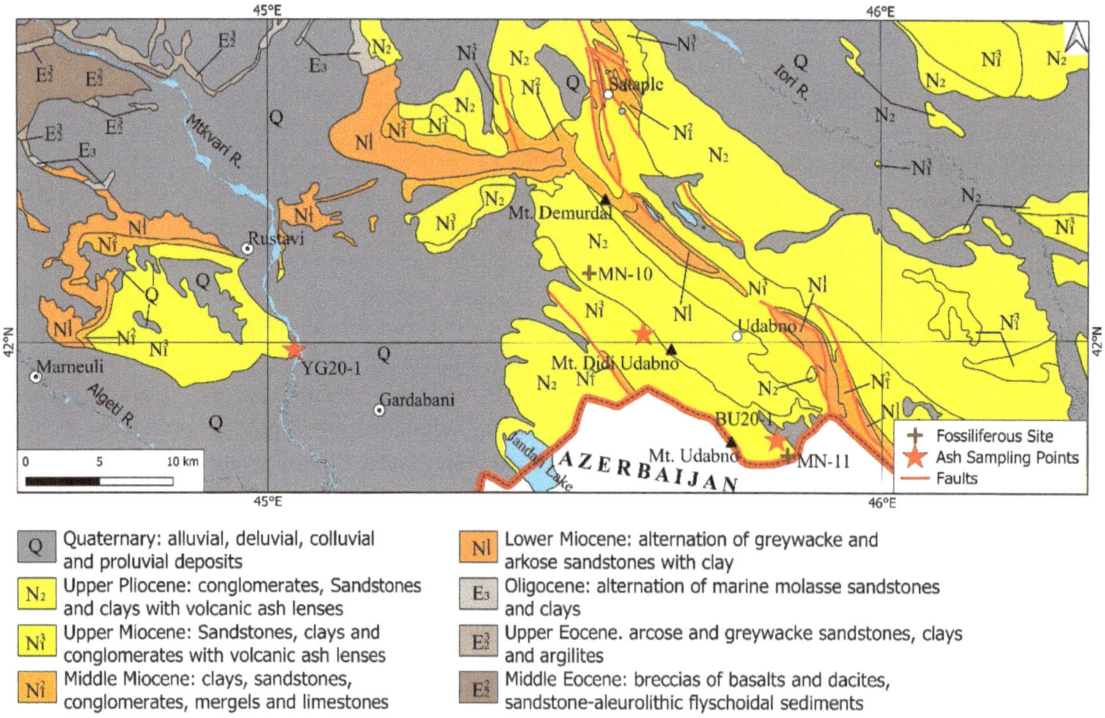

Fig. 1 Geological map of the Gareja Desert. Geological map of the sampling area from the Yaghluja (#YG20-1) and Bertubani of the Gareja (#BU20-1) volcanic ash layers. The map is modified according to geological map of the Georgian Geological Department (2019)

3 The Colorful Hills

One of the most striking geoheritage features of the Gareja Desert is the so-called "colorful hills", formed by alternating sedimentary layers of different hues that give the landscape an almost painted appearance. Comparable landscapes are known worldwide under various names, such as Rainbow Mountain in Peru or the Painted Desert in Arizona, USA (e.g., Baars, 1983). In terms of both scale and coloration, the term "colorful hills" is entirely appropriate for the Gareja Desert, and there is no doubt that they represent an important component of Georgia's geoheritage.

The vivid coloration of these hills results from weathering processes acting under semi-arid climatic conditions, controlled by a complex interplay of geological and geochemical factors. The principal agent is water, which, under elevated temperatures, promotes oxidation of metals contained in sedimentary rocks. This leads to structural disruption and argillization of the rock surface and, ultimately, the development of distinct color bands.

Under arid and semi-arid conditions, such intense coloration develops almost exclusively in sedimentary rocks, and only at or near the surface, because water and oxygen penetrate to shallow depths. Iron, one of the most abundant elements in the Earth's crust, plays the dominant role: trivalent iron imparts red, brown, and yellow tones, while divalent iron produces greenish and grayish hues. Manganese can contribute pink shades, whereas copper yields bluish to greenish tints. White horizons generally reflect the dominance of aluminum and silica within the argillized rock. The depth of this alteration depends on rock type, degree of fracturing, and the intensity of atmospheric precipitation; in most cases it does not exceed 20–30 cm, though in rare instances it may reach several metres.

In the Gareja Desert, colorful hills are exposed in the north-western sector, along the Tetri Udabno ("White Desert") Ridge, but only in two localities: near the village of Udabno (Fig. 2) and in the area of Mravaltskaro Lake. Their total area is approximately 13–15 km^2. They are composed predominantly of Upper Pliocene continental deposits: clays, sandstones, and conglomerates. The exposures typically face south, and each layer displays a characteristic and clearly separated coloration; mixing of colours and layers has not been observed. Red tones dominate, with subordinate yellow and brown bands and rare whitish-gray horizons. Field observations indicate alteration depths of 20–30 cm, nowhere exceeding 40 cm.

From this brief overview, it is evident that the colorful hills of the Gareja Desert, as in similar areas worldwide, are the product of complex geological, climatic, and geochemical processes and therefore constitute an important element of geoheritage. They are a relatively rare phenomenon on the Earth's surface, in which otherwise ordinary sedimentary rocks acquire vivid, exotic colours and effectively form a natural open-air museum. Today, the colorful hills are among the most frequently visited destinations in the Gareja region, and their long-term protection should be considered a priority. Inclusion of these sites within a future UNESCO Global Geopark would provide an effective framework for conserving their scientific and aesthetic values while enhancing their educational and geotourism potential.

Fig. 2 The colorful hills of the Gareja Desert in spring. Photo © Travel Man

4 Mud Volcanoes

Mud volcanoes are landforms of diverse morphology that intermittently erupt mud, often in association with water and hydrocarbons. Despite a long history of observation, they remain among the least fully understood geological phenomena in the Earth's crust. Some researchers emphasise their tectonic origin, linking mud volcanoes to zones of compression, faulting, and overpressured sediments; others highlight their geochemical and biogeochemical aspects, considering them as surface expressions of deep fluid–rock interaction and microbial activity (e.g., Mazzini and Etiope 2017; Wan et al. 2023). In practice, mud volcanism reflects a combination of tectonic, geochemical, and biological factors. What is well established is that mud volcanoes are most commonly associated with sedimentary basins containing hydrocarbon accumulations, where overpressured fluids find pathways to the surface along fractures and faults.

At depth, gases accumulate in hydrocarbon-bearing reservoirs, while surrounding sedimentary rocks, typically sandstones, clays, and argillites, are subjected to tectonic stress. Gas and fluid overpressure can mobilise fine-grained sediments into a plastic slurry. This mixture ascends along zones of weakness, interacting with meteoric waters that circulate in the fracture network. Chemical alteration during ascent further reduces grain size and viscosity. Upon reaching the surface, these fluids and muds are extruded to form conical or dome-shaped edifices, small craters, and gryphons. Recent work has shown that the composition and discharge style of mud volcanoes can provide valuable indirect information on reservoir characteristics and hydrocarbon potential, underlining their significance not only as geoheritage features but also as indicators of subsurface resources.

Globally, more than 2,500 mud volcanoes are known, occurring mainly in young orogenic belts and foreland basins (e.g., Azerbaijan, Italy, Trinidad, Pakistan). They are closely associated with fold-and-thrust structures and active or recently active tectonic regimes. It is noteworthy that the term "mud volcano" (Ger. *M¨udevulkan*) was first formalised in the nineteenth century after German geologist H. Abich studied and described these features in the Caucasus, including sites within present-day Georgia and Azerbaijan.

Within the Caucasus, the largest concentration of mud volcanoes is in eastern Azerbaijan, particularly in the Absheron Peninsula and adjacent Caspian regions, where hundreds of active and extinct edifices are known (Baloganov et al. 2018). On the territory of Georgia, mud volcanoes are far less numerous but still form an important element of geoheritage. They are represented both as isolated centres (e.g., Shaori, Narazeni, Kvirila, Korkhi, Patara-Klde, Tsaishi-Gvandra, Jvari-Tskali, Alajigli) and as volcanic fields (Akhtala, Kila-Kupra, Takhti-Tepha). Most of these occur on the Iori Plateau, within the Gareja Desert and adjacent areas (Akhtala, Kila-Kupra, Takhti-Tepha, Lakbe, Phoveli, Bakana, Alajigli), with a smaller number along the Tsiv-Gombori Ridge (Lakbe, Phoveli, Akhtala, Bakana), and a single occurrence in the Shida Kartli region near the town of Kaspi (Kavtiskhevi) (Fig. 3).

The mud volcanoes of the Gareja Desert are particularly interesting because of their accessibility and well-preserved morphology. They are built predominantly of fine-grained, strongly argillised material that forms smooth, grey to bluish-grey mud flows and low cones. Granulometric and chemical analyses show that these muds consist mainly of clay-sized particles and display a weakly alkaline reaction (pH 7.0–7.7). Such muds, finely dispersed, homogeneous, and slightly alkaline, are known to possess favourable therapeutic properties and have long been used in balneology (Kikvadze 2016). Thus, the Gareja mud volcanoes have both scientific and potential medical significance.

The genesis and origins of these therapeutic properties are of considerable interest. Ancient writers and physicians,

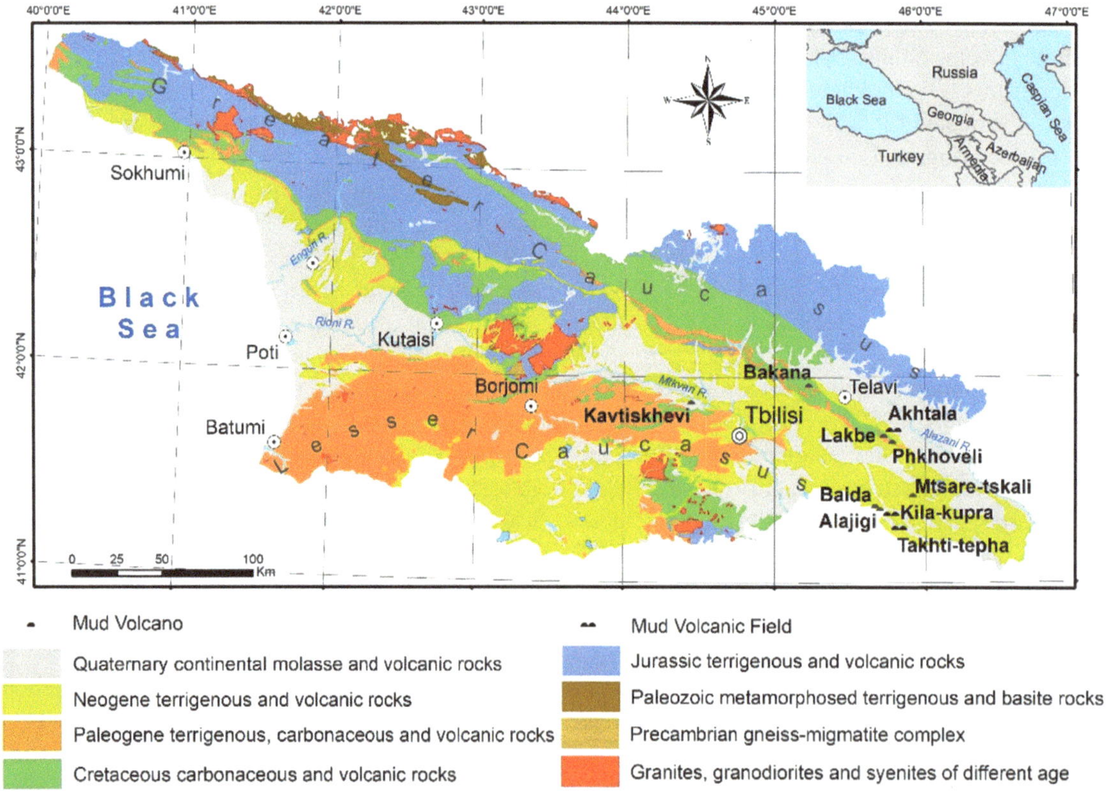

Fig. 3 Location of mud volcanoes and volcanic fields of Georgia on a simplified geological map of Georgia. The geological framework is modified after Gudjabidze et al. (2003)

including Herodotus, Pliny, and Dioscorides, described the medicinal use of mud volcano deposits in Egypt, Rome, and India. Historical accounts note that Alexander the Great treated wounded soldiers and horses with mud, while Queen Cleopatra of Egypt is said to have bathed in volcanic mud from the Nile region for therapeutic purposes. In Georgia, the use of volcanic mud in traditional medicine also has a long history, documented by Prince Vakhushti Bagrationi (1696–1758) in his "Description of the Kingdom of Georgia".

The mud volcanoes of the Gareja Desert also provide important insights into the geodynamic evolution of the region. Data reported by Kikvadze (2016) are particularly illuminating: mud volcanoes in Kakheti are characterised by a predominance of mantle-derived helium (^3He) over crustal helium (^4He), with $^3He/^4He$ ratios exceeding 3. This indicates an active inflow of mantle-derived fluids into the system. Such inflow suggests ongoing crustal stretching and thinning, which facilitates the migration of mantle ^3He. This interpretation is consistent with the earlier hypothesis of crustal thinning in the Alazani Depression proposed by Philip et al. (1989).

Among the mud volcanoes of the Gareja Desert, the Kila-Kupra and Takhti-Tepha fields are the most representative and spectacular. The active Kila-Kupra mud volcanic field is located in the southeastern part of the Iori Plateau, near the village of the same name, at an elevation of approximately 400–450 m asl (Fig. 4).

The field hosts more than 40 extinct and active cones. In its eastern sector, the volcanoes form an ellipsoidal area about 340 × 190 m in size, with cones up to 3–4 m high. Many of these edifices are actively discharging mud, gas, and minor amounts of oil, and they exhibit small gryphons and bubbling pools. On the western flank of the field, along the northern slope of a low ridge, there are numerous lower cones (1–2 m high) that emit viscous mud with persistent gas seepage. The physical properties and composition of the Kila-Kupra muds show notable similarities to those of the Akhtala balneological resort, and their therapeutic potential has long been recognised by local residents. Previous studies have even proposed establishing a balneological resort in this area (Rcheulishvili et al. 2006).

To better characterise the Kila-Kupra muds, we carried out full geochemical analyses of three samples from this field at the MSALAB Laboratory in Vancouver, Canada. Concentrations of major rock-forming elements were determined using the WRA-310 method, and trace elements were analysed by IMS-300. Chemically, the mud is mafic in composition, with SiO_2 contents of 48.4–50.1%. Al_2O_3 concentrations are relatively low (12.2–13.5%), whereas Fe_2O_3 (5.2–6.7%), CaO (7.3–7.8%), and Na_2O (4.9–5.8%) are elevated. Potassium (K_2O: 1.9–2.4%) and sulfur (SO_2: 0.56–1.88%) contents are comparatively low. The combination of high Ca and low S contents appears to be responsible for the characteristic grey coloration of mud volcano deposits in the Gareja Desert.

Radioactive elements were also measured. Uranium contents of 0.5–2.1 g/t fall within normal limits, whereas thorium concentrations are elevated, ranging from 12.53 ppm to 18.14 ppm. Zinc concentrations are particularly high, varying from 80.54 to 143.57 ppm, compared to an average of ~16 ppm for similar deposits. Recent studies of the Akhtali and Pkhoveli mud volcanoes on the Gombori Ridge (Martkopishvili 2019) show broadly comparable compositions, with similarly elevated thorium (up to 12 ppm) and zinc (up to 145 ppm) contents. These geochemical features help to explain the recognised therapeutic potential of the muds.

South of the Kila-Kupra field lies the Takhti-Tepha mud volcanic field, located partly within the Vashlovani

Fig. 4 View of one of the gryphons of the Kila-Kupra mud volcanic field. Photo © Ipove Sastumro ge

Fig. 5 General view of the Takhti-Tepa mud volcanic field. Photo © Turebi.ge

Protected Areas (Fig. 5). This field occupies elevations of 500–550 m asl and forms an ellipsoidal uplifted area with a surface densely covered by mud volcanoes and their eruptive products. More than 50 active and dormant cones and gryphons occur within the field, many of which are continuously coated by an oily film, reflecting the hydrocarbon content of the emitted fluids. The largest crater within Takhti-Tepha reaches a diameter of 3.5–4 m (Fig. 7; Gamkrelidze et al. 2021). Together, the Kila-Kupra and Takhti-Tepha fields represent outstanding examples of mud volcanism in Georgia and are among the most distinctive geoheritage sites of the Gareja Desert, combining scientific interest, aesthetic appeal, and significant balneological and geodynamic value.

South of the Kila-Kupra field lies the Takhti-Tepha mud volcanic field, located partly within the Vashlovani Protected Areas (Fig. 6). This field occupies elevations of 500–550 m a.s.l. and forms an ellipsoidal uplifted area with a surface densely covered by mud volcanoes and their eruptive products. More than 50 active and dormant cones and gryphons occur within the field, many of which are continuously coated by an oily film, reflecting the hydrocarbon content of the emitted fluids. The largest crater within Takhti-Tepha reaches a diameter of 3.5–4 m (Gamkrelidze et al. 2021). Together, the Kila-Kupra and Takhti-Tepha fields represent outstanding examples of mud volcanism in Georgia and are among the most distinctive geoheritage sites of the Gareja Desert, combining scientific interest, aesthetic appeal, and significant balneological and geodynamic value.

5 The Upper Miocene Volcanic Ash Layers

The rocks of the Neogene intermontane basin of the Eastern Caucasus consist mainly of thick molasse-type continental and shallow-marine sediments, including conglomerates, sandstones, limestones, marls, argillites, and clays. Within these successions, several subaerial volcanic ash layers of andesite–rhyolite composition have been identified. The thickest ash layer is exposed near the village of Sarabuki at the entrance to the city of Tskhinvali, where its thickness exceeds 5 m (Skhirtladze 1964). Further east, within the Upper Miocene deposits of Mount Iagludja, the thickness decreases to about 1.5 m. In the Gareja Desert it varies between 50 and 60 cm, and in the Kura Basin of Azerbaijan it reaches only about 20 cm. These systematic thickness changes, together with lithological similarity, indicate that

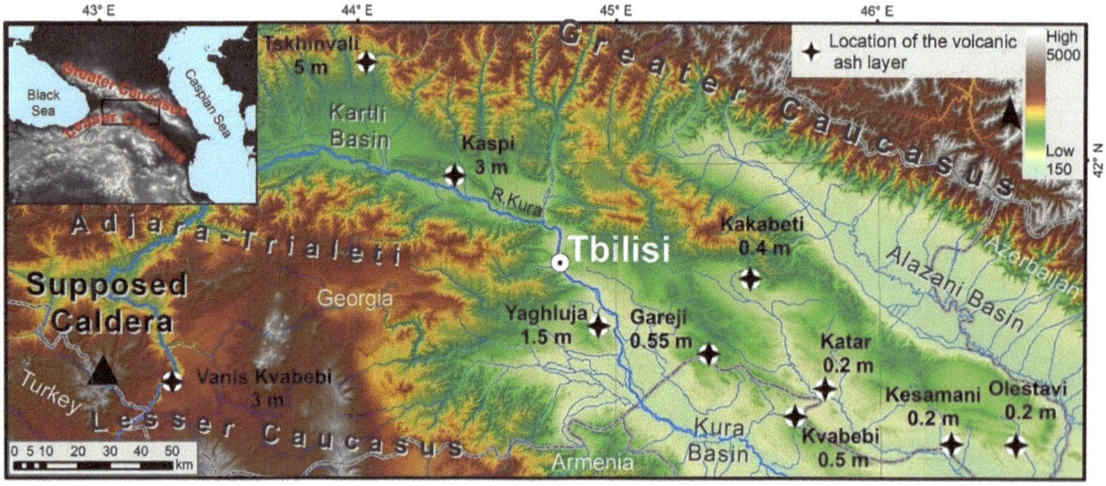

Fig. 6 Distribution of the Upper Miocene volcanic ash layers crops out in the Eastern Caucasus intermountain Neogene basin. Black color cross indicate the location of exposure, and meters—its thickness (Okrostsvaridze et al. 2024)

Fig. 7 One of the segments of the Upper Miocene volcanic ash layer (Light yellow layer) on the Gareja ridge. Photo © A. Okrostsvaridze

the ash layers belong to the same volcanic event and can be traced across much of the Eastern Caucasus intermontane basin (Fig. 6).

In the Gareja Desert, a particularly clear example of this Upper Miocene ash is exposed along the north-eastern flank of the Gareja Ridge (Fig. 7). The layer is of andesite–rhyolite composition and is interbedded within Upper Miocene continental deposits. It appears as a fine-grained, white-coloured bed with only a very small admixture of terrigenous material, suggesting deposition in a quiet, likely lagoonal setting. Petrographic observations show that the ash is composed mainly of glass shards and fine crystals of plagioclase, quartz, biotite, and hornblende, with subordinate magnetite and zircon. Major- and trace-element geochemical analyses indicate that the ash corresponds to rhyolites, with SiO_2 contents around 70%, Al_2O_3 ~17%, MgO ~0.7%, Na_2O ~3%, and K_2O ~2%, confirming its evolved, felsic character.

The ash is especially well exposed near the Bertubani monastery complex, where a continuous section allows detailed sampling. Zircon grains extracted from the ash were studied using cathodoluminescence imaging and dated by the LA-ICP-MS U–Pb method. The majority of grains, however, define a coherent Late Miocene population. The best-constrained concordant age is 7.51 ± 0.2 Ma, which is interpreted as the crystallisation age of the ash and thus of the eruption that produced it (Okrostsvaridze et al. 2024).

Zircon U–Pb dating is widely regarded as the most reliable method for determining the ages of volcanic rocks, because zircons are highly resistant to weathering and metamorphism and retain radiogenic lead over geological timescales. By providing absolute numerical ages, such geochronological data fix the depositional age of the host sedimentary unit and allow robust regional correlations. Consequently, the Upper Miocene volcanic ash layer of the Gareja Desert is not only of local interest but also serves as a key marker horizon across the Eastern Caucasus intermontane basin. Its continuity and thickness indicate that it originated from a powerful explosive eruption, most likely from a large volcanic centre in the wider Anatolia–Caucasus region. An eruption capable of dispersing ash over such a wide area would have had a significant environmental impact, with the atmosphere enriched in ash and toxic gases (e.g., CO_2, SO_2) and regional temperatures temporarily lowered by several degrees. In this sense, the Gareja ash layer records both a moment in the basin's stratigraphic history and a brief but intense climatic disturbance during the Late Miocene.

6 Terrestrial Vertebrates Fossils

Fossil remains of terrestrial vertebrates are relatively rare in Georgia; however, the Iori Plateau of the Gareja Desert represents a notable exception, being particularly rich in Late Miocene vertebrate fossils. During this epoch, the area formed part of the western coastal zone of the Eastern Paratethys. Numerous lakes and lagoons were characteristic of this environment, the climate was close to subtropical, and both flora and fauna were diverse. Their remains have been preserved within the continental-type sediments of that time (Bukhsianidze 2019).

In the Miocene sandstones of the Iori Plateau, at Udabno, Chachuna, and Eldari (the latter now located in present-day Azerbaijan), abundant remains of hipparions (the three-toed horse genus *Hipparion*) have been discovered. In addition to hipparions, numerous remains of rhinoceroses and several other ungulate taxa have been identified. These fossils are distributed across large areas of the plateau, and their study is still ongoing (Bukhsianidze 2024).

Since this volume includes a dedicated chapter, "Fossil Land Mammal Faunas of Georgia" (authored by Maia Bukhsianidze), we do not discuss this aspect of geoheritage in further detail here. Nevertheless, it is important to note that the abundance of terrestrial vertebrate remains on the Iori Plateau, distributed across several localities within a specific time interval (Late Miocene), suggests that their accumulation may have been related to a large-scale extinction event triggered by a powerful volcanic eruption. Comparable events are well documented in Earth's history and are typically associated with atmospheric enrichment in toxic gases and a sharp decrease in temperature (e.g. Bond, Wignall 2014). This hypothesis is further supported by the occurrence of Late Miocene volcanic ash layers (7.51 Ma) in the Gareja Desert, produced by a major eruption of the Keltepe–Gumbati megacaldera (Okrostsvaridze et al. 2020; Makadze et al. 2024).

Fig. 8 General view of the modern David-Gareja Lavra. Photo © Ivane Goliadze

7 The David Gareja Rock-Cut Monastery Complex

The establishment of the David Gareja rock-cut monastery complex was closely conditioned by the desert climate and geology of the Gareja region. In the sixth century, the monk David, one of the Thirteen Syrian Fathers, chose the south-facing slopes of the Gareja Ridge as a place of retreat and spiritual practice. Here he and his companions carved the first caves into the soft Upper Miocene sandstones, creating simple cells and a small church that later became known as David's Lavra (Fig. 8). Over the following centuries, the complex expanded along the ridge and its lateral spurs, so that dozens of monasteries and hermitages, together with smaller rock-hewn cells, chapels, and refectories, functioned within the broader Gareja landscape.

In the early medieval period, the David Gareja monastery complex developed into one of the most important religious and cultural centres of Georgia and the wider Caucasus. Monasteries such as Lavra, Udabno, Bertubani, Sabereebi, Dodo's Rka and many others were established along the crest and flanks of the ridge, following the natural ledges and overhangs created by the tilted sandstone beds. The main churches and refectories were cut deeply into the rock, while façades, walls, and defensive elements were constructed of local stone and mortar. This intimate integration of architecture and geology, monastic spaces literally carved into the Upper Miocene sandstones that form the desert slopes, remains one of the most distinctive features of the complex and continues to impress visitors today (Fig. 9).

Fig. 9 Fragment of the eastern part of the David's Lavra of the David-Gareja monastic complex, most of which is carved into Upper Miocene sandstones. Photo © A. Okrostsvridze

Of particular note is the ingenious water-collection and storage system developed by the monks to survive in the arid environment. Rainwater and snowmelt were captured from the sandstone slopes using carved gutters and small channels, which directed the flow into rock-cut cisterns and reservoirs hidden within the cliffs. These cisterns, often plastered to reduce leakage, formed a network that supplied drinking water and supported limited agriculture and livestock. The system demonstrates a deep practical understanding of the local geomorphology and hydrology: the monks used the natural permeability contrasts within the sandstone–clay sequence and the ridge topography to maximise water capture and minimise evaporation. This traditional knowledge is a key component of the cultural landscape and an important example of sustainable adaptation to semi-desert conditions (Fig. 10).

Fig. 10 General view of the water collection system of the David-Gareja Lavra. Photo © Salome Gogoladze

The David Gareja complex was also a major centre of creative and intellectual life. From the early Middle Ages onward, manuscripts were intensively copied and illuminated in the scriptoria of several monasteries, contributing significantly to the development and preservation of Georgian literature and theology (e.g. Skhirtladze 1985). At the same time, the walls of churches and chapels were adorned with rich fresco cycles depicting biblical scenes, portraits of kings and donors, and episodes from monastic life. Many of these frescoes, especially those at Sabereebi and Udabno, are masterpieces of medieval Georgian mural painting and display stylistic influences from both Byzantine and eastern Christian traditions. Although many images have been damaged by time, vandalism, and past iconoclasm, the surviving fragments still convey the high artistic level and spiritual intensity of the complex.

At present, only part of the original David Gareja monastery complex remains in active use, yet its cultural and spiritual significance for Georgia is undiminished. Monastic life continues at David's Lavra and several neighbouring monasteries, and the area is an important pilgrimage destination as well as a popular geotourism site. Recognising its outstanding cultural value, the David Gareja complex has been included on the UNESCO World Heritage Tentative List. At the same time, the rock-cut architecture is highly vulnerable. The soft sandstones and interbedded clays are prone to weathering, fracturing, and rockfalls, processes that are locally accelerated by climatic extremes and increasing visitor pressure. Recent geotechnical studies have documented instability in several sectors of the cliffs and recommended detailed monitoring as well as reinforcement of the most endangered sections.

The David Gareja rock-cut monastery complex is therefore best understood as a cultural landscape in which monastic architecture, traditional land use, and the semi-desert geomorphology are inseparably linked. The monasteries are carved into the same Miocene strata that host the colourful hills, mud volcanoes, and volcanic ash layers described in this chapter, and they share the same tectonic and climatic history. This tight coupling of geological and cultural features makes David Gareja a key element of Georgia's geoheritage and a central pillar of any future UNESCO Global Geopark centred on the Gareja Desert.

8 Threats and Management Considerations

Although the Gareja Desert and the David Gareja rock-cut monastery complex retain an overall sense of remoteness, signs of pressure on both geological and cultural features are increasingly evident. The soft Upper Miocene sandstones and interbedded clays that host the monasteries, colourful hills, and fossil localities are highly susceptible to weathering, slope failure, and gullying. Intense rainfall events following long dry periods accelerate erosion, particularly along informal footpaths, off-road vehicle tracks, and livestock trails. In places, fossiliferous horizons and the surfaces of mud volcano cones have been damaged by unregulated collecting and trampling.

Cultural heritage is similarly vulnerable. The rock-cut churches, cells, and cisterns of the David Gareja complex are affected by rockfalls, block detachment, and progressive opening of joints and fractures in the cliff faces. Salt crystallisation, moisture fluctuations, and biological growth further deteriorate wall paintings and plaster surfaces, particularly in caves with high visitor numbers or poor ventilation. Vandalism, graffiti, and accidental damage also occur along heavily visited routes where monitoring is limited. Without systematic geotechnical assessment and stabilisation of the most fragile sectors, there is a risk of irreversible loss of both structural integrity and mural art.

Addressing these threats will require integrated geoconservation measures. Priority actions include detailed mapping and monitoring of unstable rock faces and key geosites; the design of marked trails and viewpoints to concentrate visitor use and reduce trampling; improved interpretation that explains both the fragility and the value of the geoheritage; and stronger involvement of local communities in guiding, surveillance, and maintenance. Framing the Gareja Desert and David Gareja as a single geoheritage–geocultural landscape also opens the possibility of using a future UNESCO Global Geopark model to coordinate management across geology, biodiversity, and culture. Such an approach would not only help safeguard outstanding features but also provide a platform for sustainable local development based on geotourism, education, and the careful stewardship of this distinctive semi-desert environment.

9 Conclusion

Within a relatively small area on the Iori Plateau, the Gareja Desert concentrates a remarkable variety of geoheritage features. Gently tilted Neogene molasse strata, colourful hills formed by intense near-surface weathering, active mud volcanoes, Late Miocene volcanic ash layers, and rich terrestrial vertebrate faunas together provide an unusually complete record of the region's collisional and sedimentary history. These elements make the Gareja semi-desert a natural open-air laboratory where processes of basin infill, volcanism, climate, and erosion can be read directly from the landscape.

On this geological stage, the David Gareja rock-cut monastery complex and associated hermitages have created an equally dense cultural landscape. The monasteries are literally carved into the same Miocene sandstones that host the geoheritage sites described in this chapter, while the rock-cut churches, frescoes, manuscripts, and water-harvesting systems demonstrate a long tradition of adaptation to aridity. Spiritual practice, architecture, and geomorphology are inseparably linked, so that the cultural values of David Gareja cannot be understood without its geological setting, and vice versa.

Viewed in this light, the Gareja Desert and the David Gareja monastery complex together form a compact but highly expressive geoheritage–geocultural ensemble with clear potential for future recognition as a UNESCO Global Geopark. They offer outstanding geological and geomorphological features, a powerful overlay of spiritual and artistic heritage, distinctive semi-desert ecosystems, and an existing, if still loosely organised, geotourism base linked to nearby Tbilisi. With thoughtful, participatory management that balances access and protection, Gareja can become not only a model for small, dense geoparks in the Caucasus, but also a new facet of Georgia's national story: a desert landscape that holds deep time, deep faith, and a path toward sustainable geoheritage-based development.

References

Baars DL (1983) The colorado plateau: geologic history. University of New Mexico Press, Albuquerque pp 163–168

Baloganov EE, Abbasov OR, Akhundov RV (2018) Mud volcanoes of the world: classifications, activities and environmental hazard (informational-analytical review). Eur J Nat Hist 5:1–16

Bond DPG, Wignall PB (2014) Large igneous provinces and mass extinction: an update. In: Keller G, Kerr AC (eds) Volcanism, impacts, and mass extinctions. Geol Soc Am Spec Pap 505:32–41

Bukhsianidze M (2019) Tapirus priscus Kaup, 1833 (Tapiridae, Perissodactyla) from the Udabno site (Late Miocene, Georgia). Bull Georg Natl Acad Sci 13(2):103–107

Bukhsianidze M (2024) Revisiting the earliest terrestrial fossil record from the paleo Kura bay (Eldari and Udabno faunas, Late Miocene, South Caucasus)—diversity of bovids. EGU Gen Assem 2024, Vienna, Austria, 14–19 Apr 2024, EGU24–7327. https://doi.org/10.5194/egusphere-egu24-7327

Buleishvili DA (1960) Geology and oil and gas potential of the intermontane depression of eastern Georgia. Leningrad, 239 pp (in Russian)

Gamkrelidze I, Okrostsvaridze A, Koiava K, Maisadze F (2021) Geotourism potential of Georgia, the Caucasus. Springer, Cham. Pp 1–140. https://doi.org/10.1007/978-3-030-62966-3

Ebralidze TP, Bidzinashvili GG, Tatarashvili LI, Geladze SR, Shatirishvili NG (1975) Mud volcanoes of Georgia. VNIGRI, Moscow, pp 1–163. (in Russian)

Kikvadze O (2016) Geochemistry of mud volcanic fluids of the Caucasian region. PhD Thesis, Moscow, pp 1–23 (in Russian)

Mazzini A, Etiope J (2017) Mud volcanism: an updated review. Earth Sci Rev 168:81–112

Makadze M, Okrostsvaridze A, Keskin M, Pophkhadze N (2024) Structure and composition of the late miocene Goderdzi volcanic complex, Lesser Caucasus, Georgia: Implication for resurgent caldera existence. Abstr 1511, 37th Int Geol Congr, Busan, Republic of Korea

Margottini C, Spizzichino D, Gigli G, Frodela W, Elashvili M, Alberti S, Valagussa A, Giovani C (2020) Instability processes affecting the rupestrian monastery complex area of Davit Gareja (Georgia). In: Davit Gareja—Multidisciplinary study and development. Conf Proc, Tbilisi, pp 187–193. https://www.researchgate.net/publication/332013627

Martkopishvili A (2019) Petrogeochemical characteristics of the Pkhoveli mud volcano of the Tsiv-Gombori range and their correlation with the Akhtala mud volcano. Master's thesis, Tbilisi, pp 1–58 (in Georgian)

Okrostsvaridze A, Chung SL, Chang YH, Skhirtladze I (2020) Geology and zircon U-Pb geochronology of the Mtkvari pyroclastic flow and evaluation of destructive processes affecting Vardzia rock-cut city. Quat Int 540:137–145. https://doi.org/10.1016/j.quaint.2019.03.026

Okrostsvaridze A, Gamkrelidze I, Lee YH, Keskin M, Tormey D, Boichenko G, Gogoladze S, Makadze M (2024) U-Pb geochronology and geochemistry of volcanic ash layers in the Eastern Caucasus intermountain Neogene basin: implications for their sources. Arab J Geosci 17:129. https://doi.org/10.1007/s12517-024-11914-7

Philip H, Cisternus A, Gvishiani A, Gorshkov A (1989) The caucasus: an actual example of the initial stages of a continental collision. Tectonophysics 161:1–21

Rcheulishvili R, Geleishvili V, Salukvadze H (2006) On the use of mud volcanoes of outer Kakheti as a raw material base for balneological resorts. Min J 1–2 (16–17) (in Georgian)

The geological map (1:200 000) of Eastern Georgia. The Georgian Geological Department, 2019

Skhirtladze Z (1985) Fresco inscriptions of the Sabereebi. Sources Georg Hist 35:55–56 (in Georgian)

Skhirtladze NI (1964) On the use of volcanic ashes for dating volcanic for mation. XXII International Geological Congress, New Delhi, pp 207–214

Wan Z, Zhang J, Lin L, Zhong S, Li D, Wei J, Sun Y (2023) Formation mechanism of mud volcanoes/mud diapirs based on physical simulation. Geofluids 2021:5531957. https://doi.org/10.1155/2021/5531957

Fossil Land Mammal Faunas of Georgia

Maia Bukhsianidze

Abstract

This chapter provides an overview of several key fossil land mammal sites of Georgia, south Caucasus. It briefly discusses the fossil land mammals from famous paleontological sites across the country. These include the late Oligocene fauna of Benara -the earliest large mammalian fauna of the Caucasus with giant long necked rhino, Paraceratherium, and the late Miocene faunas–Udabno and Chachuna. It includes the earliest hipparion faunas from the Kura Basin with the last dryopithecine record in western Eurasia, and Kvabebi fauna—one of the latest hipparion faunas in Eurasia dating back to the latest Pliocene. Dmanisi—1.8 million years old, contains an extremely rich and diverse fauna with the earliest *Homo* remains outside of Africa, and Akhalkalaki—noteworthy Early Galerian fauna from the Caucasus. Paleolandscape and ecological conditions suggested by the fossil land mammals and other environmental proxies are depicted for each site.

Keywords

Benara · Chachuna · Dmanisi · Fossil land mammals · Kvabebi · Udabno

1 Introduction

The wildlife of Georgia and the Caucasus region, in general, is distinguished by its high degree of biodiversity. This biodiversity results from the region's geographic, climatic, and evolutionary history of living forms. This region serves as a crossroads for the migration of flora and fauna, a hub of evolutionary diversification with a high degree of endemism, and a refuge that boasts a variety of relict species. The Caucasus is located at the crossroads of Europe and Asia and today lies on the border between the subtropical and temperate belts. Understanding the natural history of this region is essential for grasping the evolution of the Eurasian biome as a whole, and clarifying the pathways, directions, and timelines of migrations between Asia, Europe, and Africa. The fossil land mammal sites discussed in this chapter are therefore key elements of Georgia's geoheritage: they preserve snapshots of past ecosystems and climates that underpin today's status of the Caucasus as a global biodiversity hotspot.

Georgia's paleontological record includes fossils of both marine and terrestrial organisms—flora, invertebrates, and vertebrates—spanning from the Mesozoic Era through the Cenozoic. This chapter focuses exclusively on terrestrial fossil mammals of the Cenozoic Era. Georgia has fourteen notable locations known for terrestrial fossil mammal faunas. This list does not include cave archaeological sites or locations with isolated or small numbers of fossils (which require further investigation). These fossil sites can be categorized into the following groups: (1) Sites with a high concentration and diversity of fossil forms (e.g., Iagludja, Dzidzvekhavi, Bazaleti, Kvabebi, Dmanisi, Orozmani, Akhalkalaki); (2) Complex sites that cover extensive areas and usually include several sub-sites. At such sites fossil accumulations are rare, and fossil remains are often fragmented and sporadic (e.g., Udabno; sub-sites include Mravaltskaro, Natlismtsemeli, Lavra, Hagia; Chachuna—sub-sites include Chachuna 1, 2, 3, 4; Benara); (2) Taxonomically rich sites, but with a limited number of fossil specimens (e.g., Arkneti, Melaani, Tsalka). Only one terrestrial vertebrate fossil site dates back to the Paleogene, while the others are from the Neogene and Quaternary periods. Here, we will highlight only a few exceptional sites, beginning with the oldest fauna—Benara. Many of these localities lie within landscapes that remain important today for agriculture, protected areas, and geotourism, making them

M. Bukhsianidze (✉)
Georgian National Museum, Tbilisi, Georgia
e-mail: maiabukh@gmail.com

A. Okrostsvaridze and D. Tormey (eds.), *Geoheritage of Georgia*, Geoheritage, Geoparks and Geotourism,
https://doi.org/10.1007/978-3-032-27639-1_18

were dwellers of marshes and coastal forest habitats. Typical forest inhabitants include *Colodon* and chalicotheres, while xerophilous open savanna dwellers include *paracerathers, hyracodonts, rhinoceroses,* and *Prodremotherium.*

From a biogeographical perspective, the diverse perissodactyls in Benara (e.g., *Paraceratherium, Colodon, Phyllotillon, Prothyracodon, Ardynia, Telmatodon*) have Asian origin and distribution, suggesting prolonged and close contact with Asia. Other forms are shared between Europe and Asia. The Benara fauna shows significant similarities to the Bugti fauna of Pakistan but is comparatively more archaic. It most closely resembles the Balkan-Transylvanian faunas. The Benara fauna was the first to confirm that the Balkan-Transylvanian region was connected to Asia via the Caucasus-Anatolian landmass. The Paratethys Sea acted as an impassable barrier to the west of the Balkan-Transylvanian landmass detaching the European archipelago from Asian influence.

3 Late Miocene Fossil Vertebrate Faunas from the Eldari Variegated Formation

(Fig. 2; Upper Miocene, 10–7.5 million years ago, MN10–MN11; Georgia-Azerbaijan Border, Udabno: 41°32′13.2"N 45°15′05.5″E—41°25′57.1"N 45°25′26.5″E; Chachuna: 41°13′04.7"N 45°55′37.6″E—41°11′38.3"N 45°59′56.2″E; Eldari: 41°11′21.4"N 46°07′26.0″E—41°10′47.5"N 46°14′20.8″E; Meladze 1985; Garzhiev 1996; Bukhsianidze and Koiava 2018; Bukhsianidze 2019, 2024; Vanishvili 2022).

The Iori Plateau is one of the touristic destinations in Georgia, known primarily for the David Gareji Monastery Complex, its beautiful landscape, and the diversity of avifauna. Additionally, the plateau is exceptionally rich in fossil vertebrate remains. Along the coastline of the paleo Kura Bay (a westward extension of the South Caspian Basin of the Eastern Paratethys) exposed here as a series of sandstones and overlying Variegated Eldari Formation, hipparion fauna fossil sites have been discovered. These sites include Udabno and Chachuna on Georgian territory, and Eldari in Azerbaijan. All three sites are found in penecontemporaneous sedimentary succession deposited in the same basin. Despite Eldari being located in Azerbaijan, all three sites are discussed together in this review.

Fossil remains of terrestrial vertebrates from the early and Middle Miocene are extremely rare in Georgia. Most findings from these epochs consist of fascinating fossils of marine mammals (Mchedlidze 1984). However, after a long chronological gap (more than ten million years), the terrestrial fossil record unfolds the history of the Late Miocene (11.6–5.3 Ma) Hipparion fauna. The contours of familiar landscapes and ecosystems, for those of us living in the temperate zone of Eurasia, began to take shape during the Late Miocene. At this time, the land approached its modern configuration, the climate became seasonal, modern flora took on a leading role, and the fauna began to be dominated by representatives of animal families still existing today. Extinct side branches persisted but became increasingly rare.

During the Late Miocene, the Eastern Paratethys Sea gradually receded, and by approximately eight million years ago, this once vast sea had entirely dried up, leaving only the smaller basins of the Black, Caspian, and Aral Seas. Consequently, the extent of landmass significantly increased. Late Miocene molasse basins are widely exposed in eastern Georgia, particularly in the arid region of the Iori Plateau. Such areas are conducive to discovering fossil accumulations, and indeed, most terrestrial fossil mammal sites in Georgia are found in Kakheti, on the Iori Plateau. Interestingly, terrestrial landscapes existed from the Middle Miocene (16–11.6 Ma) onwards to the west of the Iori Plateau, in Kartli, and western Georgia. Therefore, older fossil sites are expected to be discovered in these regions. However, a lack of exploratory work and dense vegetation cover, especially in western Georgia, present obstacles to identifying fossil deposits.

Udabno, Chachuna, and Eldari represent complex fossil sites. Fossils are found across large areas and at different stratigraphic levels, making stratigraphic correlation essential for data interpretation. Detailed stratigraphic studies are ongoing but not yet completed. Broadly, this sedimentary succession reflects the west-to-east retreat of the paleo Kura Bay, and thus terrestrial deposits are diachronous. In the west (Udabno), terrestrial layers (the Variegated Eldari Formation) appear at the end of the Bessarabian stage (ca. 10 Mya), while in the east (Chachuna and Eldari), variegated deposits are found later, within the Khersonian stage (9.9–7.65 Ma). Overall, this fossil record from the coastal zone of the Kura Bay spans approximately 10 to 7.6 Ma, corresponding to the Late Vallesian and Early Turolian European land mammal stages and MN10–MN11 biozones.

During this time, the Iori Plateau was a coastal zone of the Kura Bay covered with a mosaic of isolated and semi-isolated basins, i.e., lakes, lagoons, and a large river network forming deltas, transitioning further inland to drier, open landscapes. Pollen and spore spectra from Chachuna (Kokolashvili et al. 2018) reveal vegetation typical of the temperate zone, alongside subtropical and tropical elements such as *Schizaea, Gleichenia, Ginkgo, Sequoia, Keteleeria,* palms, rhododendron, magnolia, *Symplocos, Sterculia,* and members of *Icacinaceae.* Temperate flora included fir, spruce, pine, cedar, cypress, walnut, hazelnut, hornbeam, cornel, alder, oak, linden, beech, chestnut, rose family members, wormwood, and other grasses. Overall, the spectrum of plant pollen and spores indicates the presence of

Fig. 2 **a** One of the late Miocene sites—Chachuna—on the Iori Plateau exposure of the Eldari Variegated Formation; **b**. *Udabnocerus georgicus,* Skull, GNM1 32–2013/1031, Udabno, Hagia, holotype; **c**. *Miotragoceros valenciennesi eldaricus,* skull, GNM1 31–2013/70, Eldari; **d**. *Simocyon* sp. *Skull with mandible fragment, GNM1 32–2013/1503, Udabno, Hagia;* **e**. *Percrocuta gigantea,* maxilla, *GNM1 32–2013/769 (343–5) Udabno;* **f**. *Praepusa procaspica, articulated forearm, El-I 2 IPB, Eldari;* **g**. *Hipparion* sp., *ckull, 32–2013/1003, Udabno;* **h**. *Microstonyx major, maxilla, Ud. 343–4* (photos of M. Bukhsianidze)

coastal forests along Kura Bay. It is significant that in the facies farther from the seashore, the herbaceous vegetation prevailed over the tree plants and deciduous trees prevailed over tropical and subtropical species.

The dominant group in the faunas of Udabno, Chachuna, and Eldari is *Hipparion* (genus of the three-toed horses). *Hipparion* entered Eurasia from North America via the Bering Strait approximately 11–12 million years ago and in a very brief time spread all over the Old World, becoming the dominant group in the fauna (Bernor et al. 2021). Consequently, faunas from this period are referred to as *Hipparion* faunas. Two species of *Hipparion* have been identified from the sites: *H. eldaricum* and *H. garedzicum*. Among others in the odd-toed ungulates

(*Perissodactyla*), rhinoceroses are well-represented, (five species are mentioned in the faunal lists *Chilotherium aff. Zernovi*, *Chilotherium eldaricum*, *Aceratherium cf. incisivum*, *Ceratotherium gabuniai*, *C. pachygnathus*). Tapirs (*Tapirus priscus*) and chalicotheres (*Chalicotheriidae cf. Ancylotherium*, *Eldarotherium burtchaclensis*) are relatively rare.

Among the even-toed ungulates (*Artiodactyla*), bovids are dominant ruminant family, represented by the boselaphins and early gazelles. Fossil deer—muntjacs, roe deer, and tragulids—are common, while remains of spiral-horned antelopes and giraffids are rare. Fossils of suids, such as *Microstonyx erymanthius* and *M. major*, are frequently encountered. Carnivores include *Simocyon* sp., a member of the family *Ailuridae* related to the red panda, a mustelid *Plesiogulo cf. brachygnathus*, and hyenas, including durophagous forms like *Adcrocuta eximia* and *Percrocuta gigantea* and cursorial species like *Miohyaenotherium bessarabicum*. Saber-toothed cats (*Machairodontinae indet.*) are also present. Proboscideans include *Deinotherium giganteum*, *Tetralophodon cf. longirrostris* and *Choerolophodon pentelici*. Ostrich eggshell fragments and abundant remains of terrestrial and marsh turtles and castorids further enrich the faunal diversity of these sites.

Among the discovered fossil mammals, the most remarkable and celebrated fossil is the upper jaw fragment of the Miocene ape *Dryopithecus garedziensis*, first described as *Udabnopithecus garedziensis*. *Dryopithecus* was widespread in Europe between 12.5 and 9.5 Mya. *Dryopithecus garedziensis* is the easternmost and possibly the latest representative of this genus (Agustí et al. 2020). The dryopitecins were a relatively small arboreal, suspensory apes. Their diet consisted by soft vegetable foods, mainly fruits. They inhabited the moist subtropical forests of the Paratethys and Mediterranean regions. With the retreat of the seas, the general trend towards aridization, and an intensification of seasonal fluctuations the habitat of dryopithecins vanished, leading to their eventual extinction.

As a whole the fauna includes inhabitants of open spaces (rhinoceroses, giraffids, spiral-horned antelopes, hyaenas, ostriches), forest-edge biotopes (baselaphin antelopes, gazelles, roe deer), moist evergreen forests (dryopithecins, muntjacs, pigs, mustelids, simocyons), and wetlands (tapirs, marsh turtles, castorids). Such habitat diversity is because fossils cover a long period and are mostly found in river deposits, representing a wide variety of animals existing over a considerable area and time.

Kvabebi (Fig. 3.; Upper Pliocene, ca. 2.7 Ma, Early Villafranchian, MN16b; Sighnaghi municipality, Georgia; 41°27′02.6"N 45°42′54.5″E; Vekua 1972; Hemmer et al. 2004**;** Bukhsianidze and Koiava 2018**).** Vertebrate fossils were found in the Mount Kvabebi area (Sighnaghi municipality, Georgia), where the ruins of the medieval Kvabebi monastery (one of the monasteries of the Davit Gareja complex) overlook the fields at the foothills of the mountain. Kvabebi thus represents a typical Georgian geocultural setting, where a medieval monastery and a key fossil site share the same landscape, illustrating how deep time and historical heritage can be interpreted together.

Around 2.7 million years ago, this place was a coastal zone of the Akchagylian Sea with swampy, flooded banks and river tributaries. Among the fossilized remains of vertebrate animals (28 species were identified; faunal list in Bukhsianidze and Koiava 2018) there are three-toed horses (*Plesiohipparion rocinantis*, the last representative of hipparions in Western Eurasia), mastodons (*Anancus arvernensis*)—a proboscidean adapted to swampy water bodies; forest-edge and fertile meadow inhabitants such as an early representative of bovines, the hornless bull—*Ioribos aceros*, spiral-horned antelopes (*Gazellospira torticornis*). Fossils of deer and giraffids are also found. Among the predators are the remains of the extinct raccoon dog *Nyctereutes megamastoides,* possible ancestor of red fox -*Vulpes* cf. *alopecoides,* hunting hyaena—*Chasmaporthetes lunensis,* saber-toothed cat (*Homotherium davitashvili*), false saber-toothed cat (*Dinofelis cristata*), the oldest puma in Eurasia (*Puma pardoides*) and others. One of the outstanding representatives of the fauna of the Kvabebi fauna is the hyrax (*Kvabebihyrax kachethicus*)—its upturned eye-sockets and backward shifted nasal opening suggest a semi-aquatic life style. The remains of ostriches add an even more exotic touch to the late Pliocene wildlife of the South Caucasus.

Dmanisi. (Fig. 4.; Lower Pleistocene, 1.8 Ma, Late Villafranchian/Villanian fauna, MN17–MQ1, Village Patara Dmanisi, Dmanisi Municipality, Georgia, 41°19′23.4"N 44°20′59.5″E, Vekua 1995**;** Lordkipanidze et al. 2007, 2013; Bartolini-Lucenti et al. 2022). Along with early human stone tools and skeletal remains, thousands of bones of prehistoric animals are buried under the ruins of the medieval city of Dmanisi on a promontory at the confluence of the Mashavera and Pinesauri rivers. The animals and the people, who made these stone tools in the Dmanisi area, lived at the same time 1.8 million years ago. Today, Dmanisi is recognized not only as a cornerstone of human evolutionary research, but also as one of Georgia's most important geoheritage sites, where volcanism, river incision, and the fossil record of early Homo and associated fauna can be understood as a single, integrated story.

The wildlife of prehistoric Dmanisi was incredibly diverse compared to today. Early humans (*Homo erectus ergaster georgicus*) were an integral part of the fauna. They were omnivores and often competed with predators over animal carcasses. Early humans lived in well-coordinated small groups. While resting and processing prey, they probably took shelter in trees and rocks. The difficult terrain

Fig. 3 **a.** View on Kvabebi mountain; **b.** *Homotherium davitashvili*, skull, GNM1 29–2013/453 (K14), holotype; **c.** *Kvabebihyrax kachethicus*, skull and mandible K3010; **d.** *Puma pardoides*, mandible, GNM1 29–2013/576 (K3035); **e.** *Gazellospira torticornis*, frontlet with horn cores, GNM1 29–2013/204 (K597); **f.** *Eosyncerus iivericus*, frontlet with horn cores, GNM1 29–2013/534 (K1051), holotype (photos of M. Bukhsianidze)

of the Dmanisi area, where it was easy to hide places, must have been a favorable living environment for them, as food, water, and raw materials for making stone tools were abundant.

The leading role in the accumulation of fossils undoubtedly belongs to carnivores– giant short-faced hyaenas and big cats. Traces left by predators, tooth-marks and gnawed areas, are clearly visible on many remains of large mammals. The nature of the bone-cracking, the ratio of the different bones of the skeleton, and the parts of specific bones that are preserved suggest extensive hunting, dismembering of the animal carcasses, carnivore competition over

Fig. 4 **a** View on Dmanisi promontory (photo of G. Tsibakhashvili); **b**. *Bison (Eobison) georgicus*, skull fragment, D354, holotype; **c**. *Capra dalii*, horn core dex., D75, holotype; **d**. *Gallogoral meneghini sickenbergii*, skull, Dm.4/154.B1.79; **e**. *Pontoceros surprine*, skull fragment D5552, holotype; **f**. *Homotherium crenatidens*, skull, Dm.5/154.A1/87; **g**. *Panthera onca georgica*, mandible dex., D2027, holotype; **h**. *Lynx issiodorensis*, skull, D3497; **i**. *Pachycrocuta brevirostris*, juvenile mandible dex., Dm.65/63.B1x.153; j. *Canis borjgali*, skull with mandible, D3420, holotype; **k**. *Megantereon whitei*, skull, D1341; **l**. *Homo erectus ergaster georgicus,* skull and mandible, skull 5, D4500 and D2600 (photo of G. Bumbiashvili); **m**. *Praemegaceros obscurus*, antler fragment dex., D430; **n**. Alcini, antler fragment dex., D1949; **o**. *Arvernoceros insolitus*, antler dex., D2747, holotype; **p**. *Mammuthus meridionalis taribanensis*, articulated hind leg fragment, Dm.64/68.B1x.338; **q**. *Stephanorhinus etruscus*, skull, D3270; **r**. *Pseudodama nestii*, antler sin., D1495 (fossil specimens photos others than *Homo* of M. Bukhsianidze)

carcasses, and carnivores hiding their prey in protected locations.

The human contribution to the accumulation of animal bones at Dmanisi promontory was negligible, as revealed by the detailed spatial and taphonomic studies despite the thousands of stone tools and cobbles left by early humans at this site (Coil et al. 2020; Tappen et al. 2022b). The number of artifacts clearly indicate that early *Homo* frequented this place. The carcasses of hunted animals were waiting for them on the Dmanisi promontory. Dmanisi men threw cobbles in order to repel predators, and tried to slice meat and dismember the carcasses with sharp-edged stone tools.

Numerous scenes of competition between predators and early *Homo* are revealed by the fossil animal bones, on which both, tooth-marks made by carnivores and cutmarks made by stone tools are preserved; in some cases, the tooth-marks preceded cutmarks, sometimes traces of stone tools are found on such bones that predators consume first of all, e.g., the thigh bones, and shoulder bones, indicating early access to the carcasses by the humans (Tappen et al. 2022b). The fossil remains of the early *Homo* from Dmanisi are the result of some of the many unlucky episodes, when humans were the losers in a battle with predators.

In terms of faunal diversity, Dmanisi is one of the richest Early Pleistocene sites in the world. The remains of vertebrate animals, besides *Homo*, found in Dmanisi belong up to fifty-four species (faunal list in Bartolini-Lucenti et al. 2022). Large mammals are particularly diverse, comprising fourteen carnivores, thirteen even-toed ungulates, four odd-toed ungulates, and one proboscidean. These are Eurasian animals (all large mammals expect *Homo* and possibly *Megantereon*) or taxa that have migrated out of Africa a long time before the *Homo* has reached Eurasia.

Elephants, rhinos, horses, short-necked giraffids, early moose, early species of giant deers, mountain and spiral-horned antelopes, early bison, and giant ostriches interspersed in herds of herbivores added a special touch to the Dmanisi landscape. Among the predators, there were large short-faced bone-cracking hyaenas, felines that were actively hunting for their prey (saber-toothed tigers—homotherium and megantereon, European jaguars, giant cheetah, large size), and pack hunting early wolves and xenocyons.

The fossil fauna of Dmanisi reflects the presence of a diverse, mosaic-like landscape. This environment provided suitable living conditions for both closed forest biotopes and open savannas, aquatic habitats, and dry, rocky slopes. Broadleaf forests covered the mountains, while riparian zones along river valleys likely resembled those of today. However, 1.8 million years ago, the river canyons were likely less deeply incised than they are now. The areas covered by lava flows, such as the Dmanisi promontory, would have been open plains interspersed with shrubs. The volcanism also contributed to the creation of dry, rocky habitats. This variety of habitats made Dmanisi a uniquely rich and dynamic environment, supporting a wide array of fauna and human activity during the Early Pleistocene.

4 Akhalkalaki Fauna

On the Javakheti Plateau, near the eastern edge of the town Akhalkalaki, next to a local cemetery, a large accumulation of fossilized vertebrate remains, known as the Akhalkalaki Fauna (Vekua 1962, 1987), was discovered (Fig. 5).

Javakheti is a land of extinct volcanoes. The combination of Pliocene volcanic domes, high-altitude plateaus, and rich fossil accumulations makes the Javakheti region a prime example of how geological evolution, paleontological heritage, and present-day landscapes intersect in Georgia. The Amiranisgora Mountain, at the base of which the fossils were found, is a prominent volcanic dome that emerged ca. 3.1 million years ago (Lebedev et al. 2007, 2008). The fossil-bearing sediments are the erosional product (colluvium) of this dome. The age of the Akhalkalaki fauna, based on paleomagnetic data, geological correlations, and fossil assemblage, is estimated to be around 0.8–0.9 Ma, predating Brunhes–Matuyama reversal at 0.78 Ma (Tappen et al. 2002a).

The accumulation of large mammal remains of the Akhalkalaki fauna is likely due to predator activities (Tappen et al. 2002a). Evidence supporting this includes the scarcity of axial skeleton elements (e.g., ribs and vertebrae) and the dominance of teeth and fragmented lower jaws among cranial remains. Additionally, the minimum number of individuals calculated from teeth is significantly lower than estimates based on limb bones. This suggests that predators consumed the softer, more nutritious parts of the carcasses elsewhere and brought limbs to their dens. The age distribution of the animals, dominated by juvenile and old individuals as inferred from teeth, also supports predator-driven accumulation. Many bones bear bite marks and fractures, typical of predator activity. However, such traces are not overly abundant due to the dominance of distal limb bones in the assemblage—remains often left behind after consuming the upper, fleshier parts of limbs.

Half of the Akhalkalaki fossil collection consists of the large horse, *Equus suessenbornensis*, remains, with some fossils of the smaller horse *E. hipparionoides*. Other herbivores include rhinoceroses (*Stephanorhinus aff. hundsheimensis*), elephants (*Mammuthus (Mammuthus) aff. trogontherii*), bison (*Bison* sp.), spiral-horned antelopes (*Pontoceros ambiguus*), mountain-dwelling caprins (*Caprini indet.*), deer (*Eucladoceros* sp.), and a hippopotamus (*Hippopotamus amphibius antiquus*). The latter is arguably the most spectacular animal in the Akhalkalaki fauna. Among carnivores, the collection includes fossils of the European jaguar (*Panthera onca gombaszoegensis*), the short-faced hyena (*Pachycrocuta brevirostris*), badgers (*Meles* sp.), marbled polecats (*Vormela cf. peregusna*), otters (*Lutra* sp.), bears (*Ursus* sp.), *Xenocyon lycaonoides*–ancestral form of red and African hunting dogs and the wolf (*Canis mosbachensis*).

Taking into account the habitat of the mentioned animals, one can visualize the landscape of the Javakheti Plateau as it was 0.8–0.9 million years ago: vast open plains interspersed with rivers and creeks, shallow lakes and wetlands, riparian forests on the background of towering volcanic mountain

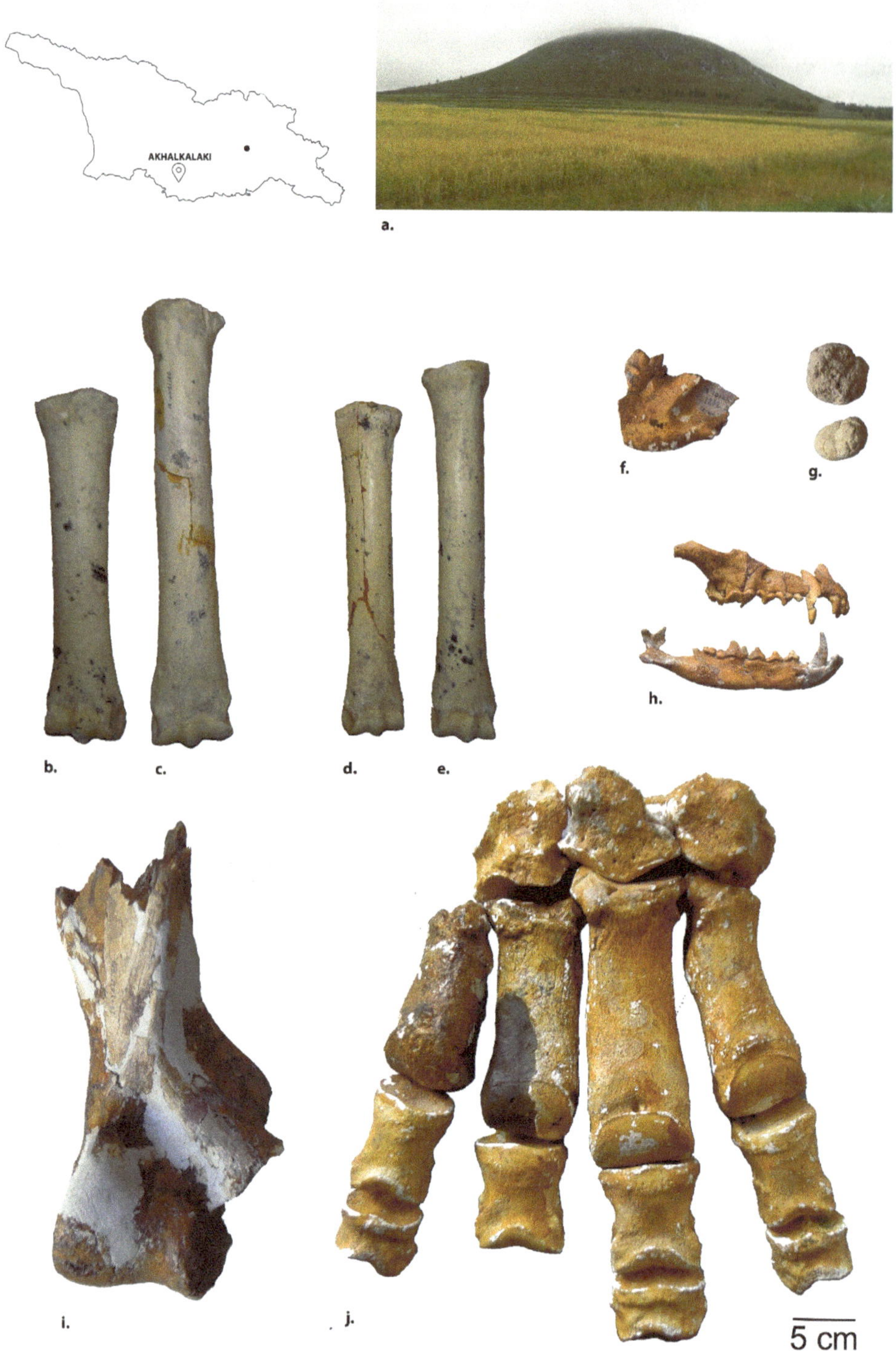

Fig. 5 **a** View on Amiranisgora (photo of R. Chagelishvili); **b.** *Equus suessenbornensis,* metacarpal sin. GNM1 18–2012/132 (Akh 1); **c.** *E sueasenbornensis,* metatarsal dex. GNM1 18–2012/180 (Akh 2); **d.** *E hipparionoides,* metacarpal dex. GNM1 18–2012/155; **e.** *E. hipparionoides,* metatarsal sin. GNM1 18–2012/164; **f.** *Pachycrocuta brevirostris,* mandible fragment with m1 sin. GNM1 18–2012/1262 (Akh 784); **g.** *P. brevirostris* coprolits GNM1 18–2012/1307 and GNM1 18–2012/1308; **h.** *Canis mosbachensis,* maxilla (GNM1 18–2012/1215 (Akh. 681, 684)) and mandible (GNM1 18–2012/1216 (Akh. 214, 883, 907)) with full toothrows (holotype of *Canis tengizii*); **i.** *Stephanorhinus aff. Hundsheimensis,* humerus, GNM1 18–2012/603 (Akh. 26)*;* **j.** *Hippopotamus amphibius antiquus,* articulated manus, GNM1 18–2012/586–588, 590–593, 594–600, *(holotype of Hippopotamus georgicus) (fossil specimens photos of M. Bukhsianidze)*

chains. The remarkably symmetrical dome of the Amiranisgora Mountain would have been a benchmark of the landscape as it is today. The aquatic habitats and wetlands are supported species like hippopotamuses and otters. The mountainous terrain covered with shrubs and grasses, their rocky slopes, and detritus at the mountain's base forming well-drained soil, likely constituted excellent habitats for caprins, marmots, ground squirrels, and marbled polecats. Lightly wooded areas and meadows were home to hedgehogs, hares, badgers, deer, antelopes, bison, bears, wolves, and jaguars. Elephants, horses, rhinoceroses, short-faced hyenas, and *xenocyon* thrived in the open landscapes.

5 Summary

The fossil record of land mammals of Georgia is significant, spanning from the end of the Paleogene epoch to the Holocene, with a gap in the Lower and Middle Miocene. This brief overview highlighted several key fossil sites—Benara—the earliest large mammalian fauna of the Caucasus with *Paraceratherium*; Udabno, Chachuna (in Georiga) and Eldari (in Azerbaijan)—the earliest hipparion faunas from the Kura Basin with the last dryopithecine record in western Eurasia; Kvabebi—one of the latest hipparion faunas in Eurasia; Dmanisi—the Early to Middle Villafranchian fauna with the earliest *Homo* remains outside of Africa, and Akhalkalaki—noteworthy Early Galerian fauna from the Caucasus. Every fossil is unique and each prehistoric species holds valuable evolutionary information, yet, primates add special value to this faunal record.

There is an enormous paleontological potential in Georgia. The molasse basins of Western and Eastern Georgia show a wide distribution of shallow, coastal marine facies and a well-developed hydrological network. These are the favoring factors for the accumulation of terrestrial fossils. With more intensive exploration and scientific study, there is a potential to bridge existing chronological gaps in land fossil record. This would provide deeper insights into the biodiversity of prehistoric wildlife. In addition to their scientific importance, these fossil localities form an integral part of Georgia's geoheritage portfolio, complementing the country's modern biodiversity and geosites and providing deep-time context for the landscapes highlighted elsewhere in this volume.

References

Agustí J, Oms O, Piñero P, Chochisvili G, Bukhsianidze M, Lordkipanidze D (2020) Late survival of dryopithecine hominoids in southern Caucasus. J Hum Evol 138:102690. https://doi.org/10.1016/j.jhevol.2019.102690

Bartolini-Lucenti A, Cirilli O, Pandolfi L, Bernor RL, Bukhsianidze M, Carotenuto F, Lordkipanidze D, Tsikaridze N, Rook L (2022) Zoogeographic significance of Dmanisi large mammal assemblage. J Hum Evol 163:103125. https://doi.org/10.1016/j.jhevol.2021.103125

Bernor RL, Kaya F, Kaakinen A, Saarinen J, Fortelius M (2021) Old world hipparion evolution, biogeography, climatology and ecology. Earth Sci Rev 221:103784. https://doi.org/10.1016/j.earscirev.2021.103784

Bukhsianidze M (2019) *Tapirus priscus* Kaup, 1833 (Tapiridae, Perissodactyla) from the Udabno site (late Miocene, Georgia). Bull Georg Natl Acad Sci 13(2):103–107

Bukhsianidze M (2024) Revisiting the earliest terrestrial fossil record from the paleo Kura bay (Eldari and Udabno faunas, late Miocene, South Caucasus)—diversity of bovids. EGU general assembly 2024, Vienna, Austria, Apr 2024, EGU24-7327 pp 14–19. https://doi.org/10.5194/egusphere-egu24-7327

Bukhsianidze M, Koiava K (2018) Synopsis of the terrestrial vertebrate faunas from the middle Kura Basin (eastern Georgia and Western Azerbaijan, South Caucasus). Acta Palaeontol Pol 63(3):441–461

Coil R, Tappen M, Ferring R, Bukhsianidze M, Nioradze M, Lordkipanidze D (2020) Spatial patterning of the archaeological and paleontological assemblage at Dmanisi, Georgia: an analysis of site formation and carnivore-hominin interaction in block 2. J Hum Evol 143:102773. https://doi.org/10.1016/j.jhevol.2020.102773

Gabunia LK (1964) Benara fauna of oligocene vertebrates. Metsniereba press, Tbilisi, pp 1–267. (in Russian)

Gadzhiev DV (1996) Eldari upper Sarmatian Hipparion Fauna. ELM press, Baku, pp 1–344. (in Russian)

Hemmer H, Kahlke RD, Vekua AK (2004) The Old World puma – *puma pardoides* (Owen, 1846) (Carnivora: Felidae)—in the lower Villafranchian (upper Pliocene) of Kvabebi (East Georgia, Transcaucasia) and its evolutionary and biogeographical significance. Neues Jb Geol Paläontol Abh 233:197–231

Kokolashvili I, Shatilova I, Bukhsianidze M (2018) The first pollen date from the upper Sarmatian deposits of the Chachuna 2 section (eastern Georgia). Bull Georg Acad Sci 12(4):93–99

Lebedev VA, Bubnov SN, Chernyshev IV, Chugaev AV, Dudauri OZ, Vashakidze GT (2007) Geochronology and genesis of subalkaline basaltic lava flows at the Javakheti Highland, Lesser Caucasus: K-Ar and Sr-Nd isotopic data. Geochem Int 45(3):211–225. https://doi.org/10.1134/S0016702907030019

Lebedev VA, Bubnov SN, Dudauri OZ, Vashakidze GT (2008) Geochronology of Pliocene volcanism in the Dzhavakheti Highland (the lesser Caucasus). Part 2: eastern part of the Dzhavakheti Highland. Regional geological correlation. Stratigr Geol Correl 16(5):553–574. https://doi.org/10.1134/s0869593808050080

Lordkipanidze D, Jashashvili T, Vekua A, de León MSP, Zollikofer CPE, Rightmire GP, Rook L (2007) Postcranial evidence from early homo from Dmanisi. Georgia Nature 449(7160):305–310. https://doi.org/10.1038/nature06134

Lordkipanidze D, de León MSP, Margvelashvili A, Rak Y, Rightmire GP, Vekua A, Zollikofer CPE (2013) A complete skull from Dmanisi, Georgia, and the evolutionary biology of early Homo. Science 342(6156):326–331. https://doi.org/10.1126/science.1238484

Mchedlidze GA (1984) General features of the paleobiological evolution of Cetacea. Russian Translations Series 23. A.A. Balkema, Rotterdam, 180 pp. ISBN 90-6191-438-8.

Meladze GK (1985) Review of the Hipparion faunas of the Caucasus. Metsniereba press, Tbilisi pp 1–75. (in Russian)

Shatilova I, Mchedlishvili N, Rukhadze L, Kvavadze E (2011) The history of the flora and vegetation of Georgia (South Caucasus). Georgian National Museum press, Tbilisi, pp 1–200

Tappen M, Adler DS, Ferring CR, Gabunia M, Vekua A, Swisher CC (2002a) Akhalkalaki: the taphonomy of an Early Pleistocene locality in the Republic of Georgia. J Archaeol Sci 29(12):1367–1391. https://doi.org/10.1006/jasc.2001.0797

Tappen M, Bukhsianidze M, Ferring R, Coil R, Lordkipanidze D (2022b) Life and death at Dmanisi, Georgia: Taphonomic signals from the fossil mammals. J Hum Evol 171:103249. https://doi.org/10.1016/j.jhevol.2022.103249

Vanishvili N (2022) The smallest late Miocene phocine from the southern Caucasus and the eastern Paratethys seal community crisis. Hist Biol 36(1):165–176. https://doi.org/10.1080/08912963.2022.2151365

Vekua A (1962) Early Pleistocene mammalian Fauna of Akhalkalaki. Academy of Sciences of GSSR press, Tbilisi, pp 1–207

Vekua A (1972) Kvabebi fauna of Akchagylian vertebrates. Nauka, Moscow, pp 1–351. (in Russian)

Vekua A (1987) The lower Pleistocene mammalian fauna of Akhalkalaki (southern Georgia, USSR). Palaeontogr Ital 74:63–96

Vekua A (1995) Die Wirbeltierfauna des Villafranchium von Dmanisi und ihre biostratigraphische Bedeutung. Jahrb Römisch-Ger Zent-Museums Mainz 42(1):77–180

The Unique Early Cretaceous Dinosaur Footprints of Sataplia, Georgia

Khatuna Mikadze and Nino Lapachishvili

Abstract

Outstanding dinosaur footprints were found in exposed Lower Cretaceous sediments on Mount Sataplia near the city of Kutaisi in Georgia, in the 1930s. These footprints were preserved in two stratigraphic layers, with a vertical distance of 2 m between them. The distinctive feature of Sataplia is the unique arrangement of dinosaur footprints. Around 150 traces of predatory dinosaurs were preserved in the marl limestone of the lower layer, while only about 15 traces of herbivorous dinosaurs were fossilized in the upper layer of the same rocks. The footprints in both layers were made by dinosaurs on two legs. The footprint area was located in the coastal region of the Lower Cretaceous Sea. This area preserved traces of two distinct ages and species of dinosaurs under conditions of two periods of transgression and regression. It is clear that the fossilized dinosaur footprints in Sataplia are the result of a very rare geological event, and they represent a significant monument to geological history of global significance. Currently, dinosaur footprints at Sataplia are protected by the state and are one of the most popular tourist attractions in Georgia, along with the nearby karst cave.

Keywords

Sataplia · Lower cretaceous · Limestones · Dinosaurs · Footprints

K. Mikadze (✉)
Georgian National Museum, Institute of Paleobiology, Tbilisi, Georgia
e-mail: mkhatuna21@gmail.com

N. Lapachishvili
Paleontological Museum, Javakhishvili State University, Tbilisi, Georgia

1 Introduction

Characteristic dinosaur footprints can be found in Sataplia, located in western Georgia, about 5 km from the city of Kutaisi, in the Tskaltubo municipality, on Mount Sataplia (42°18′58.2″N; 42°40′21.6″E; 190 m esl). This mountain forms part of a Lower Cretaceous limestone massif that covers an area of approximately 300 ha. According to legend, there are wild bees that nest in the limestone ledges on the southern slope of the mountain. Because of this, the locals gave this place the name "Sataplia", which means "place of honey" in Georgian. 95% of the area of Mount Sataplia is occupied by relic Colchian forests, which harbour some of the most significant paleontological and speleological geoheritage are located here. The Ogaskuri River has carved a stunning 600 m. long karstic cave into the Lower Cretaceous urgonian limestone, with unique traces of dinosaurs from the Lower Cretaceous period visible just 100 metres from its northern opening.

The karst cave was discovered in 1925 by Peter Chabukiani, a museum employee, teacher, and researcher in Kutaisi. In 1933, he also found the tracks of predatory dinosaurs near the cave. Later, he discovered the tracks of herbivorous dinosaurs next to these tracks. Because of the traces of dinosaur activity and the presence of the Colchis forest, the entire Sataplia area was declared a nature preserve in 1935.

The Sataplia Nature Reserve houses a protected karst cave, which has become an important geological heritage in Georgia due to the presence of dinosaur footprints preserved in two distinct stratigraphic layers. However, in this chapter, our focus will be exclusively on the dinosaur footprints themselves, while karst caves will be explored in greater detail in a separate section of the same publication.

Within Georgia's emerging Key Geoheritage Areas (KGA) framework (see Chap. 2), Sataplia is one of the core geosites of the Imereti karst and cave KGA, together

A. Okrostsvaridze and D. Tormey (eds.), *Geoheritage of Georgia*, Geoheritage, Geoparks and Geotourism,
https://doi.org/10.1007/978-3-032-27639-1_19

with nearby caves and mineral springs in the Tskaltubo–Prometheus area. In this compact territory, dinosaur footprints, karst caves, Colchian relic forest and cultural traditions such as the "place of honey" toponym all come together in a single landscape.

2 A Brief Geological Description of Sataplia Area

The Sataplia area is located on the boundary of the southern slope of the Greater Caucasus and the Rioni basin, on the uplift between the valleys of the Rioni and Tskhenistskali rivers. These deposits are composed of Cretaceous and Jurassic rocks (Fig. 1).

During the Mesozoic period, this region represented an active continental margin, where Paratethys oceanic crust subduction occurred below the Greater Caucasus island arc. In the Middle Jurassic Epoch, active volcanic processes took place in the region, while in the Cretaceous period, volcanic activity weakened, and the deposition of terrigenous and carbonate sediments began.

During this period, the coverage area represented a regressive marine coastal zone. The Greater Caucasus, including its southern slope, began rapid uplift over the last 7–5 million years and the contemporary location of the coverage area was formed during this period.

The fossil remains of dinosaurs from Sataplia are preserved in the limestone layers of the littoral sea. This is confirmed by the sedimentological analysis of these deposits and the presence of bivalve shells. The strata with traces are overlain by a meter-high layer of clay marls. According to the general geological assessment, the dinosaur footprints in Sataplia are estimated to be from the Early Cretaceous period and are approximately 135 million years old (Gabunia 1951, 1958).

The first studies of dinosaur footprints in Sataplia were published in the works of (Kandelaki and Dzotsenidze 1938). Later, L.Gabunia identified a new species, "*Satapliasaurus tschabukiani* Gabunia", and in 1958 he assigned two more species to the genus: "*Satapliasaurus kandelaki* Gabunia" and "*Satapliasaurus dsocenidze* Gabunia". At the same time, he noted the similarity of these species with Columbosauripus stenberg found in the Upper Cretaceous deposits of Canada.

In the layers located 2 m above the tracks of predatory dinosaurs in Sataplia, traces of short-legged, three-toed herbivorous dinosaurs have been found, which have been identified as belonging to the family Camptosauridae (Nesov 1995). Unfortunately, research and publications on

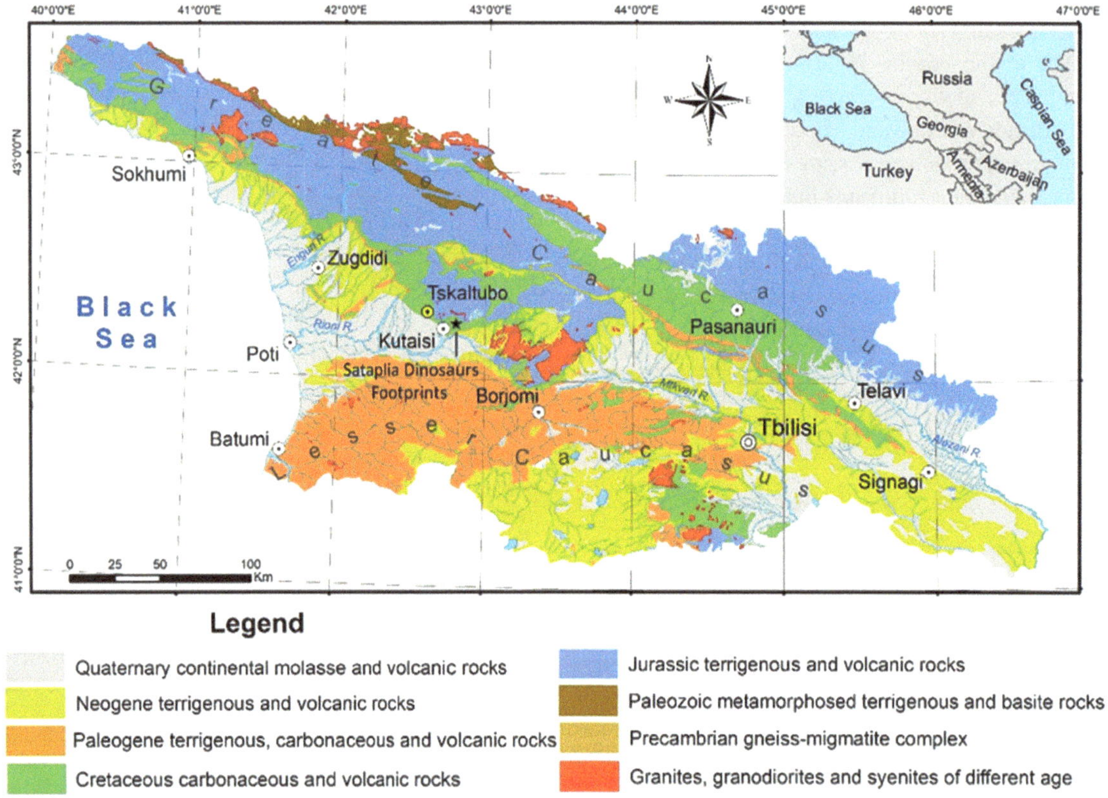

Fig. 1 Location of the Sataplia Early Cretaceous Dinosaurs footprints on a geological map of Georgia. The geological framework is modified after Gudjabidze et al. (2003)

dinosaur footprints in Sataplia in the modern scientific community are extremely limited (e.g. Gabunia et al. 1998, Gamkrelidze et al. 2021).

3 Analysis of Dinosaur Footprints in Sataplia

The area of the lower layer of Sataplia exceeds 300 ha., on which up to 150 mostly three-toed, rather deep footprints have been preserved. These tracks vary in size, shape and features of fossilization. All these traces belong to bipedal dinosaurs.

The prints in the lower layer of the fossil footprint belong only to predatory dinosaurs with highly developed claws. Ungulates have relatively elongated paws, and the front paw is more commonly represented among the prints (Fig. 2). There are also smaller prints, likely belonging to younger individuals. Therefore, the prints found in the lower layer can be attributed to both adult and juvenile predatory dinosaurs. Studies of ungulate footprints have shown that most dinosaurs left theirs at approximately the same time (Fig. 3).

Fig. 3 One of the segments of the tracks of the predatory dinosaurs from Sataplia. Photo© Khatuna Mikadze

The tracks of the upper layer, with footprints, belong to herbivorous dinosaurs. These tracks lack claws and have webbed paws, which is characteristic of herbivores. In addition, these tracks are very large, with some exceeding 50 cm in length. One of the prints shows the last phalanges of the toes and the imprint of the "heel" (Fig. 4).

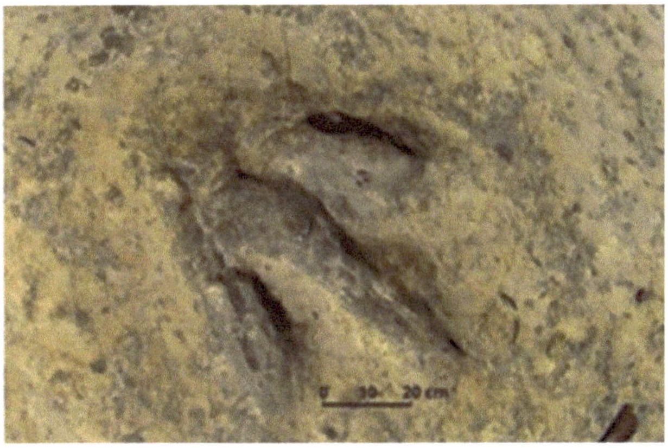

Fig. 4 One of the footprints of herbivorous dinosaurs from Sataplia. Photo© Khatuna Mikadze

A study of the direction of the dinosaur tracks showed that up to 25 different predators passed through this territory. Most of them were moving from north to southwest, but some were changing directions to the south or southeast. Only two tracks were pointing north. The tracks of herbivores in the upper layer have the opposite direction to the tracks in the lower layer, that is, they are oriented mainly to the north.

Fig. 2 Footprints of predatory dinosaurs from the lower layer of Sataplia. Photo© Khatuna Mikadze

4 Discussion

The footprints of dinosaurs in Sataplia are imprinted on the limestone layers in the littoral area of the sea. This is confirmed by the presence of bivalve shells. The age of the marl limestone with these imprints is estimated to be from the Early Cretaceous period, based on both geological and paleontological evidence. After the dinosaurs moved away, the sea level began to rise, covering the footprints with new carbonate deposits over a long period of time, creating layers up to 2 m thick. After this, the sea once again receded, exposing the coast where predatory dinosaurs had walked and left their tracks several thousand years earlier. Herbivorous dinosaurs also walked along this area, and favorable geological and climatic conditions led to the preservation of their footprints in the limestone. Soon, this coastline started to sink, being covered with seawater again. These hardened dinosaur footprints were then covered with new layers of sediment.

As a result, the Sataplia tracksite can be regarded as a reference locality for Early Cretaceous coastal environments in the Caucasus region. The two superposed track horizons record successive phases of shoreline occupation by different dinosaur communities and provide an unusually clear snapshot of how vertebrate behaviour, substrate conditions and sea-level change interacted along the active margin of the Neotethys Ocean. Within the Imereti karst and cave KGA, this makes Sataplia not only a spectacular visitor attraction but also a key scientific anchor point for understanding the evolution of western Georgia's Mesozoic basins.

5 Conclusion

The Lower Cretaceous dinosaur footprints of Sataplia represent a unique geoheritage preserved in two stratigraphic levels. It is noteworthy that these footprints belong to dinosaurs of two different ages and species, although both species walked on two legs. This layers stratigraphic levels are spaced vertically by ~ 2 m; the lower layer contains up to 150 footprints of predatory dinosaurs, while the upper layer contains only 15 footprints of herbivorous dinosaurs. No similar case has been described in the world's scientific literature, which determines the uniqueness and global significance of these footprints. Thus, the footprint of the Sataplia has significant scientific, educational, and geotourism significance, which obliges us to take care of it and its unwavering protection.

Within the national geoheritage framework outlined in Chap. 2, the Sataplia dinosaur tracksite is therefore recognised as a key geosite in the Imereti karst and cave Key Geoheritage Area and a priority element in any future Sataplia–Prometheus UNESCO Global Geopark proposal or regional geotourism route. In this wider context, its global uniqueness can support not only research, but also education, conservation and sustainable development in western Georgia.

References

Gabunia L (1951) Dinosaur tracks from the Lower Cretaceous outcrop at Sataplia (Kutaisi Region) // Collection of works of the Institute of Geology and Mineralogy of the Academy of Sciences of Georgia SSR, pp 209–222

Gabunia L (1958) Dinosaur tracks (based on material from the Satapliya Mountains and literature data) // publishing house of the Academy of Sciences, Moskow, pp 1–72

Gabunia L, Mchedlidze G, Chkhikvadze V, Lucas S (1998) Jurassic sauropod dinosaur from the republic of Georgia. J Vertebr Paleontol 18:233–236. https://doi.org/10.1080/02724634.1998.10011047

Gamkrelidze I, Okrostsvaridze A, Koiava K, Maisadze F (2021) Geotourism potential of Georgia, the Caucasus. Springer International Publishing, pp 1–81. https://doi.org/10.1007/978-3-030-629 66-3

Gudjabidze, G., Gamkrelidze, I., Abesadze, G. et al. (2003). Geological map of Georgia: Georgian State Department of geology and National oil Company, Tbilisi, Georgia, scale 1:500,000

Kandelaki N, Dzocenidze G (1938) The karst caves of the Tskhaltubo and Sataplia // nature reserves and monuments of the Georgian SSR. Tbilisi, pp 121–131. (in Georgia)

Nesov L (1995) Dinosaurs of northern Eurasia //new data on the composition of complexes, ecology and paleobiogeography. St Petersburg, pp 118–119

The Goderdzi Petrified Wood Site of Georgia

Mirian Makadze, George Mustoe, and Davit Makadze

Abstract

The Goderdzi Volcanic Formation of southwestern Georgia hosts one of the richest Upper Miocene paleofloras in the Caucasus, preserved within thick dacitic–andesitic pyroclastics and overlying lava flows. Abundant silicified wood and fossil leaves occur in laharic, tuffaceous, and fine-grained laminated deposits, reflecting varied depositional settings including small lacustrine basins. Radiometric K–Ar and zircon U–Pb ages (6.5–9.8 Ma) confirm a Late Miocene age for the sequence. More than 200 plant species from 120 genera and 60 families are known from the formation, with new taxa reported in 2019 reinforcing its subtropical affinities. Paleoecological interpretations indicate mixed evergreen and deciduous forest communities, whose modern analogues span Southeast Asia, Northeastern India, North America, the Antilles, the Canary Islands, the Mediterranean, and the Caucasus. Petrified wood displays multi-stage silicification—from opal-A to opal-CT to quartz—driven by hydrothermal fluid circulation during and after burial. Together, these geological and paleobotanical records provide important insights into Late Miocene environments and fossilization processes in the Lesser Caucasus. Within Georgia's emerging Key Geoheritage Areas framework, the Goderdzi petrified wood and fossil flora represent a major geoheritage resource of the Lesser Caucasus volcanic highlands, with high potential for geotourism, education and inclusion in future UNESCO Global Geopark proposals and national georoutes.

Keywords

Goderdzi petrified woods · Opal-A · Opal-CT · Silica · Goderdzi pass

1 Introduction

Neogene volcanic terrains frequently preserve exceptionally detailed fossil records, providing valuable insights into ancient environments, climate regimes, and vegetational dynamics. In southwestern Georgia, the Goderdzi Volcanic Formation represents one of the most significant of these archives. Extensive Upper Miocene volcanic activity produced thick pyroclastic deposits, lahar flows, and localized sedimentary basins that facilitated the rapid burial and preservation of diverse plant remains, including silicified wood and well-preserved fossil leaves.

The fossil flora of the Goderdzi region is known for its remarkable taxonomic diversity, reflecting subtropical and warm-temperate forests that once occupied the Lesser Caucasus. These assemblages are crucial for understanding Miocene paleoclimate and biogeographic links between the Caucasus and other regions across Eurasia and North America.

A particularly important aspect of this research involves the mineralogical characterization of petrified wood. High-resolution Scanning Electron Microscopy (SEM) and X-ray Diffraction (XRD) analyses of Goderdzi fossil wood were conducted by George Mustoe at Western Washington University, providing detailed evidence for multi-stage silicification processes and the mineral transformations from opal-A to opal-CT and microcrystalline quartz. These analytical results offer a deeper understanding of

M. Makadze (✉)
Faculty of Exact and Natural Sciences, Tbilisi State University, Tbilisi, Georgia
e-mail: mirian.makadze@tsu.ge

G. Mustoe
Geology Department, Western Washington University, Bellingham, USA

D. Makadze
The Caucasian Institute of Mineral Resources, Tbilisi State University, Tbilisi, Georgia

A. Okrostsvaridze and D. Tormey (eds.), *Geoheritage of Georgia*, Geoheritage, Geoparks and Geotourism,
https://doi.org/10.1007/978-3-032-27639-1_20

the diagenetic and hydrothermal conditions that governed fossilization within this volcanogenic environment.

Together, the geological, paleobotanical, and mineralogical records of the Goderdzi Formation form an essential foundation for reconstructing environmental evolution in the Lesser Caucasus during the Late Miocene.

2 Geoheritage Context and Key Geoheritage Area Significance

Within Georgia's national geoheritage framework (see Chap. 2), the Goderdzi Formation is recognised as a core component of a Lesser Caucasus volcanic highlands Key Geoheritage Area, encompassing the Goderdzi Pass, Javakheti and adjacent uplands. This KGA records the post-collisional volcanic history of southwestern Georgia and preserves an exceptional archive of Late Miocene vegetation in the form of petrified wood and fossil leaves. As such, it links directly to other volcanic geoheritage chapters in this volume, including those on ignimbrites, the Rock-Cut City of Vardzia, and Neogene basins of the Lesser Caucasus. The extraordinary abundance of silicified trunks, stumps and leaf beds exposed along roads and in natural outcrops around Goderdzi Pass makes this area a natural geoheritage landscape: a place where visitors can directly see how volcanic processes, ecosystems and fossilisation interact.

Recognising the Goderdzi Formation as a KGA helps to integrate its scientific importance with practical considerations for conservation and geotourism. It provides a clear basis for treating the main petrified wood localities and leaf floras as priority geosites within any future UNESCO Global Geopark or geotourism route in the Lesser Caucasus volcanic highlands, and for developing interpretive material that links their fossils to wider stories of climate change, vegetation history and volcanic hazards.

3 Geological Setting

The widespread abundance of silicified wood and fossil leaves in southwestern Georgia is associated with the Upper Miocene volcanic deposits of the Goderdzi Formation (Makadze et al. 2019; Djigauri and Petrosov 1989). The Goderdzi Formation represents a post-collisional, subaerial volcanic sequence characterized predominantly by andesitic and dacitic compositions (Tutberidze 2004). Classic exposures are located in the Goderdzi Pass area, particularly in the headwaters of the Dzindze and Adjaristskali rivers, where the Goderdzi Formation unconformably overlies Middle Eocene volcanic units (Fig. 1).

The Goderdzi Formation is divided into two major units: a lower pyroclastic sequence and an upper lava sequence. The lower unit, with a thickness of approximately 500–800 m, is dominated by volcaniclastic rocks of dacitic and andesitic composition, locally containing intraformational basaltic lava sheets in its basal levels (Makadze et al. 2019). These pyroclastic units contain numerous localities with fossil wood and fossil leaves. It should be emphasized that this thick pyroclastic succession is not uniform; rather, it consists of alternating pyroclastic sequences displaying various types of grading that are commonly associated with laharic flow processes. In addition, in some areas the fossil leaves occur within laminated, fine-grained pyroclastic deposits, which may reflect sedimentation in small lacustrine basins developed within the volcanic landscape. This substantial pyroclastic succession is overlain by the upper unit, which comprises laminated to massive andesite and dacite lavas. Radiometric K–Ar dating of these lavas yields ages ranging from ~7 Ma (Lebedev et al. 2004) to 9.4–9.8 Ma (Aslanyan et al. 1984), placing the formation firmly within the Upper Miocene. Parts of the Goderdzi Formation—such as the Mtkvari Ignimbrites and the andesitic lava flows of the Niala caldera structure—have been dated using zircon U–Pb geochronology, yielding ages of ~7.51 Ma for ignimbrites and 6.4 Ma for andesitic lava flows (Okrostsvaridze et al. 2020; Okrostsvaridze et al. 2024).

4 The Biodiversity of the Goderdzi Fossil Flora and its Paleoecological Significance

The pyroclastic deposits of Goderdzi Formation preserve a significant fossil record of wood and leaves, majorly silicified through complex mineralization processes driven by volcanic and hydrothermal activity. Fossilized plant remains at the Goderdzi site include well-preserved leaves with detailed venation, found in diverse lithological settings that suggest varied depositional environments. According to Uznadze and Tsagareli (1979), four primary burial types have been distinguished: (1) Massive, unstratified microbreccias containing tough-leaved plants, palms, and Osmanthus; (2) Sandy, poorly stratified tuffs, rich in Magnolia dzundzeana and ferns; (3) Fine-grained, laminated tuffs, preserving abundant Castanopsis adjarica leaves in leaf-fall deposits; and (4) Clay-rich tuffs dominated by Myrica lignitum and Populus populina, indicative of a riparian environment. Petrified wood trunks frequently found within the massive pyroclastic lahar sequences indicate that catastrophic volcanic events played a key role in the rapid burial and preservation of ancient trees.

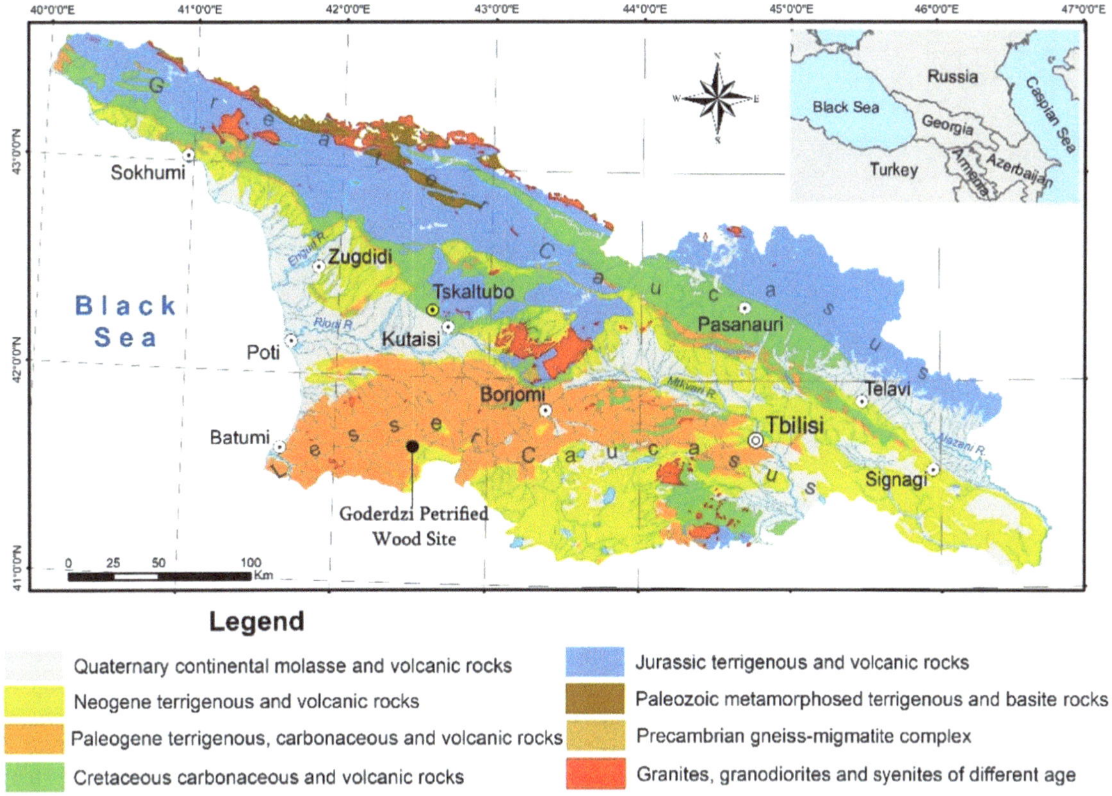

Fig. 1 Location of the Goderdzi petrified wood site on the geological map of Georgia. The geological framework is modified after Gudjabidze et al. (2003)

The Goderdzi fossil flora represents one of the most significant and varied paleobotanical assemblages in the Caucasus included more than 200 identified fossil plant species from 120 genera and 60 families (Djigauri and Petrosov 1989). However, the number of known taxa increased with the discovery of additional fossil leaf species in 2019 (Makadze et al. 2019). These newly identified species further reinforce previous interpretations that the Goderdzi fossil flora represents a distinctly subtropical vegetation assemblage. Based on the numerous studies (Uznadze 1946; Uznadze 1949; Uznadze 1965; Uznadze and Tsagareli 1979; Makadze et al. 2019), the fossil flora consists of three ecological groups: (1) Subtropical flora–the dominant group with Lauraceae, Myrtaceae, Apocynaceae, Sapindaceae, and Verbenaceae, indicating a warm and humid climate; (2) Warm-temperate flora with Platanaceae and Fagaceae, suggesting seasonal temperature variations; and (3) Temperate flora with Betulaceae and additional Platanaceae members, indicative of moderate climate conditions. The Goderdzi Formation lacks boreal elements but contains numerous warm-temperate deciduous and subtropical species. The co-occurrence of evergreen and deciduous species in the Goderdzi formation suggests a diverse and dynamic forest ecosystem influenced by volcanic activity.

The modern counterparts of the plants represented in the Goderdzi fossil flora are presently distributed across a wide geographic range, including Southeast Asia, northeastern India, North America, the Antilles, the Canary Islands, the Mediterranean region, and the Caucasus (Uznadze and Tsagareli 1979).

5 Silicification Processes of Petrified Wood

The study of petrified woods within the Goderdzi Formation integrates geological, mineralogical, and paleobotanical data to reconstruct the stages of silicification, the variability in silica mineralization, and the paleoenvironmental conditions that facilitated exceptional preservation.

The study of petrified wood samples and wood petrification processes around the world provides crucial insights into the interaction between biological materials and geochemical environments (Mustoe 2001; Mustoe 2003; Mustoe 2015; Scurfield and Segnit 1984; Stein 1982). Fossilized wood from the Goderdzi Formation offers a unique opportunity to investigate silicification processes in a volcanogenic setting.

Fig. 2 Transported log embedded within breccia layers, indicating deposition by volcanic flows or flood events. Photo© A. Okrostsvaridze

Fig. 3 Upright petrified tree stump preserved in its original growth position, indicating in situ fossilization. The enclosing sediments suggest rapid burial, which promoted silicification. Photo© Mirian Makadze

Fossilized trees in the Goderdzi Formation are primarily preserved as silicified wood through different stages of silicification by hydrothermal fluids. These specimens often exhibit mineralization with different type of opal, chalcedony, and quartz, replacing original plant tissues while maintaining anatomical structures.

The preservation of fossilized leaves and trees varies depending on taphonomic conditions, volcanic activity, and mineralization processes. Silicified trunks and branches, transported logs and fragmented deformed wood in breccia layers are resulted of volcanic flows or floods associated with lahar deposits. Indicating rapid burial process during volcanic mudflows. Upright fossil tree stumps, sometimes preserved with root systems, suggest in-situ preservation, implying rapid burial with minimal transport, allowing petrification to occur in place (Figs. 2, 3 and 4). Charred and carbonized wood specimens indicate exposure to pyroclastic surges during volcanic eruptions, with prolonged carbonization in anaerobic environments (Figs. 4b and 5).

The transformation of organic wood into petrified material follows a complex, multi-stage silicification process (Mustoe 2003). Initially, amorphous silica (opal-A) precipitates within cell walls, preserving anatomical details. Over time, diagenetic processes lead to the transformation of opal-A into opal-CT, a transitional silica phase containing cristobalite (Fig. 5). With continued burial, opal-CT gradually converts into microcrystalline quartz, yielding highly mineralized fossil wood. In some specimens, Ca–K zeolite crystals and, more commonly, pyrite crystals are observed filling the vessel lumina (Fig. 6).

Thin-section and XRD analyses reveal significant variability in silica composition within the fossil wood specimens. Some samples contain high proportions of opal-A, indicative of early-stage silicification, while others exhibit a predominance of quartz, representing more advanced diagenetic alteration. The heterogeneous mineralization within a single fossilized log suggests episodic infiltration of silica-rich hydrothermal fluids, which played a critical role in the preservation of fossil wood in the Goderdzi Formation. The circulation of silica-saturated water through buried wood facilitated the replacement of organic tissues with silica minerals. The presence of opal-A in many specimens suggests rapid mineral precipitation, possibly linked to volcanic fumarolic activity. Some samples exhibit secondary mineralization phases, where fractures and voids were filled with later-stage silica deposits. The coexistence of multiple silica phases within a single sample highlights the prolonged and dynamic nature of silicification in this region (Fig. 7).

6 Discussion and Conclusions

Intense Late Miocene volcanic activity in the Goderdzi Pass area produced thick pyroclastic deposits, lahars, and localized ash-derived basins that rapidly buried vegetation. These volcanic processes created ideal conditions for preserving and silicifying wood, making the Goderdzi Formation one of the most significant petrified wood localities in the region. The Goderdzi paleoflora is extraordinarily diverse, with over 200 fossil plant species representing ~120 genera and 60 families. These fossils reflect mixed evergreen–deciduous subtropical to warm-temperate forests, notably lacking any boreal elements. Many taxa have modern analogues in Asia, the Americas, and the Mediterranean, indicative of a humid subtropical climate and broad biogeographic links. Such diversity underscores the importance of the Goderdzi paleoflora for reconstructing

Fig. 4 Petrified wood trunks preserved within volcanogenic lahar deposits. **a** Upright silicified tree stump in growth position. **b** Carbonized tree stump. **c** Intensely silicified green-colored tree stump enclosed in massive lithic breccia. **d** Silicified tree stump within massive lithic breccia. Photo© Mirian Makadze

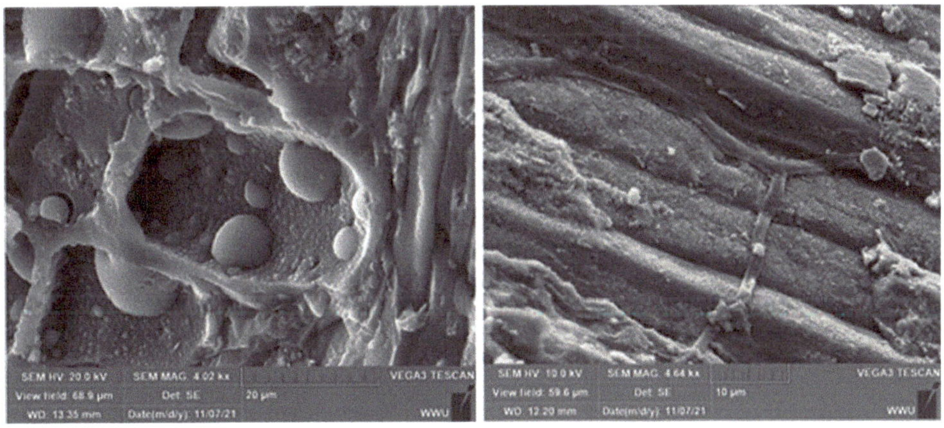

Fig. 5 SEM pictures shows opal—A, possible transition to opal-CT, and fungal hyphae that are evidence that early biological degradation (early decay) of the plant tissue occurred prior to full silicification

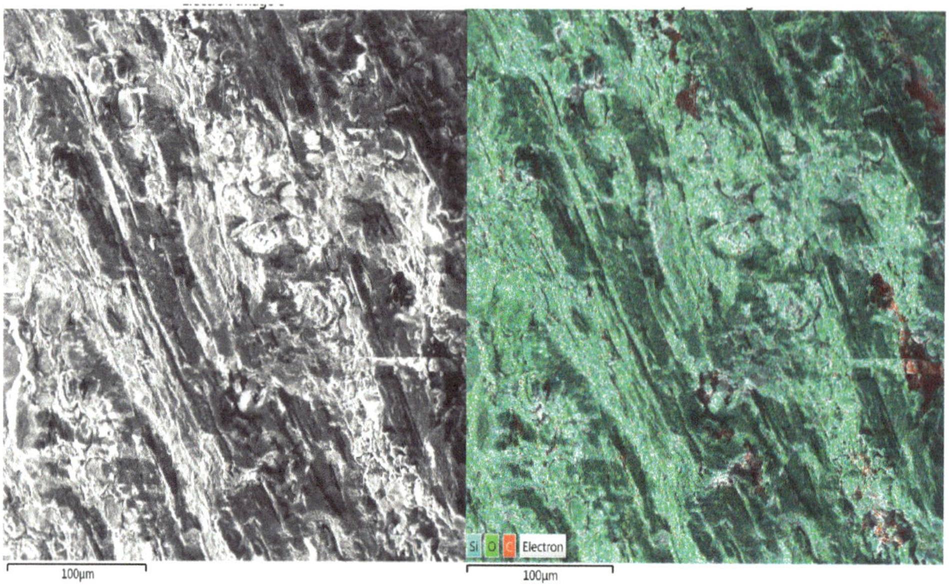

Fig. 6 SEM and elemental mapping images showing petrified wood composed mainly of opal-CT, with minor quartz and localized carbon-rich remnants preserving fragments of the original wood structure

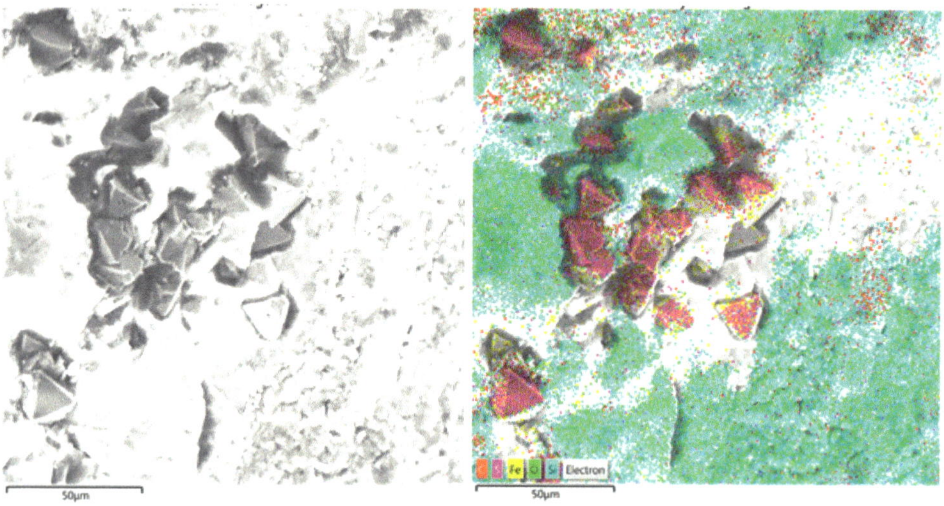

Fig. 7 SEM image and elemental map showing euhedral pyrite crystals dispersed within a fine-grained opal-CT matrix, accompanied by minor silica phases and localized carbon-rich areas

Miocene paleoclimate and biogeographic exchanges in the Lesser Caucasus.

Petrified wood from the Goderdzi Formation reveals a complex multi-phase silicification sequence: initial opal-A permineralization, followed by opal-CT formation, and final recrystallization to stable quartz. High-resolution SEM and XRD analyses (conducted by G. Mustoe, WWU) confirm this opal-A → opal-CT → quartz progression and reveal diagenetic features indicative of hydrothermal silicification. Circulation of silica-rich hydrothermal fluids during and after burial permeated the wood and preserved cellular anatomy in remarkable detail. This multi-stage fossilization pathway, coupled with rapid entombment by volcanic lahars enabled exceptional preservation of the Goderdzi biota. Together, the geological, paleobotanical, and mineralogical records of the Goderdzi Formation provide an integrative window into Late Miocene paleoenvironmental evolution in the Lesser Caucasus.

Within the broader *Geoheritage of Georgia* framework, the Goderdzi petrified wood and fossil flora thus have significance that extends beyond scientific interpretation. As a flagship locality of the Lesser Caucasus volcanic highlands KGA, Goderdzi offers outstanding opportunities for geotourism, outdoor education and public engagement with themes of volcanism, climate change and biodiversity through time. Road-cut exposures of petrified trunks and leaf-bearing tuffs around Goderdzi Pass can be developed as key geosites in future geopark proposals and national georoutes, provided that appropriate conservation measures are adopted to manage erosion, infrastructure development and uncontrolled collecting.

References

Aslanyan A, Baghdasaryan G, Gabunia L, Rubinstein M, Skhirtladze N (1984) Radiometric age of Neogene volcanic formations of the Georgian SSR and Nakhchevan ASSR. Bulletin of the Academy of Sciences of the Armenian SSR, Sciences of the Earth 35(1):3–24. (in Russian)

Djigauri D, Petrosov N (1989) Compilation and correlation of lithological-stratigraphic maps of continental effusion formations in southern Georgia: report of the mapping team on the results of thematic work conducted in 1981–1985. (in Russian)

Gudjabidze G, Gamkrelidze I, Abesadze G (2003) Geological map of Georgia: Georgian State Department of geology and National oil Company, Tbilisi, Georgia, scale 1:500,000

Lebedev V, Chernyshev I, Dudauri O, Arakelyants M, Chugaev A, Goltsman Y, Vashakidze G (2004) Geochronology of Neogene-quaternary dacitic volcanism of the Javakheti Highland (lesser Caucasus, southern Georgia). In: Proceedings of the Geological Institute of the Georgian Academy of Sciences, new series, vol 119, pp 535–544. (in Russian)

Makadze M, Tutberidze B, Akhalkatsishvili M, Kobakhidze N, Makadze D (2019) New species of the fossil flora in the tuffs of the Goderdzi suite (Goderdzi pass area). Bull Georg Natl Acad Sci 13(3):103–109

Mustoe GE (2001) Washington's fossil forests. Wash Geol 29(1–2):10–20

Mustoe GE (2003) Microscopy of Silicified Wood. Microsc Today 11(6):34–37

Mustoe GE (2015) Late tertiary petrified wood from Nevada, USA: evidence of multiple silicification pathways. Geosciences 5:286–309

Okrostsvaridze A, Chung SL, Chang YH, Skhirtladze I (2020) Geology and zircon U-Pb geochronology of the Mtkvari pyroclastic flow and evaluation of destructive processes affecting Vardzia rock-cut city. Quaternary Int 540:137–145. https://doi.org/10.1016/j.quaint.2019.03.026

Okrostsvaridze A, Gamkrelidze I, Yuan Hsi Lee YH, Keskin M, Tormey D, Boichenko D, Gogoladze S, Makadze M (2024) U-Pb geochronology and geochemistry of volcanic ash layers in the Eastern Caucasus intermountain Neogene Basin: implications for their sources. Arab J Geosci 17:129

Scurfield G, Segnit ER (1984) Petrifaction of wood by silica minerals. Sediment Geol 39:149–167

Stein CL (1982) Silica recrystallization of petrified wood. J Sediment Petrol 52:1277–1282

Tutberidze B (2004) Geology and petrology of Alpine late orogenic magmatism of the central part of Caucasian segment. Tbilisi University Publisher, Tbilisi, pp 1–362. (in Russian)

Uznadze M (1946) Flora of the Goderdzi suite. Bull Acad Sci Georg SSR 7:427–434. (in Russian)

Uznadze M (1949) Description of the Goderdzi suite Flora, Tbilisi. (in Georgian)

Uznadze M (1965) The Neogene Flora of Georgia. "Metsniereba" Publ.house, Tbilisi, pp 1–180. (in Russian)

Uznadze M, Tsagareli A (1979) The Sarmatian Flora of gorge of river Dzindza. "Metsniereba" Publ.house, Tbilisi, pp 1–112. (in Russian)

The Kldeisi Gryphaea Mass Fossil Site, Lesser Caucasus, Georgia

Avtandil Okrostsvaridze, Mevlud Sharikadze, and Salome Gogoladze

Abstract

Gryphaea is an extinct bivalve genus that is typically found as isolated fossils in sedimentary rocks ranging in age from the Jurassic to the Upper Eocene. However, in the Lesser Caucasus of Georgia, on the eastern margin of the pre-Jurassic crystalline Khrami Massif (~500 km^2), an exceptional mass occurrence site of Gryphaea is exposed. This site, located near the village of Kldeisi and referred to as the Kldeisi Gryphaea locality, preserves a dense accumulation of well-preserved Gryphaea within Maastrichtian carbonate sediments. The exposure extends for approximately 55 m in length with an average thickness of ~5 m. It is crosscut in its central part by a ~ 1.5-meter-thick subvolcanic intrusion and is overlain by a ~ 2-meter-thick volcanic–carbonate unit. Geochemical analyses indicate SiO_2 contents of approximately 47% in the subvolcanic intrusion, ~ 25% in the volcanic–sedimentary unit, and ~ 79% in the silicified Gryphaea fossils. The section records multiple phases of volcanic activity, with episodic fossil accumulation occurring between eruptive events, followed by subsequent mortality likely related to renewed volcanic activity. The Kldeisi Gryphaea site is therefore interpreted as a unique record of the interaction between biological communities and active volcanic processes. Its exceptional preservation and stratigraphic context make it a significant geoheritage locality with high potential for scientific research, education, and geotourism.

Keywords

Gryphaea · Mass fossil · Kldeisi · Khrami massif · Lesser Caucasus · Geoheritage

1 Introduction

Gryphaea one of the marine bivalve mollusks in the family Cypraeidae, a genus of extinct oysters, known as "devil's toenails". They lived in large colonies in warm, shallow waters and sat on the sea-floor with their shells partially buried in the mud, typically measuring approximately 5–7 cm. These fossils range from the Triassic period to the middle Paleogene period but are primarily found to the Triassic and Jurassic periods. Its distribution is common in various areas of Europe. The complete fossils consist of two articulated valves: a larger gnarly-shaped shell (the "toenail") and a smaller, flattened shell, the "lid". The soft parts of the animal occupied the cavity between the two shells, just like modern oysters (Duff 2018).

We discovered an extensive burial colony of Gryphaea within Upper Cretaceous volcanic–carbonate sediments on the eastern periphery of the Khrami Massif, near the village of Kldeisi, in the Lesser Caucasus. The destruction of the colony was caused by contemporaneous volcanic activity during the Late Cretaceous period. Subvolcanic magma intruded into the central part of the living colony, while the overlying effusive material subsequently buried it, inevitably leading to its extinction. Analysis of the exposure indicates that such volcanic activity, although of relatively low intensity, occurred repeatedly. During the post-volcanic stage, the Gryphaea shells, originally composed of aragonitic calcium carbonate ($CaCO_3$) (Pakhnevich et al. 2022), were replaced by amorphous silica (SiO_2). As is evident, this exposure records a series of highly informative biogeological

A. Okrostsvaridze (✉) · S. Gogoladze
Ilia State University, Institute of Earth Sciences, Tbilisi, Georgia
e-mail: avtandil.okrostsvaridze@iliauni.edu.ge

M. Sharikadze
Faculty of Geology and Mines, Georgian Technical University, Tbilisi, Georgia

A. Okrostsvaridze and D. Tormey (eds.), *Geoheritage of Georgia*, Geoheritage, Geoparks and Geotourism,
https://doi.org/10.1007/978-3-032-27639-1_21

processes and represents an important element of geological heritage. For this reason, the present publication was prepared.

2 Geoheritage Significance and Key Geoheritage Area Context

Within Georgia's emerging Key Geoheritage Areas (KGA) framework, the Kldeisi Gryphaea site forms part of the broader Khrami–Bolnisi volcanic and sedimentary belt, a KGA that also includes major ore districts and volcanic centres described elsewhere in this volume. In this chapter, however, the focus is not mining but the intimate interaction between life and volcanism at the end of the Cretaceous, recorded in an exceptional shallow-marine fossil bed overrun by volcanic products.

The Kldeisi exposure is globally rare as a mass burial of Gryphaea preserved in three dimensions within volcanogenic sediments. It captures repeated low-intensity volcanic events, silicification and the subsequent destabilisation of the host slope, providing a "Pompeii of Gryphaea" that documents how a marine community responded to pulses of volcanism in a tectonically active basin. Few comparable examples are known from the literature, which gives this site particular international scientific value.

At the same time, the site has clear geoheritage and geotourism potential. It lies directly beside the Tetritskaro–Tsalka motorway and can be visited safely from the road verge, making it an accessible geosite for school groups, university excursions and independent travellers. The combination of a striking outcrop, easily visible fossils and a compelling interpretive story, shallow sea life, volcanic disturbance, silicification and slope failure, make Kldeisi a natural candidate geosite within any future geopark or geotourism route in the Khrami—Bolnisi region.

3 Brief Description of the Kldeisi Gryphaea Site

The Geology of Georgia is discussed in Chap. 5. For this site, the Khrami pre-Jurassic crystalline massif (~500 km^2) is situated in southern Georgia, approximately 100 km southwest of Tbilisi, within the Lesser Caucasus orogenic belt. It represents a horst-like uplift of the Artvin–Bolnisi zone and is primarily composed of a pre-Jurassic gneiss–migmatite complex and late Variscan granitoids (Gamkrelidze et al. 2020). Its eastern periphery is covered by Upper Cretaceous sedimentary rocks (Tsagareli 1954). Here near the village of Kldeisi, a large colony of well-preserved Gryphaea fossils is exposed within Cenomanian volcanogenic–sedimentary rocks (Okrostsvaaridze et al. 2023) (Fig. 1).

The Kldeisi Gryphaea mass fossil site is exposed on the right side of the Tetritskaro–Tsalka motorway, near the village Kldeisi, where it forms an outcrop approximately 55 m long and 5–7 m high. In its central part, the exposure is intruded by a 1.5 m thick mafic body, which is covered by a ~ 2 m thick, dark-colored, fine-grained volcanogenic–sedimentary layer enriched in carbonate material. In some parts of the outcrop, shrubs and small trees grow, while the upper surface is completely covered by soil and dense deciduous forest. On the eastern periphery of the site, in a small forested area dark-colored pillow lavas are exposed with individual pillows ranging in diameter from 25 cm to 50 cm.

Beneath these pillow lavas, the exposure of the silicified Gryphaea colony is disrupted into large blocks (approximately 1.2 × 2.3 × 4.5 m), which have detached from the parent rock and slipped downslope due to erosion. Because of these conditions, it is not possible to observe the Kldeisi Gryphaea site as a single, continuous exposure. Therefore, a schematic sketch is provided here to illustrate the overall structure and the geo-biological processes that developed at the site (Fig. 2).

Fossilized Gryphaea at the Kldeisi site exposure are unevenly distributed. They are particularly abundant in the eastern part of the subvolcanic intrusion, where the rocks are almost entirely composed of these mollusks (Figs. 3 and 4). The Gryphaea specimens from Kldeisi exhibit characteristic elongated shapes, with lengths reaching 7–8 cm and widths ranging from 3 to 5 cm (Fig. 5).

These silicified mollusks are closely spaced, with the small voids between them filled by volcanic products. Notably, the outcrop exhibits an alternation of volcanic and Gryphaea layers, indicating multiple volcanic eruptions. We consider the Kldeisi Gryphaea site to be unique due to the exceptional preservation of this colony in its original form, attributed to volcanic activity.

We determined the contents of major and minor chemical elements in the fossilized Gryphaea, subvolcanic intrusive rocks, volcanic-sedimentary layer, and pillow lava at the MSALAB laboratory in Vancouver, Canada (Tables 1 and Tables 2).

The numbering of the samples is the same as in Fig. 1.

As shown in these tables, the main component of the fossilized Gryphaea is amorphous quartz (SiO_2), which constitutes more than 78% of its composition. It also contains a relatively small amount of Al_2O_3 (9.80%) and an even smaller amount of Na_2O (3.48%). The original principal component of the Gryphaea, CaO, has been almost completely replaced by SiO_2, and its content does not exceed 0.65%.

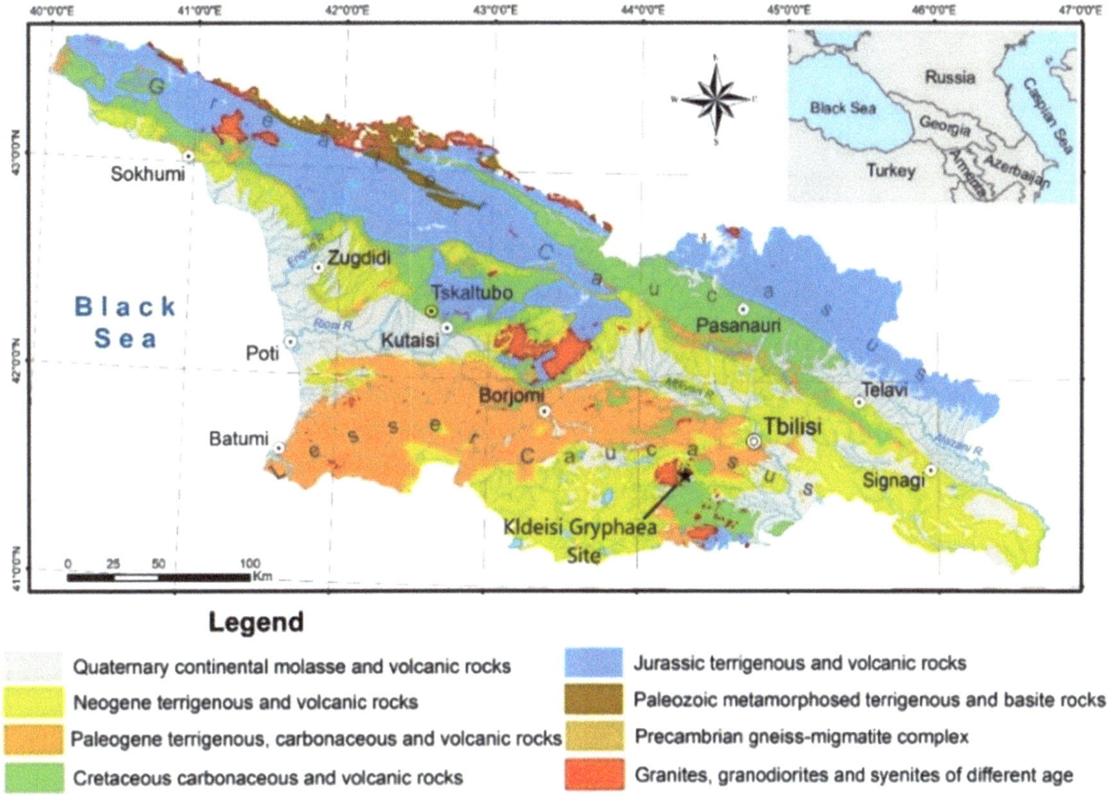

Fig. 1 Location of the Kldeisi Gryphaea mass fossil site on a simplified geological map of Georgia. Geological framework modified after Gudjabidze et al. (2003)

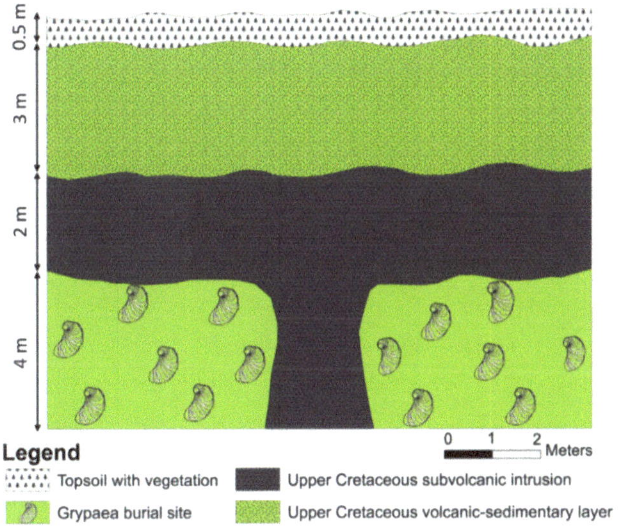

Fig. 2 Conceptual geological sketch of the Kldeisi Gryphaea site

The chemical composition of the subvolcanic intrusive rock is typical of basaltic magma, with a SiO_2 content of 47.78% and a total iron (Fe_2O_3) content of 10.21%, as well as a relatively high Na_2O content of 3.48%. The pillow lava exhibits a nearly identical composition, with SiO_2 at 48.28% and total iron at 9.37%. This similarity indicates that both originated from a common magma source, with the intrusive body likely representing the feeder channel for the pillow lava.

4 Discussion

This study of the Kldeisi Gryphaea mass fossil site is pioneering; therefore, many aspects remain unclear and open to discussion. Unfortunately, no records of a comparable Gryphaea mass burial have been found in the scientific literature. Nevertheless, our data confidently indicate that this Gryphaea colony inhabited a shallow marine environment, as evidenced by the composition of the associated terrigenous material.

It is established that rifting processes began in the Transcaucasus microplate during the Maastrichtian, gradually separating its southern provinces, specifically the Artvin-Bolnisi and Loki-Karabakh blocks (Gamkrelidze et al. 2020). The volcanic activity recorded at the Kldeisi site was most likely part of this tectonomagmatic phase and played a key role in preserving the life conditions and burial of a Maastrichtian-age Gryphaea colony.

Fig. 3 Fragments of displaced blocks from the Kldeisi Gryphaea mass fossil site outcrop, mainly composed of Gryphaea shells. Photo © A. Okrostsvaridze

Fig. 4 Close-up view of the Kldeisi Gryphaea mass fossil site. Photo © A. Okrostsvaridze

Fig. 5 Silicified Gryphaea shells from the Kldeisi Gryphaea mass fossil site. Photo © Rabi Gabrielashvili

Table 1 Major chemical element contents (wt%) of the Kldeisi Gryphaea mass fossil site

Sample	SiO_2	TiO_2	Al_2O_3	Fe_2O_3	MnO	MgO	CaO	Na_2O	K_2O	P_2O_5	BaO	Cr_2O_3	SrO
23KI1	78.72	0.42	9.80	2.90	0.05	0.83	0.65	3.48	0.83	0.05	0.03	<0.01	0.02
23KI2	47.78	1.03	19.21	10.21	0.18	4.21	5.11	4.22	0.75	0.13	0.04	<0.01	0.03
23KI3	24.69	0.21	4.69	2.13	0.17	0.95	35.96	0.81	0.52	0.06	0.02	<0.01	0.04
23Kl4	48.28	1.13	18.11	9.37	0.25	4.12	5.57	5.11	0.83	0.15	0.03	<0.01	0.03

Sample: 23KI1-fossilized Gryphaea; 23KI2-Subvolcanic intrusive; 23KI3-Volcanic-sedimentary layer; 23Kl4-Pillow lava.

Table 2 Trace chemical element contents (ppm) in the Kldeisi Gryphaea mass fossil site

Sample	Sc	Y	La	Ce	T	U	Cr	Cd	Be	Bi	Nb	Li	Hf
23KI1	8.4	10.9	16.7	31.00	4.8	1.1	37	0.04	0.69	0.13	5.9	11.6	1.6
23KI2	42.4	22.1	23.9	15.77	1.0	0.3	4	0.04	0.66	0.06	1.9	41.1	1.6
23KI3	3.8	18.2	6.6	43.34	3.2	1.0	11	0.08	0.80	0.04	5.9	12.9	1.3
23Kl4	40.7	25.8	21.7	21.44	1.3	0.4	6	0.06	0.58	0.05	2.3	38.9	1.4

Within Georgia's national geoheritage framework, this outcrop can be read not only as a palaeontological and volcanological case study, but also as a key element of the Khrami–Artvin–Bolnisi Key Geoheritage Area. The Kldeisi Gryphaea mass burial complements nearby sites that record Cretaceous–Palaeogene volcanism, sedimentation and metallogeny, and together these geosites illustrate how the evolution of the Transcaucasus microplate shaped both geological resources and living communities. As such, Kldeisi should be treated as a priority geosite for documentation, conservation and future inclusion in regional geotourism routes and any proposed UNESCO Global Geopark in this belt.

5 Conclusions

On the eastern periphery of the Khrami Massif, near the village of Kldeisi, a large colony of well-preserved mollusks is exposed within Maastrichtian carbonate sediments. The fossils belong to Gryphaea, an extinct bivalve genus of the family Gryphaeidae. These shallow-marine, warm-water organisms existed from the Triassic to the early Paleogene.

The Gryphaea-bearing outcrop extends for approximately 55 m in length with an average thickness of about 5 m. In its central part, it is crosscut by a subvolcanic gabbroic intrusion approximately 1.5 m thick and overlain by a volcanic unit about 2 m thick.

The Kldeisi Gryphaea mass fossil exposure is interpreted as a unique example of fossil preservation associated with episodic volcanic activity. It represents an exceptional case of interaction between active volcanic processes and the biosphere, sometimes referred to as the "Pompeii of Gryphaea." This mass occurrence is considered a significant geoheritage site with high scientific, educational, and geotourism value. Within the broader context of the Geoheritage of Georgia volume, the Kldeisi Gryphaea site illustrates how a single outcrop can integrate multiple layers of value, including scientific, educational, and geotourism significance. Its recognition within the Khrami–Bolnisi geoheritage framework provides a basis for potential inclusion in future geopark and georoute initiatives, ensuring that this unique locality is appropriately documented, interpreted, and integrated into Georgia's geoheritage inventory.

References

Duff J (2018) My favorite fossil: the extinct oyster Gryphaea (AKA devil's toenails). Biologos. https://biologos.org

Gamkrelidze I, Okrostsvaridze A, Maisadze F, Boichenco G, Skhirtladze I (2020) Main features of geological structure and geotourism potential of Georgia, the Caucasus. Mod. Environ. Sci. Eng. 5:422–442. https://doi.org/10.15341/mese(2333-2581)/05.05.2020/007

Gudjabidze G, Gamkrelidze I, Abesadze G et al. (2003) Geological map of Georgia: Georgian State Department of geology and National oil Company, Tbilisi, Georgia, scale 1:500:000

Okrostsvaridze A, Sharikadze M, Gogoladze S, Gabrielashvili R (2023) Mass burial site of the Gryphaea in upper Cretaceous sedimentary cover of the Khrami massif, lesser Caucasus, Georgia. In: a extended abstract of MEDGU23 international conference, Istanbul, Turkey

Pakhnevich A, Nikolayev D, Lychagina TA (2022) Crystallographic texture of the mineral matter in the bivalve shells of Gryphaea dilatata Sowerby. Biology 11:1300. https://doi.org/10.3390/biology11091300

Tsagareli A (1954) Upper cretaceous fauna of Georgia. Monograph, Tbilisi (in Russian)

on global markets. If this trend continues, additional metal deposits are likely to be added to Georgia's exploitable resource base, with the potential to make a substantial contribution to national economic growth.

At the same time, these ore deposits and mining districts are not only sources of raw materials; they also form part of Georgia's geoheritage. They record the action of Tethyan tectonic and magmatic processes, and they underpin a long history of mining, metallurgy and cultural narratives, from the gold sands of Colchis to the industrial landscapes of Chiatura and Bolnisi. In this chapter we therefore focus on ore deposits and mining landscapes that are both economically important and of clear geoheritage interest, and we consider how they can be integrated into Georgia's emerging Key Geoheritage Area inventory and future geopark and georoute concepts.

2 Mining History

The ancient Georgian states, particularly the early kingdoms of Colchis and Iberia, are considered by many researchers to be among the first places where mining and metal-extraction technologies were developed. Numerous scholars argue that the Kartvelian tribes, including the Khalibs (Chaldees), Mosiniks and Tubals, represent one of the homelands of mining and metallurgy (e.g. Richardson 1934; Forbes 1950). Richardson (1934) suggested that iron mining and steel-making technology were first developed in the fourteenth century BC by the Chaldeans, a Kartvelian tribe living along the Halis River in the valley of present-day Trabzon Province, Turkey. The French scholar Dussaud (1930) proposed that the Greek word *chalkos* (copper) derives from the name of the Colchian tribe of Chaldea, with the Greek suffix *-kos* (indicating origin) added to this root.

Artifacts recovered from Early Bronze Age burials in the region suggest that the mining and processing of ferrous, non-ferrous and precious metals were already well developed during this period (Courcier et al. 2008). These finds, together with the classical textual evidence, indicate that metallurgy was a central technological and cultural practice in ancient Georgia.

The first written accounts of gold extraction from river sands in ancient Georgia appear in the works of Greek and Roman authors. The Greek historian Strabo (I AD-I BC) noted that rich gold deposits were the reason the Argonauts journeyed to the Kingdom of Colchis. The Roman scholar Pliny the Elder (I AD-I BC) also described gold mining in Colchis: in his *Natural History*, he wrote that "in Colchis, the land of the Svanetian people, large quantities of gold and silver are mined from the ground" (Book VI). In the second century ad, the historian Appian of Alexandria (II AD) reported that "many rivers carry gold-bearing sands from the Caucasus Mountains, and the local people extract gold from these sands using sheepskins." In addition to these examples, there is a wealth of other written evidence on gold mining and processing in ancient Georgia, many of which are explored in more detail in the Golden Fleece/Enguri chapter of this volume.

At present, the Bolnisi mining district is the only major hard-rock mining centre in Georgia. Underground workings, furnaces and historical records indicate that copper mining in the Bolnisi area began in the 4th–3rd millennium BC, while iron mining dates back to the 2nd–1st millennium BC. In later periods, beginning in the Middle Ages, non-ferrous and noble metals (Dambludi, Bnelikhevi) as well as iron (Balidara, Chatakhi, Madnistskaro, Demursu) were mined in the region (Okrostsvaridze 2014). The Bolnisi district thus represents a long-lived metallogenic and cultural landscape where ancient, medieval and modern mining overlap.

Many artefacts related to mining and metal processing in Georgia are preserved in the National Museum of Georgia. Among them is a unique gold-washing pan made of ash wood (Fig. 1), originating from the village of Sgurishi in Upper Svaneti. While gold-washing pans around the world are typically round, this example has an unusually precise rectangular shape. The skill required to produce such a tool, and its preservation, provide indirect evidence that alluvial gold mining in Georgia has a long and sophisticated history (Okrostsvaridze et al 2014; Okrostsvaridze 2017).

The foundations of Georgia's modern mining industry were laid in the nineteenth century, beginning with the Tkibuli coal mine in 1849, followed by the development of the Chiatura manganese mine in 1879. From the 1930s, during the Soviet period, intensive industrial development of Georgia's mining sector took place, although much of the revenue was redirected to the central budget of the Soviet Union. The most important mineral resource exported from Georgia in this era was manganese from the Chiatura mine. Until the 1990s, Chiatura accounted for about 40% of global manganese production, and the annual export value of manganese concentrate represented 75–80% of Georgia's national budget (Tvalchrelidze 2003).

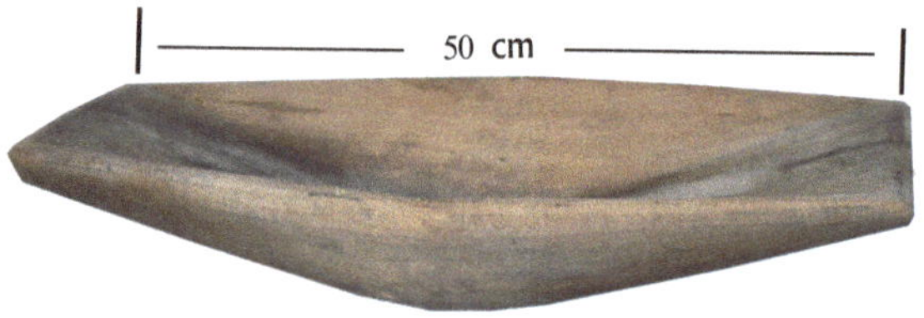

Fig. 1 Svan gold-washing wooden pan made of ash wood. The village of Sgurishi, Chuberi community. Preserved in the Georgian National Museum

In 1975, the Madneuli Mining and Beneficiation Complex was put into operation. It processed copper and gold–silver-bearing quartzites, polymetallic ores and barite ores from the Madneuli deposit, with a throughput of 1.5 million tonnes of ore and production of about 70,000 tonnes of copper concentrate per year (Report of GGMC 2004). In addition to these large operations, around 140 smaller mining enterprises were active in Georgia during this period, producing coal, barite, arsenic, zeolites, diatomite, marble and other materials.

After the collapse of the Soviet Union in the 1990s, Georgia's mining industry suffered a severe decline. The shift in the economic system and the loss of the Soviet market led to the closure of many enterprises, with production in some cases falling to just 5–10% of previous levels (Report of GGMC 2004). Following independence, however, new investments in mineral resources increased, particularly in copper, gold, manganese, oil, barite and mineral waters. Today, revenues from the Madneuli gold–polymetallic and Chiatura manganese deposits make a substantial contribution to the national budget through the mining enterprises currently operating in Georgia. From a geoheritage perspective, this long arc—from Bronze Age mining and the Argonauts' gold sands, through Bolnisi's early furnaces, to the industrial landscapes of Chiatura and Madneuli—means that ore deposits in Georgia are not only sources of raw materials, but also key elements of the country's cultural and industrial heritage. Ancient workings, artefacts, mining towns and environmental legacies form part of a mining landscape that is central to Georgia's geoheritage stor.

3 Ore Deposits of Georgia

Georgia is rich in manganese, gold, arsenic, antimony, non-ferrous metals, iron and other minerals (Fig. 2). Among these, two mining regions stand out for both their economic importance and their geoheritage significance: the Chiatura manganese deposit and the Bolnisi gold–copper–polymetallic mining district. Together, they represent major metallogenic centres and industrial landscapes that are real wealth for Georgia in both economic and geoheritage terms.

Beyond these flagship areas, numerous other ore deposits and occurrences are distributed across the country, including gold–arsenic–antimony systems in Svaneti and Racha (e.g. Tsana, Lukhumi, Zofkhito, Sakeni), polymetallic deposits and prospects in central and eastern Georgia (e.g. Dzama–Gharta, Begtaqari, Sakdrisi, Madneuli, Tsitelsopeli), and younger hydrothermal and placer systems along the southern slopes of the Greater Caucasus and in Adjara–Guria (e.g. Stori, Vakijvari, Merisi, Shkhara).

3.1 Chiatura Manganese Deposit

The Chiatura Oligocene sedimentary manganese deposit is located in central Georgia on the northeastern edge of the Dzirula Paleozoic crystalline massif. It is exposed along the Kvirila River and its tributaries and represents an important geoheritage asset and real wealth of Georgia.

The deposit formed on the stable crystalline basement during a regressive–transgressive cycle. An Oligocene basal conglomerate is overlain by sandstone, which in turn is overlain by manganese-bearing strata. The Chiatura sedimentary sequence forms nearly horizontal beds 1.5–4 m thick (average ~2 m), of which 1–1.5 m is high-grade ore, covering an area of about 150 km^2. The deposit contains three main ore types: primary oxide ores, carbonate ores and oxidised carbonate ores. Oxide ores are volumetrically dominant. The common manganese minerals in these ores are manganite, pyrolusite, braunite and psilomelane, with bixbyite, rhodochrosite and vernadite occurring less frequently. The main oxide ores contain on average 36.6 wt% MnO, 2.45 wt% Fe_2O_3, 23.5 wt% SiO_2, 5.88 wt% Al_2O_3 and 3.84 wt% CaO. The carbonate and mixed ores, in contrast, formed in somewhat deeper parts of the shallow sea. Manganese in the oxide ores was syngenetically transported from deeper-water depositional environments into the main mineralisation area, which lay shallower than the sites of carbonate ore formation (Fig. 3; Sasmaz et al. 2021).

The Chiatura manganese deposit is a world-class formation with a long history of development. Exploitation began in 1879, and by the 1890s Chiatura accounted for 30–40% of global manganese production and became a primary supplier to Western European steel plants. At the beginning of the twentieth century, the influx of Western European capital into the region increased significantly but was halted by Georgia's annexation by Russia in 1921.

During the Soviet era, Chiatura supplied manganese to steel factories across the Soviet Union and Eastern Europe. Since 2006, the Chiatura deposit has been operated by "Georgian Manganese Holding Limited", a mining corporation that processes ore in the city of Chiatura. Beneficiation and waste disposal in and around the town have created major ecological problems for the local population (Fig. 4). Manganese ore is smelted at the Zestafoni ferroalloy plant, and the resulting products are primarily exported to Western European countries.

Over the history of manganese mining in Chiatura, up to 260 million tonnes of raw ore have been extracted. Nevertheless, according to various expert estimates, Chiatura still holds the largest manganese reserves in Europe, at approximately 215 million tonnes (Report of EMC 2016). From a geoheritage perspective, Chiatura is thus both a world-class stratiform manganese deposit and a

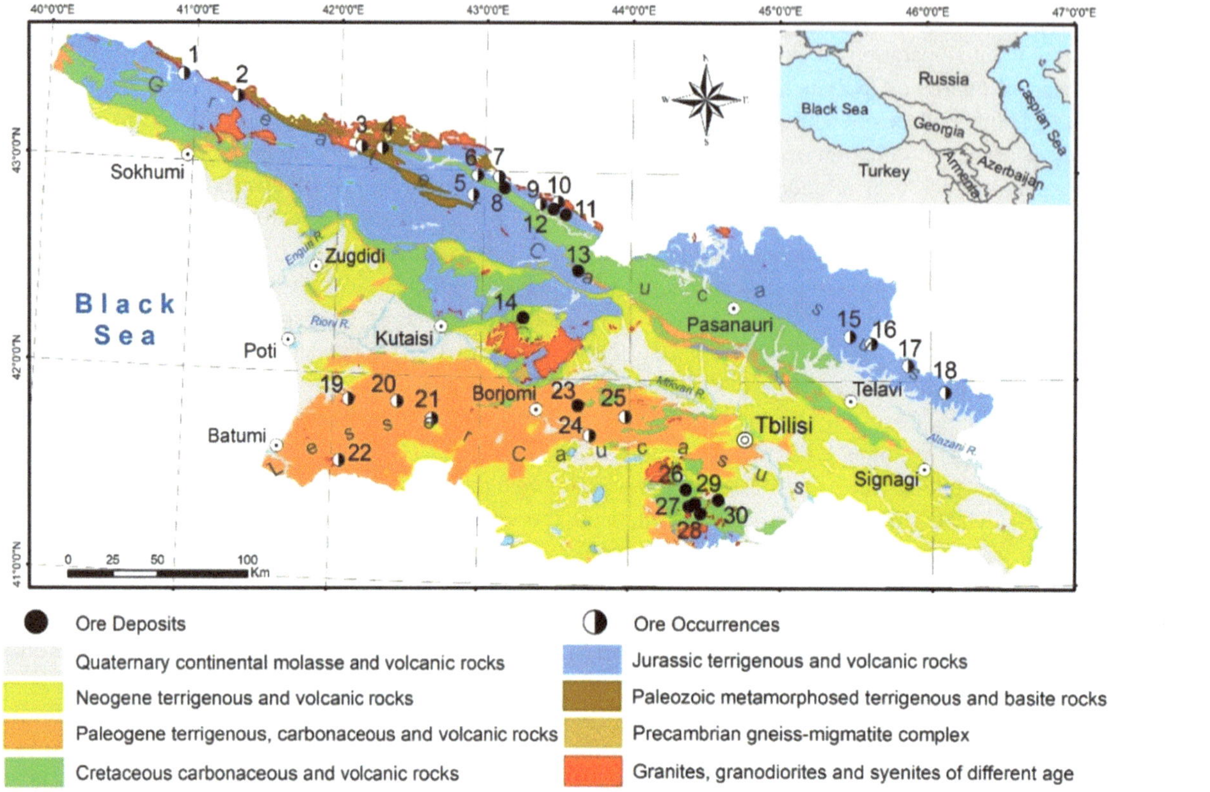

Fig. 2 Location of main ore deposits and main ore occurrences of Georgia, on a simplified geological map of Georgia. The geological framework is modified after Gudjabidze et al. (2003). **The ore deposits**: 8. Tsana (Au, As); 11. Lukhumi (Au, As); 12. Karobi (Mo); 13. Kvaisa (Pb-Zn); 14. Chiatura (Mn); 23. Dzama-Gharta Au, Ag, Cu, Mo, Fe); 26. Begtaqari (Au, Ag, Zn, Pb), 27); Sakdrisi (Au, Ag, Cu, Zn, Pb); 28. Madneuli (Au, Ag, Cu, Zn, Pb); 29. Davit-Gareji (Au, Ag, Cu, Zn, Pb); 30. Tsitelsopeli (Au, Ag, Cu, Zn, Pb). **The ore occurrences**: 1. Akhei (Hg); 2. Adange (Cu); 3. Sakeni (Au, Sb, As, W); 4. Dolra (Au, Sb, W, As); 5. Luji (Au); 6. Kvebi (Au, Cu, Zn, Pb, As); 7. Shkhara (U, Th); 9. Zofkhito (Au, Sb, As); 10. Zeskho Au, As, Sb); 15. Stori (Cu, Fe, Au, Th); 16. Artana (Au, Cu, Co); 17. Chelti (Cu, Co); 18. Areshi (Cu, Co); 19. Vakijvari (Au, Cu, Pb, Zn, Fe, Th); 20. Zoti (Au, Ag, Zn, Cu, Fe); 21. Ghaghvi (Au, Ag, Zn, Pb, Cu, Mo); 22. Merisi (Au, Ag, Zn, Pb, Cu); 24. Gujareti (Au, Ag, Cu, Mo); 25. Khachkovi (Au, Ag, Pb, Cu, Mo)

striking industrial landscape, exemplifying the tension and potential coexistence between resource extraction, environmental impact and the conservation of mining-related geoheritage. It is a natural candidate for recognition as a metallogenic KGA and as a key node in any future "mining and industry" georoute in central Georgia.

3.2 Gold, Arsenic and Antimony Deposits

Gold mining in Georgia has an ancient history, demonstrating the country's richness in gold deposits. Nearly all major genetic types of primary gold mineralisation are represented, together with numerous placer deposits of secondary origin. As of 2019, Georgia's estimated proven gold reserves were about 340 tonnes, with total resource estimates of approximately 600 tonnes (Report of NSOG 2022). The most famous gold deposits in Georgia are the Bolnisi group deposits, which are currently being mined and are described separately in this volume.

In addition to the Bolnisi group, the Merisi ore field in the Adjara region of the Adjara–Trialeti fold-and-thrust belt contains significant gold reserves. This ore field is an epithermal system genetically linked to the Eocene syenitic pluton of Merisi, with gold mineralisation associated with quartz vein systems in the host rocks. The total gold content of the Merisi ore field is estimated at about 40 tonnes (Report of GGMC 2004).

Two ore fields in the Guria region of the same belt, Zoti and Vakijvari, are genetically similar to Merisi. Both are associated with middle Eocene syenitic plutons, and ore deposition is primarily represented by copper–porphyry and contact–hydrothermal types. Calculated gold reserves are estimated at 20 tonnes in Zoti and 44 tonnes in Vakijvari (Chkhikvishvili et al. 1992).

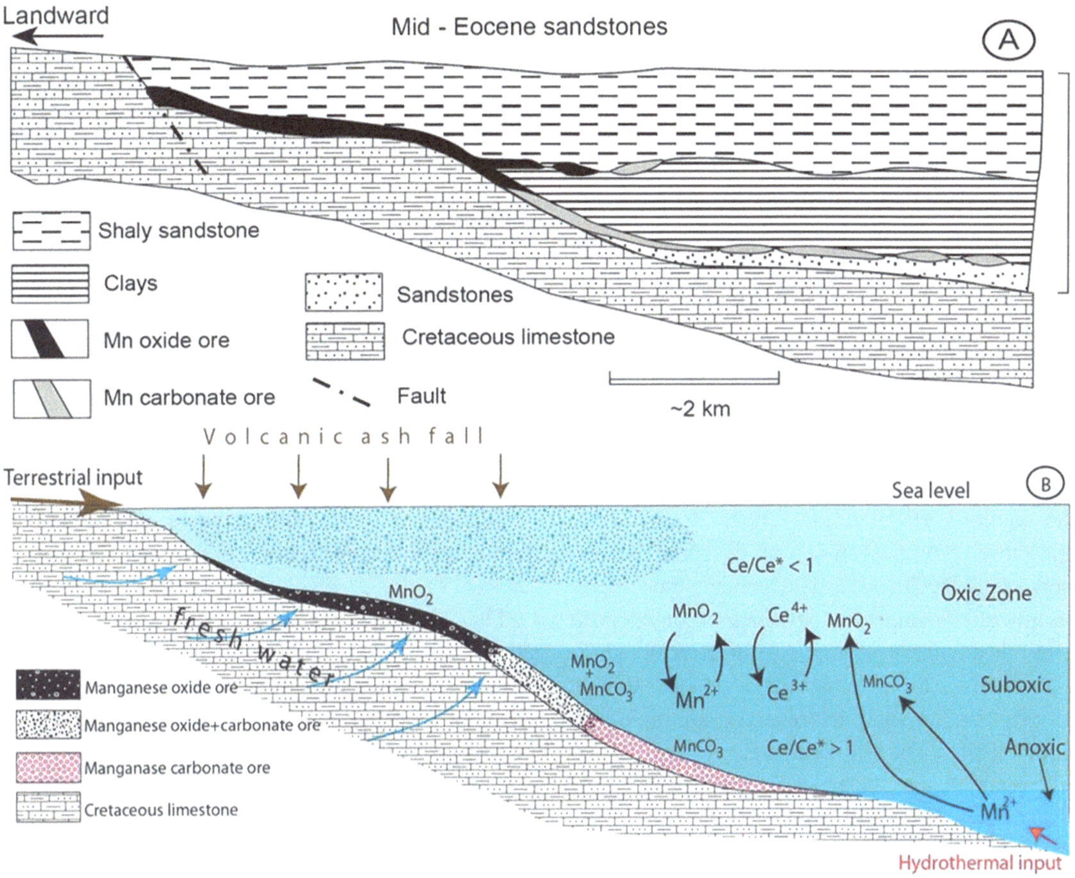

Fig. 3 The genetic model of the formation of the Chaitura Mn deposit (Sasmaz et al. 2021)

Fig. 4 General view of the Soviet-era manganese quarry in Chiatura. *Photo* © David Kvavadze

Significant gold reserves also occur in the Sakeni ore field, hosted in the crystalline basement of the Caucasus. Structurally, this field is associated with the northern contact zone of the late Paleozoic Sakeni quartz–diorite intrusive and is characterised by gold–antimony orogenic mineralisation. The field encompasses four districts—Achapara, Hokrila, Memuli and Kakrinachkuri—with combined gold reserves estimated at 60–65 tonnes (Okrostsvaridze and Bluashvili 2009).

A second group of gold deposits is associated with the Lower Jurassic black-shale series of Racha–Svaneti. Important occurrences include Lukhra and Arshira in Svaneti, and Zofkhito in Racha. In all three localities, gold mineralisation is associated with quartz veins; at Zofkhito, gold is commonly found in quartz–antimony veins. Estimated gold reserves are approximately 60 tonnes at Zofkhito, 30 tonnes at Lukhra and 40 tonnes at Arshira (Tvalchrelidze 2003).

The Lukhumi deposit in Racha hosts significant gold and arsenic reserves. Gold grades there range from 3 to 5 g/t, with similar concentrations in carbonate rocks hosting the mineralisation. Total gold reserves at Lukhumi are estimated at about 20 tonnes. The gold polymetallic deposits of Dzama (Kareli region), genetically associated with the Shuano dioritic intrusive, are estimated to contain 10–15 tonnes of gold. Comparable gold resources are reported at the Zeskho–Lafuri Cu–VMS deposit. Smaller gold reserves, about 7 tonnes, are estimated at the Shkenashi gold–polymetallic deposit in the Mestia district, and 4–5 tonnes are associated with the Karobi (Racha) Mo–Cu porphyry deposit (Report of GGMC 2004).

As noted above, arsenic and antimony are genetically closely associated with gold and often occur together in composite ores. In Georgia, such ore types are represented by the arsenic deposits of Tsana (Lower Svaneti) and Lukhumi (Upper Racha). At Tsana, arsenic is mainly present as arsenopyrite; at Lukhumi as realgar and orpiment. During the twentieth century, particularly in the Soviet era, intensive industrial extraction of arsenic took place from these deposits, although mining has now ceased. Remaining arsenic reserves are estimated at ~11,000 tonnes at Lukhumi and ~55,000 tonnes at Tsana (Report of GGMC 2004).

Georgia also possesses significant antimony reserves. The antimony veins at Lukhumi, formed in Liassic clay shales on the southern slope of the Greater Caucasus in Upper Racha, have estimated reserves of about 41,000 tonnes (Report of GGMC 2004). The genetic type represented at Zofkhito also includes the Adishi antimony deposit (Upper Svaneti, headwaters of the Adishi River), which is similarly hosted in Lower Jurassic clay shales, though it remains poorly studied and its reserves are unknown (Okrostsvaridze 2014). The Hokrila area of the Sakeni ore field in Upper Svaneti, hosted in Paleozoic biotite migmatites, also contains substantial antimony reserves. This hydrothermal deposit is estimated to hold >30,000 tonnes of antimony (Okrostsvaridze and Bluashvili 2009).

Taken together, these gold, arsenic and antimony deposits define a broad metallogenic belt extending across Svaneti, Racha and Adjara–Trialeti. From a geoheritage perspective, this belt underpins both the ancient gold-sands narratives of Colchis and the modern Bolnisi mining district, and constitutes a natural candidate metallogenic KGA representing orogenic and epithermal gold, as well as associated As–Sb mineralisation, in the Caucasus segment of the Tethys.

3.3 Alluvial Gold Deposits of Georgia

There are numerous rivers in Georgia, although only a few contain placer gold. These alluvial deposits undoubtedly represent geoheritages in their own right.

The highest concentrations of placer gold are found in Upper Svaneti, particularly in the upper reaches of the Enguri River and its tributaries. General estimates suggest that the Enguri and its tributaries contain approximately 20–25 tonnes of gold. In Lower Svaneti, the sands of the Tskhenistskali River and its tributaries are estimated to hold 3–4 tonnes of gold. Significant placer gold also occurs in the Bolnisi district, in the sands of the Khrami and Mashavera rivers, where reserves are estimated at 5 tonnes and 4 tonnes respectively (Geleishvili 1990).

3.4 Iron Deposits

The Black Sea coastal iron deposit of Supsa–Natanebi extends over a ~14 km stretch between the Supsa and Natanebi rivers. It formed by erosion of iron ores from the Vakijvari ore field, with magnetite-rich material transported by rivers and reworked into coastal placer sands. The magnetite fraction in these sands ranges from 3% to 5%, and total iron reserves exceed 15 million tonnes. This coastal black-sand belt is both an iron resource and a striking geomorphological feature, forming part of Georgia's coastal geoheritage and linking metallogeny directly to beach landscapes and tourism along the Black Sea (Fig. 5).

The industrially significant iron deposits in Georgia include Tkibuli–Shaori, Supsa–Natanebi, Dzama and Foladauri, with total ore reserves of about 770 million tonnes and contained iron reserves of around 95 million tonnes.

The Tkibuli–Shaori iron deposit comprises two thick siderite layers extending over about 5 km. In the Tkibuli area these layers crop out at the surface, whereas in the Shaori depression they are overlain by Jurassic sediments. The upper layer reaches a thickness of ~4 m, with an average iron content of 34.3%; the lower layer averages ~10 m in thickness with an iron content of about 34.6%. National reserves at Tkibuli–Shaori are estimated at 72.24 million tonnes (Report of GGMC 2004).

The Dzama magnetite deposit, located in the Kareli district, is a skarn-type iron deposit formed in the exocontact zone of the middle Eocene Rkviana pluton. It contains ore with an average Fe content of ~32%, and reserves estimated at 5.4 million tonnes. Smaller iron reserves occur at the Foladauri skarn-type deposit in the Bolnisi region, with total reserves of about 2.3 million tonnes and an average Fe content of 36% (Report of GGMC 2004).

Fig. 5 View of the Ureki magnetite sands. *Photo* © A. Okrostsvaridze

3.5 Thorium and Uranium Deposits

In recent years, several thorium–and uranium-bearing occurrences have been identified in Georgia. From the perspective of thorium and rare-element mineralisation, the Stori ore field in the Kakheti segment of the southern slope of the Greater Caucasus is especially promising. It is hosted in hydrothermally altered Lower Jurassic clay shales exposed along the Stori River canyon. Over a stretch of nearly 4 km, these shales have been intensely fractured and altered, with almost complete loss of their primary appearance (Fig. 6). The altered rocks are dominated by quartz–sericite–chlorite and quartz–chlorite–epidote–albite associations.

In 2011, significant thorium mineralisation was discovered in these altered rocks within 20–40 cm thick carbonated quartz veins that cut the Pliensbachian clay shales (Okrostsvaridze et al. 2011). Thorium contents in these veins range from 40 g/t to 3882 g/t, bismuth from 112 ppm to 4806 ppm, and yttrium from 247 ppm to 6307 ppm, indicating a highly enriched Th–Bi–Y system of potential economic and scientific interest.

In addition to Stori, a second thorium-bearing occurrence has been documented in the Merisi ore field (Adjara). East of Merisi Falls, Eocene trachytic volcanic rocks have undergone intense hydrothermal alteration. Along an exposure of nearly 800 m, thorium concentrations range from 52.3 ppm to 147.7 ppm, while uranium reaches up to 17.5 ppm.

Until recently, mineral deposits with industrial uranium concentrations were not known in Georgia; however, new work has revealed significant mineral occurrences of this element (Okrostsvaridze et al. 2022a, b). Uraninite mineralization occurrence was discovered along the Main Thrust of the Greater Caucasus, within the Paleozoic Shkhara crystalline massif. The massif forms a ~15 km long and ~5 km high ridge and is composed of Lower to Middle Palaeozoic biotite schists, gneisses, and migmatites. The Lower to Middle Palaeozoic rocks of the Shkhara Massif, almost along the entire length, are cut by a granitoid pluton of the Variscan generation. The pluton is predominantly composed of granodiorites, with smaller amounts of granites and quartz-diorites. The entire Shkhara massif is directed south, along the Main Thrust of the Greater Caucasus thrust over the Lower Jurassic black shales (Fig. 7).

Fig. 6 Hydrothermal albite-chlorite-epidote associations formed in Lower Jurassic clay shales in the Stori ore field. *Photo* © A. Okrostsvaridze

Zircons in biotite granodiorites of the Shkhara pluton and enclosing biotite gneisses were dated by the LA-ICP-MS method. A weighted mean 206Pb/238 U age of 488.5 ± 8.5 Ma was found for zircons of the biotite-gneiss and a weighted mean 206Pb/238 U age of 316.9 ± 8.8 Ma for zircons of the cutting granodiorites This rock ensemble is intersected by biotite plagiogranite intrusions of the Late Variscan generation, the 206Pb/238 U age of which corresponds to 310.2±7.5 Ma (Okrostsvaridze et al. 2023).

A high radiation background was noted on the surface of biotite plagiogranite intrusions, ranging from 1.5 μSv/h up to 2.0 μSv/h, which is 10–15 times higher than normal (Fig. 8). The plagiogranite intrusions are slightly fragmented and fractured, and fractures and nests are filled with Th-enriched uraninite veins and impregnations. The whole-rock X-ray fluorescence (XRF) analyses were conducted by MS Analytical Laboratory (MSALABS), Vancouver, Canada, and in the US Geological Survey, Denver Regional Center Laboratory. Analyses have shown that the content on this rocks of Th ranges from 25 ppm to 90 ppm and that of U from 55 to 390 ppm (Okrostsvaridze et al. 2022a, b).

Investigations of the uranium veins with a JXA-8230 electron probe microanalyzer show that the ore is represented by Th-bearing uraninite. Chemical analyses indicate that uraninite grains are compositionally homogeneous both within individual crystals and across the vein. The main components vary within the following ranges: UO_2 from 82.6 wt% to 71.7 wt%, ThO_2 from 5.7 wt% to 9.5 wt%, PbO from 2.95 wt% to 3.9 wt%, and Y_2O_3 from 0.40 wt% to 4.6 wt%. Th-U-total Pb chemical ages range between 275 and 305 Ma, with an average age of 291±14 Ma (Fig. 9). (Wilke et al. 2025). The chemical age of the uraninite mineralization and the U-Pb isotopic ages of the Shkhara variscan orogenic granitoids, which were dated to the Late Carboniferous, indicate the rise of hot melts and fluids along reactivated faults caused by relaxation during post-orogenic expansion.

According to the type of U mineralization, composition, geodynamic setting, tectonic localization, and average Th-U-Total Pb chemical age (291±14 Ma) Shkhara uraninite mineralization is in full correlation with the same

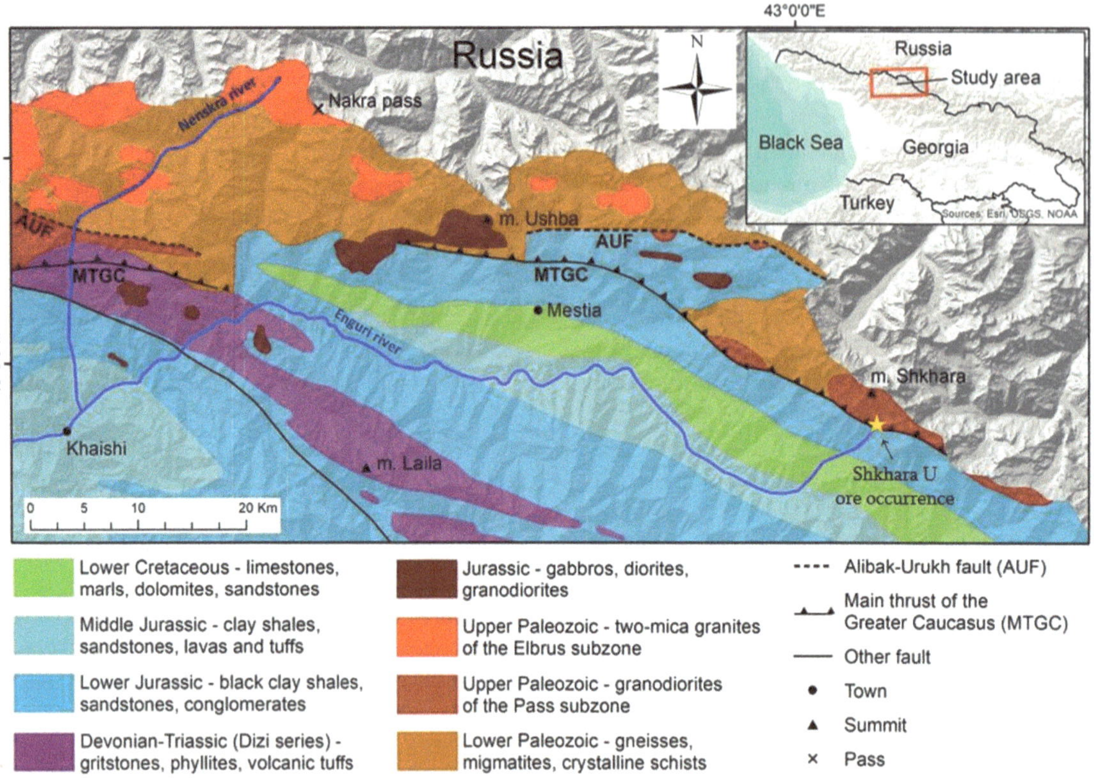

Fig. 7 Geological map of the Greater Caucasus Svaneti segment, with the location of the Shkhara uraninite mineralization occurrence (The map was modified after Gudjabidze et al. 2003)

Fig. 8 Sowing one of the U-Th baring plagiogranite vein (A) crossed by a diabase dike, folded biotite-plagiomigmatite with hait concentration of U-Th (B), and a microscopic image of a uraninite vein in plagiogranite (plane polarized light) (Wilke et al. 2025)

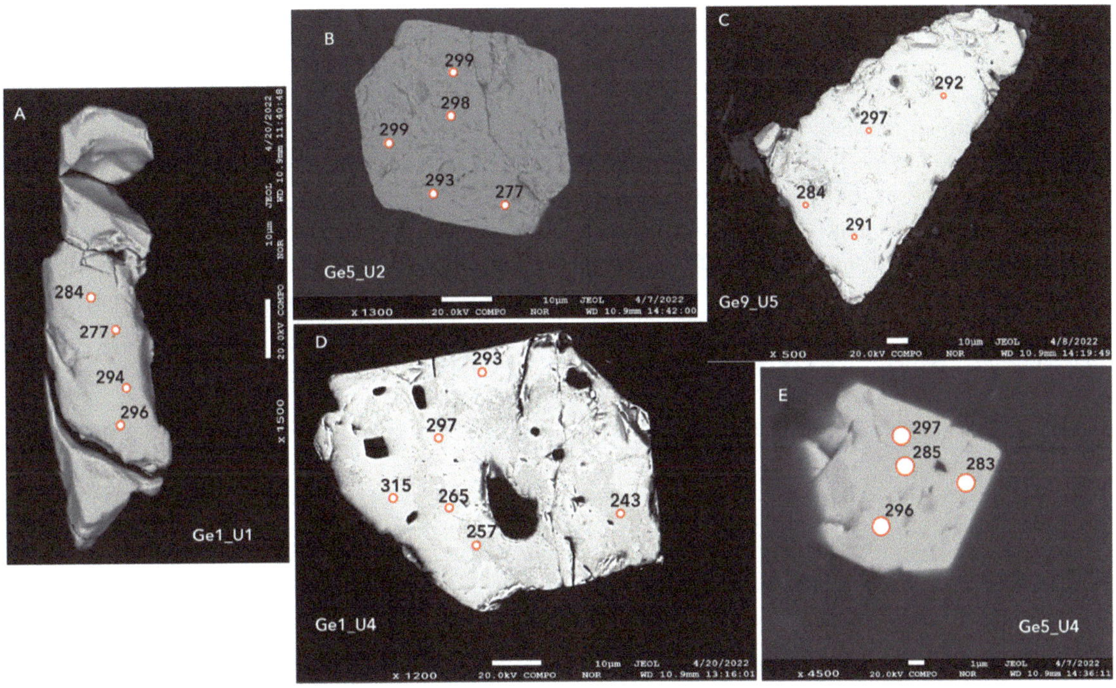

Fig. 9 Backscatter electron images of uraninite grains from the three veins dated. **a–c** No zonation is visible. Spots are ca. 1 μm, as during EPMA measurements, indicating the location of measurement. Numbers show spot ages in Ma, with uncertainties of ±15 Ma for Ge1 and ±7 Ma for Ge5 and Ge9. **d** Ge1-U4 showing inhomogeneous patterns within a grain. Darker areas have higher porosity and, when looking at the chemical composition, show increased incorporation of Er, Dy and Y at the expense of U and Th. Points placed on darker areas yielded younger ages of 243–265 Ma, while lighter areas show ages of 293–315 Ma. **e** The Ge5-U4 grain exhibits a homogeneous age and chemical distribution, while the chemical composition of this single grain differs from the other uraninites examined here in that it has higher ThO_2 and lower UO_2 concentrations (Wilke et al. 2025)

type of U-bearing granitic veins in different regions of the world. This finding is particularly interesting in the European case, as it indicates an eastward continuation of the Late Variscan U metallogenic belt from the Iberian Massif in the west, via central Armorica and the Massif Central, the Erzgebirge and the Bohemian Massif in Central Europe, to the Greater Caucasus at the easternmost border of Europe.

From a geoheritage and critical-minerals perspective, the Stori, Merisi and Shkhara occurrences together define a new metallogenic theme for Georgia: Th–U–rare-element mineralisation in hydrothermal systems of the Greater Caucasus and Adjara–Trialeti belts. These sites are important both for understanding the long-term evolution of Tethyan magmatism and for exploring the potential role of Georgia in future thorium–and uranium-related energy debates. They are natural candidates for recognition as a specialised metallogenic KGA, although any future development or geotourism interpretation will need to be carefully balanced with radiation safety, environmental protection and national policy considerations.

3.6 Deposits of Non-ferrous Metals

In Georgia, the extraction of non-ferrous metals involves complex associations that include barite, coppe lead and zinc deposits. Many of these deposits also contain significant gold, which is why some were discussed above in the context of gold mineralisation. There are also Cu–VMS (copper–colchedan) deposits where gold content is minor or absent, such as Zeskho in Lower Svaneti, Adange in Abkhazia, Artana in Kakheti and the Pb–Zn deposit at Kvaisa (Okrostsvaridze 2014).

Among the non-ferrous metal deposits in Georgia, the gold–copper porphyry systems of the Bolnisi region are the largest. These belong to the epithermal–porphyry family and include both sulphide and non-sulphide gold ores. A separate subsection in this volume is devoted to Bolnisi, so details are not repeated here. From a geoheritage perspective, however, Bolnisi represents a key metallogenic district and an industrial landscape of national significance, and is already central to current geopark discussions in eastern Georgia.

4 Mining Landscapes as Geoheritage of Georgia

The ore deposits described above are not only sources of raw materials; they also define some of Georgia's most distinctive geoheritage landscapes. They record the action of Tethyan magmatic and tectonic processes, the evolution of metallogenic belts, and the long human history of mining, metallurgy and settlement. From the gold sands of Colchis and Svaneti through the stratiform manganese of Chiatura to the coastal magnetite placers of Supsa–Natanebi and the Th–U occurrences of Stori, Merisi and Shkhara, these deposits form a network of mining landscapes that are central to Georgia's geoheritage story.

Several regions stand out as potential metallogenic Key Geoheritage Areas (KGAs) within Georgia's emerging national inventory (see Chap. Geoheritage Pathways for Georgia: Key Geoheritage Areas and The UNESCO Global Geopark Framework). The Chiatura manganese district is a world-class Oligocene stratiform Mn deposit and a striking industrial landscape, combining scientific importance, exceptional scale and a complex environmental legacy. The Bolnisi district represents a major cluster of epithermal–porphyry gold–copper deposits with a continuous mining history from the Bronze Age to the present. The Svaneti–Racha–Adjara–Trialeti belts, with their orogenic and epithermal gold, arsenic and antimony deposits and rich placer systems in rivers such as the Enguri, Tskhenistskali, Khrami and Mashavera, link ancient alluvial gold washing and the Golden Fleece tradition to modern mining and exploration.

These mining landscapes illustrate the paradox and potential of ore deposits within geoheritage: mining can damage natural systems and communities, but it also creates industrial and cultural heritage and reveals the subsurface processes that formed the ores. Recognising Chiatura, Bolnisi, Svaneti–Racha, Supsa–Natanebi and Stori–Merisi–Shkhara as metallogenic KGAs would make this dual character explicit and encourage management approaches that balance resource use, environmental restoration and heritage conservation. At the same time, these areas provide a natural backbone for future georoutes and geopark proposals: for example, a "gold sands and mountains" route linking Svaneti and Bolnisi, a "manganese and industry" route centred on Chiatura, or a coastal route that connects Supsa–Natanebi with other Black Sea geosites. In this way, ore deposits and mining districts can be integrated into Georgia's geoheritage framework not as exceptions, but as key components of a national geotourism and conservation strategy.

5 Discussion and Conclusion

This chapter has necessarily offered only a brief overview of Georgia's ore deposits and mining districts; a complete treatment would require a separate volume. Even so, a number of themes emerge clearly. Georgia occupies a key position in the Tethyan orogenic system, with a metallogenic endowment that includes stratiform manganese, epithermal–porphyry gold–copper systems, orogenic gold, arsenic and antimony belts, coastal iron placers and emerging Th–U–rare-element occurrences. These deposits have supported Bronze Age metallurgy, fuelled Soviet-era industry and continue to contribute significantly to the modern economy.

At the same time, the ore deposits and mining landscapes described here are integral parts of Georgia's geoheritage. They record the deep processes that built the Caucasus, and they underpin cultural narratives that range from the Argonauts' quest for the Golden Fleece to the industrial townscapes of Chiatura and Bolnisi. Ancient underground workings, Early Bronze Age artefacts, mining towns, transport systems, waste piles and reclaimed sites all form layers of a mining landscape that is both scientific and cultural heritage. In the language of the Key Geoheritage Area framework (Chap. Geoheritage Pathways for Georgia: Key Geoheritage Areas and The UNESCO Global Geopark Framework), regions such as Chiatura, Bolnisi, Merisi, Vakijvari, Shkara and Stori are natural candidates for metallogenic KGAs that represent the ore-forming history of the Caucasus segment of the Tethys.

These landscapes also embody the paradox at the heart of mining geoheritage. Ore extraction can degrade ecosystems, pollute rivers and damage communities, as seen in parts of Chiatura and other districts. Yet mining also reveals geologic structures, exposes ore bodies, and creates industrial and social histories that are now part of what we seek to understand and, in some cases, to conserve. The task for Georgia is not to pretend that mining and geoheritage are separate domains, but to bring them into a single management conversation: identifying where extraction should be limited or avoided, where restoration and remediation are urgent, and where mining legacies can be interpreted as part of a wider story of mountains, metals and people.

In this sense, Georgia's ore deposits offer both opportunities and responsibilities. As global demand for industrial and critical metals continues, Georgia may see renewed interest in many of the deposits outlined here. If guided by a geoheritage-informed approach—one that recognizes metallogenic KGAs, integrates mining landscapes into potential geoparks, and requires strong environmental safeguards—

this next phase of development can help ensure that the country's mineral wealth strengthens, rather than undermines, the conservation and celebration of its outstanding geoheritage.

References

Appian of Alexandria (II AD) Books XIII-XVII (Civil Wars) (1996) Translation: John Carter. Penguin books Ltd, Harmondsworth, Middlesex, England, pp 1–436

Chkhikvishvili Z, Bandzeladze T, Tokhadze G, Kakabadze R (1992) Preparation of 1:25000 scale geological-geochemical-geophysical activities on of gold in Guria ore field. Report of Georgian geological department, Tbilisi, Georgia, pp 1–173 (in Georgian).

Courcier A, Kufaradze D, Pataridze D (2008) Archeometallurgical researches on the early beginnings of metallurgy (IV-III millennia BC) in the Caucasus: an example of interdisciplinary studies. Meta 15:35–50

Dussaud R (1930) La Lidie et ses voisins. Geuthner, France, Paris, pp 1–374

Forbes R (1950) Metallurgy in antiquity: a notebook for archaeologists and technologists. E. J. Brill, Leiden, Netherlands, pp 1–489

Gudjabidze G, Gamkrelidze I, Abesadze G et al (2003) Geological Map of Georgia. Tbilisi, Georgia, scale 1:500,000

Geleishvili V (1990) Gold-bearing capacity of Georgia. Doctoral dissertation, Tbilisi, pp 1–325 (in Russian)

Moritz R, Melkonyan R, Selby D, Popkhadze N, Gugushvili V, Tayan R, Ramazanov V (2016) Metallogeny of the Lesser Caucasus: from arc construction to post-collision evolution, vol 19. Society of Economic Geologists Special Publication, pp 157–192

Okrostsvaridze A, Bluashvili D (2009) The Sakeni ore field—geological evidence and perspectives (greater Caucasus). Bull Georg Nat Acad Sci 4(1):127–131

Okrostsvaridze A, Akimidze K, Bluashvili D (2011) The first data on industrial concentration of thorium and bismuth in hydrothermally altered lower-Jurassic clay-shale of the Stori Canyon, Greater Caucasus. Bull. Georg. Nat. Acad. Sci. 5(2):76–82

Okrostsvaridze A (2014) The dynamics of the development of the mining industry of Georgia and the current situation. Report of the Caucasus Environmental Non-Government Network (CENN), Tbilisi, pp 1–72

Okrostsvaridze A, Bluashvili D, Gagnidze N (2014) Field investigation of the mythical "Gold Sands" of the ancient Colchis kingdom and modern discussion on the Argonauts' mission. Episodes 41(1):27–35

Okrostsvaridze A (2017) The Argonauts: a modern investigation of the mythical "Gold Sands". Lambert Academic Publishing, Saarbrucken, pp 1–92

Okrostsvaridze A, Bluashvili D, Gogoladze S, Gabrielashvili R (2022a) The discovery of U mineralization in the late Variscan plagiogranite vein of the Shkhara crystalline massif (Greater Caucasus, Georgia). Geologica Balcanica 51(1):3–14

Okrostsvaridze A, Bluashvili D, Gogoladze S, Boychenko G, Gabrielashvili R (2022b) Thorium—an energy of the future and its ore occurrences in Georgia. Universal, Tbilisi, pp 1–222 (in Georgian)

Pliny The Elder (I AD–I BC) (1898) The natural history, vol 6. G. Bell and Sons, London, UK, pp 1–228

Report of GGMC (2004) Report of the Georgian geology and mining company (GGMC). In: Buadze V (ed) Mineral-resource complex of Georgia, Tbilisi, pp 1–378

Report of EMC (2016) Report of human rights education and monitoring center (EMC). In: Spotlight on heavy industry: the case of Chiatura, Tbilisi, pp 1–47

Report of NSOG (2022) Statistical report of the national statistics office of Georgia (NSOG). In: Natural resources of Georgia and environmental protection, Tbilisi, pp 1–59

Richardson R (1934) Iron, prehistoric and ancient. Am J Archaeol 38(4):555–583

Richards JP (2015) Tectonic, magmatic, and metallogenic evolution of the Tethyan orogen: from subduction to collision. Ore Geol Rev 70:323–345

Sasmaz Ah, Sasmaz B, Hein GH (2021) Geochemical approach to the genesis of the oligocene-stratiform manganese-oxide deposit, Chiatura (Georgia). Ore Geol Rev 128(4):35–47 103910. https://doi.org/10.1016/j.oregeorev.2020.103910

Strabo (I AD–I BC) (2010) In: Radt S (ed) Geography. Vandenhoeck and Ruprecht, pp 1–321

Tvalchrelidze AG (2003) Mineral resource base of Georgia in the XXI century. In: Mineral resource base of the southern Caucasus and system for its management in the XXI century. Kluwer Academic Publishers, pp 19–20

Wilke F, Okrostsvaridze A, Bluashvili D, Gabrielashvili R (2025) Geochemistry and Th-U-Total Pb chemical ages of the Late Variscan uranium mineralization at Shkhara, Greater Caucasus. Minerals 15(9):960. https://doi.org/10.3390/min15090960

The Bolnisi Gold-Copper-Polymetallic Mining District of Georgia: Key Geosite of National Importance

Nino Popkhadze, Malkhaz Natsvlishvili, and Johny Shubitidze

Abstract

Bolnisi ore district is disposed in the Upper Cretaceous volcanic series of the Artvin—Bolnisi zone (Lesser Caucasus) of the Tethys—Eurasian metallogenic belt. The generation of gold–copper epithermal mineralization of the Bolnisi ore district is related to the late subduction stage of development of the Eurasian active margin. Numerous ongoing projects, major exploration and mining activities (including the Sakdrisi, Madneuli, Kazreti, Beqtakari, Bnelikhevi, Mushevani, Tsitelisopeli, Dambludi) in the Bolnisi region are active. Late Cretaceous rocks predominate in the Bolnisi district, which represents the products of an explosive volcanic event and the host rocks for the ore deposits and manifestations.

Keywords

Late cretaceous · Ore deposits · Exploration · Ignimbrite

1 Introduction

The Bolnisi ore district is located in southeastern Georgia, about 80 km south of the capital Tbilisi. It hosts Georgia's first major non-ferrous metal production centre: a gold–copper–polymetallic mining and processing complex that has played a central role in the country's modern mining industry. Several epithermal deposits and prospects are known in this region. Systematic study of the field began in the 1940s with the exploration of the Madneuli deposit, followed by staged drilling and mapping programmes prior to industrial development.

In 1959, construction began on one of Georgia's largest non-ferrous metal mining and processing plants, "Madneuli", and the industrial settlement of Kazreti was established to support its operations. Since 1975, the plant has extracted copper, barite and gold–silver-bearing silica rocks, and has carried out primary processing, beneficiation and sale of gold–copper and barite ores. As discussed in the national ore chapter, mining in the Bolnisi area has much deeper roots, with evidence of copper and iron production extending back to the Bronze Age. In this chapter we focus on the geological setting, ore deposits and geoheritage significance of the Bolnisi district, and consider its role as a key metallogenic area within Georgia's emerging Key Geoheritage Area inventory and future geopark and georoute network.

2 Ore Deposits and Occurrences in the Bolnisi District

The Bolnisi ore district includes the following deposits and prospects: Madneuli, Tsiteli Sopeli, Kazreti, Kvemo Bolnisi, Sakdrisi, Darbazi, Imedi, Beqtakari, Samgereti, Bneli Khevi, Balichi and Grma Khevi (Fig. 1). The Late Cretaceous Bolnisi district (~87–71 Ma) in southeastern Georgia forms a distinct tectonic zone along the Tethyan orogenic belt, at the junction between the Lesser Caucasus and the Eastern Pontides (Moritz et al. 2020). Late Cretaceous volcanic and volcanogenic–sedimentary rocks predominate in the Bolnisi area, cropping out between the Transcaucasian crystalline basement massifs of Khrami and Loki.

Generation of gold–copper porphyry and epithermal non-sulphide gold mineralisation in the Bolnisi district is linked to the late subduction stage of development

N. Popkhadze (✉)
Al. Janelidze Institute of Geology of I. Javakhishvili Tbilisi State University, Tbilisi, Georgia
e-mail: nino_popkhadze@yahoo.com

N. Popkhadze · M. Natsvlishvili · J. Shubitidze
RMG Rich Metals Group, Tbilisi, Georgia

A. Okrostsvaridze and D. Tormey (eds.), *Geoheritage of Georgia*, Geoheritage, Geoparks and Geotourism,
https://doi.org/10.1007/978-3-032-27639-1_23

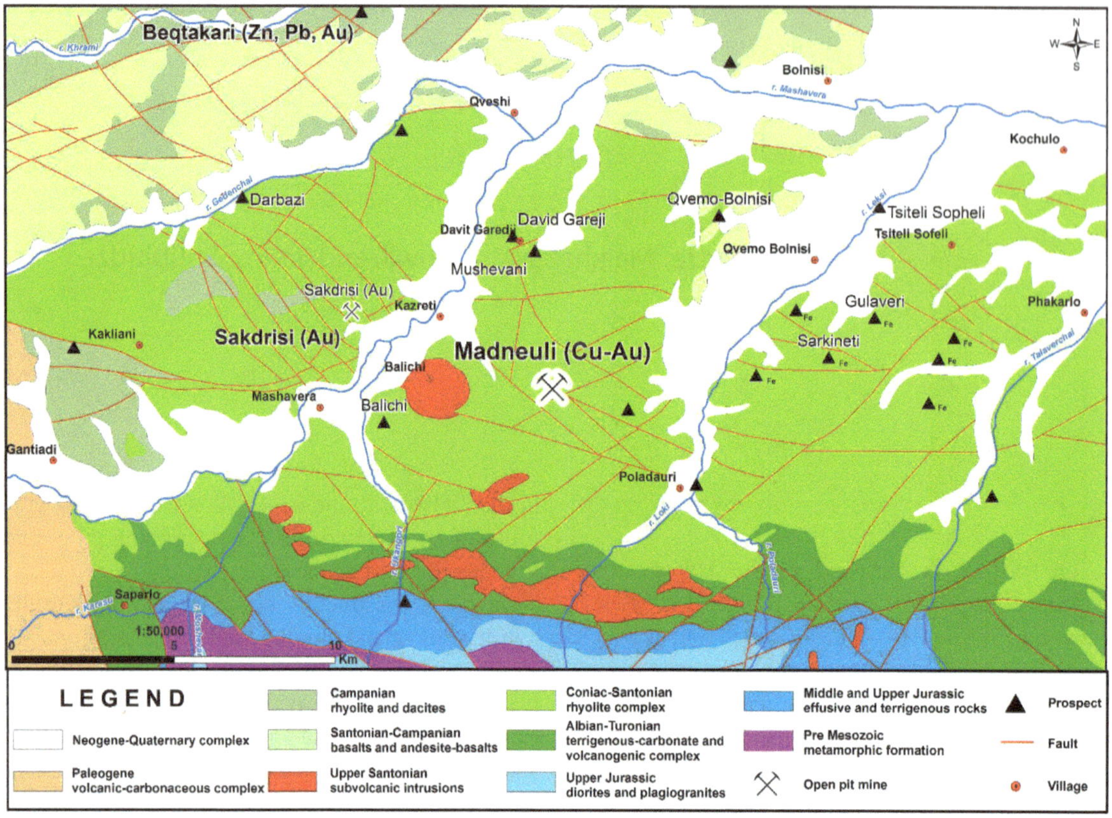

Fig. 1 Geological map of the Bolnisi district (The map modified after Vashakidze and Gugushvili 2006)

of the Eurasian active margin. The Late Cretaceous host rocks for the main ore deposits are subdivided into six volcanogenic suites, interpreted as Cenomanian to Campanian in age and overlain by Maastrichtian limestones (Apkhazava 1988). Abundant pyroclastic rocks are exposed in the region, including ignimbrites, ash-fall tuffs and pumice-rich deposits (Popkhadze et al. 2014), cross-cut by extrusive domes and subvolcanic intrusions and dykes predominantly of rhyolitic, dacitic and andesitic composition. In contrast, the Tandzia suite comprises aphyric to porphyritic basalt, basaltic andesite and trachybasalt lava flows, tuffs and dykes with interlayered sedimentary rocks, whereas the Shorsholeti suite consists mainly of porphyritic trachyandesite and basaltic trachyandesite lavas interlayered with pyroclastic rocks and limestone (Moritz et al. 2016, 2020) (Fig. 2).

Fig. 2 General view of the Madneuli open pit as of 2020. Photo © Nino Popkhadze

The Madneuli deposit is hosted in an Upper Cretaceous volcanogenic and volcanogenic–sedimentary sequence known as the Mashavera sequence: medium-to acidic tuffs, tuff breccias and ignimbrites, cross-cut by various dacitic intrusions, rhyodacites and dykes of andesitic–basaltic composition. Detailed mapping at Madneuli has distinguished three main lithostratigraphic and structural units; Cu–Au ore of economic importance is associated with the lower (Madneuli) complex. Hydrothermally altered and strongly oxidised, silicified rocks are widespread within the deposit and host gold and copper–polymetallic mineralisation. The main ore minerals are pyrite, chalcopyrite, sphalerite, galena, covellite, chalcocite and barite.

On the basis of explored reserves and historical production, the most important ore types at Madneuli are gold–copper ores, followed by gold–silver-bearing secondary quartzites, barite and barite–gold-bearing quartzite ores, and copper–zinc and barite–polymetallic ores. Ore textures include massive, breccia, vein and disseminated types, reflecting variations in ore-forming processes and host-rock permeability. Current reserve–grade figures for Madneuli (Popkhadze et al. 2025) are approximately 6.5 Mt. @ 0.5 g/t Au and 0.4% Cu (proven and probable).

The Sakdrisi deposit, like Madneuli, lies within the Kazreti area, about 11 km northwest of Madneuli and~7 km from the processing plant. Sakdrisi consists of five ore zones arranged along a northeast-trending fault zone of approximately 2.5 km in length: Sakdrisi I (Kachagiani), Sakdrisi II (Mamulisi), Sakdrisi III (Eastern Postiskhedi), Sakdrisi IV (Western Postiskhevi) and Sakdrisi V (Eastern and Western Kviratskhoveli). These are segmented by a set of transverse northwest-trending faults. In terms of mineral composition and hydrothermal alteration, Sakdrisi belongs to the epithermal type of deposits. Some authors have suggested that it is related to the southern peripheral part of the complex Late Cretaceous Darbazi–Abulmulk caldera (Apkhazava 1988).

The volcano–sedimentary architecture at Sakdrisi is subdivided into a Lower Volcano–Sedimentary Complex (LVSC) and an Upper Volcano–Sedimentary Complex (UVSC), separated by a NE-trending thrust fault zone. The upper complex is characterised by barren ignimbrites and late cross-cutting dykes, whereas the lower complex hosts the main mineralisation. Vertical zoning of gold grades has been observed in quartz / quartz–barite / quartz–sulphide veins, with low-to medium-grade stockworks in the lower complex, and disseminated ore described in permeable host rocks and explosive breccias in the Sakdrisi IV and V zones (Popkhadze et al. 2023). Reserve–grade estimates for Sakdrisi (Popkhadze et al. 2025) are about 10 Mt. @ 1.6 g/t Au and 0.14% Cu (proven and probable).

The Beqtakari epithermal prospect comprises two distinct ore zones. One is a silicified zone exposed at surface, with local barite and gold enrichment but essentially devoid of base metals. The second, intersected by drilling, consists of a lithologically controlled, folded breccia sequence mineralised with both base and precious metals. In the latter zone, the main opaque minerals are sphalerite, chalcopyrite, pyrite, barite and galena, together with tennantite–tetrahedrite that cements breccia clasts (Lavoie 2015). Reserve–grade figures for Beqtakari (Popkhadze et al. 2025) are about 1.5 Mt. @ 4 g/t Au, 0.14% Cu, 15 g/t Ag, 1.14% Pb and 2.5% Zn (proven and probable).

The Mushevani 2 gold–copper polymetallic deposit is located between the Madneuli and Kvemo Bolnisi deposits, on the southeastern slope of a ridge at an elevation of 1026.6 m. Host rocks are volcanics and volcanogenic–sedimentary rocks of the Upper Cretaceous (Cenomanian) Mashavera Formation. Ores at Mushevani 2 are of vein–disseminated type, with ore-bearing veins of several orientations. The industrial ore type is low-sulphide, non-oxidised and partially oxidised (mixed) gold–copper ore near the surface, with low-sulphide non-oxidised ores predominating at depth. The main ore minerals are chalcopyrite and pyrite, with lesser sphalerite, galena and melnikovite; chalcocite, covellite, bornite and smithsonite are present in minor amounts. Reserve–grade estimates for Mushevani 2 (Popkhadze et al. 2025) are about 4 Mt. @ 1.6 g/t Au and 0.3% Cu (probable).

Fig. 3 General view Heap leach operation site on the Madneuli plant. Photo © Nino Popkhadze

Rich Metals Group (RMG) companies have been involved in mining activities in the Bolnisi region since 2012. RMG produces polymetallic concentrates and gold dore alloy from surface and underground mining and processing of copper-, gold- and polymetal-bearing ores. RMG Gold extracts precious metals from gold-bearing silica rocks by heap-leaching methods (Fig. 3). The heap-leach production cycle, in operation since 1997, is described as complying with international standards of mining practice. Collectively, the open pits, underground workings, processing facilities, heap-leach pads and the town of Kazreti form a characteristic mining landscape that is central to the geoheritage significance of the Bolnisi district.

3 Geoheritage Significance of the Bolnisi District

The ore deposits of the Bolnisi district are not only of economic importance; they also define one of Georgia's most distinctive mining-related geoheritage landscapes. From a geoheritage perspective, Bolnisi has several complementary values. Scientifically, it is one of the best-studied epithermal–porphyry districts in the Lesser Caucasus, representative of Late Cretaceous metallogeny in the central Tethyan belt. Educationally, its open pits, underground workings and natural outcrops offer accessible sections through host rocks, alteration halos and ore bodies that are ideal for university field courses and professional training. Culturally and industrially, the creation of the Kazreti settlement, the construction and evolution of the Madneuli plant, and the long continuity of mining activity in the area form a characteristic industrial landscape with deep social and economic roots.

Within Georgia's emerging Key Geoheritage Area inventory (see Chap. 2), Bolnisi is therefore a natural candidate metallogenic KGA representing Late Cretaceous porphyry–epithermal gold–copper systems in the Caucasus segment of the Tethys. At the same time, Bolnisi exemplifies the broader paradox of mining geoheritage: extraction has altered the landscape and generated environmental pressures, yet the resulting exposures, industrial structures and community histories are now part of what we seek to understand and, in some cases, to conserve and interpret. As discussions on an eastern Georgian geopark advance, the Bolnisi district is likely to serve as a core area or anchor node in a future geopark and georoute network that integrates metallogeny, mining heritage and surrounding cultural sites.

4 Conclusions

The Bolnisi gold–copper–polymetallic mining district is a distinctive and significant component of Georgia's geoheritage. It preserves a complete example of a Late Cretaceous porphyry–epithermal system in the Lesser Caucasus and has been the focus of continuous gold and copper mining from Bronze Age workings to modern industrial operations. The district's unique mineral deposits illustrate the long-term geological evolution of the central Tethyan belt, while its mining histories show how natural resources have shaped settlement, industry and culture in southeastern Georgia.

As such, Bolnisi stands as a key geosite of national importance, embodying both the natural and cultural dimensions of Georgia's geological wealth. Within the framework of the emerging Key Geoheritage Area inventory (see Chap. 2), the district is a natural candidate metallogenic KGA representing Late Cretaceous porphyry–epithermal Au–Cu systems in the Lesser Caucasus. At the same time, Bolnisi exemplifies the wider mining–geoheritage paradox: active and historical extraction have transformed the landscape and created environmental pressures, yet the resulting exposures, industrial structures and community histories are themselves part of the geoheritage we now seek to document and, in appropriate ways, to conserve and interpret.

As Georgia develops its geoheritage strategy, the Bolnisi district is likely to play a central role in future georoutes and geopark concepts in eastern Georgia, linking ore deposits and industrial landscapes with nearby cultural sites and the wider tectonic story of the Tethyan belt. Managed with strong environmental safeguards and an explicit geoheritage lens, Bolnisi can continue to contribute to national development while also serving as a living laboratory and showcase for the coexistence of metallogeny, mining and geoheritage.

References

Apkhazava M (1988) Late cretaceous volcanism structures of Bolnisi volcano-tectonic depression. Ph.D. thesis, Caucasus Institute of Mineral Resources, pp 1–269

Lavoie J (2015) Genetic constraints of the late cretaceous epithermal Beqtakari prospect, Bolnisi mining district. Lesser Caucasus, Georgia. M.Sc. thesis, Switzerland, University of Geneva, pp 1–119

Moritz R, Melkonyan R, Selby D, Popkhadze N, Gugushvili V, Tayan R, Ramazanov V (2016) Metallogeny of the lesser Caucasus: from arc construction to postcollision evolution. Soc Econ Geol Spec Publ 19:157–192

Moritz R, Popkhadze N, Hassig M, Golay T, Lavoie J, Gugushvili V, Ulianov A, Ovtcharova M, Grosjean M, Chiaradia M, Dumitrica P (2020) At the crossroads of the lesser Caucasus and the eastern Pontides: late to early Eocene magmatic and geodynamic evolution of the Bolnisi district, Georgia. Lithos 105872:378–379

Popkhadze N, Gogia B, Tagviashvili Sh, Ugrekhelidze G, Natsvlishvili M, Shubitidze J, Sonmez SU, Moritz R (2023) 3D Lithological and structural model and prospectivity of the late cretaceous sakdrisi copper-gold epithermal deposit, Lesser Caucasus, Georgia. In: Proceedings of the 17th SGA biennial meeting: mineral resources in a changing world, August 28–September 1 2023/ Proceedings 1:184–188

Popkhadze N, Moritz R, Gugushvili V (2014) Architecture of upper cretaceous rhyodacite hyaloclastite at the polymetallic Madneuli deposit, lesser Caucasus, Georgia. CentL Eur J Geosci 6:308–326

Popkhadze N, Tagviashvili SH, Natsvlishvili M, Shubitidze J, Ugrekhelidze G, Gogia B, Sonmez SU, Moritz R (2025) Ore deposits and prospects in the Bolnisi mining district: general overview and future exploration prospectivity, Lesser Caucasus, Georgia. In: Proceedings of the 18th SGA biennial meeting. pp 531–535

Vashakidze S, Gugushvili V (2006) Geological map of Bolnisi district (1:50 000). Caucasian Institute of Mineral Resources, Tbilisi, Georgia

Mineral and Thermal Water Riches of Georgia: The Phenomenon of "Immortality Water"

Daniel Tormey, Avtandil Okrostsvaridze, and Giorgi Boichenko

Abstract

Georgia lies within the Caucasus collision zone, where long-lived convergence between the Arabian Plate and Eurasia has produced a complex mosaic of terranes, multistage magmatic activity, and deep-seated fault systems. In addition, Georgia's position between the Greater and Lesser Caucasus Mountains and along the warm Black Sea supports abundant precipitation (~760 mm/yr) and substantial recharge of both groundwater and surface-water systems. This combination, active tectonics, pervasive faulting, inherited heat from magmatic centers, and high recharge, creates favorable conditions for the formation and circulation of mineral and thermal waters. As a result, within a small territory (69,700 km^2), Georgia hosts ~2000 natural and engineered mineral and thermal springs. Over centuries, these waters have shaped culturally anchored patterns of human use and meaning, expressed today through internationally distributed mineral-water brands (e.g. Borjomi, Nabeglavi, Sairme) and through historic thermal-water traditions such as the Tbilisi sulfur baths and spa landscapes including Tskaltubo. In Georgian folklore, their healing reputation is captured in the enduring motif of "immortality water." In geoheritage terms, mineral and thermal waters represent a living expression of Georgia's geodiversity at the intersection of geology, culture, and well-being, an asset whose sustainable use and protection are essential for future generations.

D. Tormey
Catalyst Environmental Solutions, Santa Monica, USA

D. Tormey (✉)
IUCN World Commission on Protected Areas, Gland, Switzerland
e-mail: dtormey@ce.solutions

A. Okrostsvaridze · G. Boichenko
Institute of Earth Sciences, Ilia State University, Tbilisi, Georgia

Keywords

Mineral waters · Thermal waters · Geoheritage · Georgia · "Water of immortality"

1 Introduction

The formation of mineral and thermal waters is among the most visible expressions of complex geological processes occurring within the Earth's crust. Their origin, chemical composition, and spatial distribution are controlled by interacting factors that include post magmatic activity and residual heat, the thermal effects associated with radioactive decay in rocks, and tectonic structures that provide pathways for fluid circulation. Water's physical and chemical properties make it an active solvent and transport medium, which further shapes mineralization as fluids move through heterogeneous rocks at depth and, in many cases, discharge at the surface. These waters have long held multifaceted cultural and economic value for local communities. Their circulation mechanisms, temperature regimes, and chemical signatures also provide scientific information about crustal structure and subsurface processes. For these reasons, mineral and thermal waters in Georgia should be treated not only as resources, but also as important elements of national geoheritage (e.g. Tsulukidze 2004).

The study of mineral waters in Georgia has a long history, spanning both the historical past and the modern scientific period (e.g. Javakhishvili 1939; Buachidze 1987; Tsulukidze 2004; Gvakharia 2006; Tsertsvadze 2017; Zautashvili and Mkheidze 2022). A notable early contribution comes from the Georgian scholar Vakhushti Batonishvili, who described the therapeutic properties and geographic distribution of these waters in his foundational work, Description of the Kingdom of Georgia (Batonishvili 1745).

A. Okrostsvaridze and D. Tormey (eds.), *Geoheritage of Georgia*, Geoheritage, Geoparks and Geotourism,
https://doi.org/10.1007/978-3-032-27639-1_24

Today, Georgia hosts approximately 2000 natural and engineered mineral and thermal springs of diverse origin, composition, and temperature. Common water types include CO_2 bearing hydrocarbonate sodium waters, chloride sodium waters, sulfide waters, radon bearing waters, and various thermal waters (Gvakharia 2006; Tsertsvadze 2017). The distribution of spring groups reflects Georgia's complex geological structure, with many springs clustered along the margins of intermontane basins, within deeply incised river gorges, and near structural junctions between the Greater and Lesser Caucasus.

This hydrogeologic richness has also shaped a distinctive Georgian culture of healing waters. Balneological resorts such as Borjomi, Likani, Sairme, Nabeglavi, Tskaltubo, and the Tbilisi sulfur baths have long traditions of therapeutic use that combine local beliefs with formal medical practice. On a smaller scale, countless village springs, monastery adjacent holy waters, and sulfur pools are woven into everyday life, pilgrimage, and regional identity. Together, these sites form a national landscape of water-based wellness in which geology, climate, and culture are tightly intertwined.

Belief in the healing properties of mineral waters has been strong enough in Georgia that settlements, villages, and cities developed around them. The most prominent example is the capital, Tbilisi, founded around hot sulfur springs, with the name Tbilisi derived from the Georgian word for warm. These traditions also appear in Georgian folklore through the enduring motif of the water of immortality. The motif has deep historical roots and is linked in popular retellings to the Argonaut legend, in which Medea, daughter of Aeetes of ancient Colchis, is said to have healed the sick using warm mineral waters and medicinal herbs. Some authors have even suggested a linguistic association between Medea and medicine, although this remains interpretive and should be treated cautiously.

It is not possible in a single chapter to provide a complete characterization of all mineral and thermal waters of Georgia or to address all associated social and economic issues. Accordingly, we present a concise overview of the national context and then highlight several representative mineral and thermal water sites that are especially important for understanding Georgia's geoheritage at the intersection of geology and culture.

2 Geological Overview: Parent Rocks and Tectonic Framework Factors for Mineral and Thermal Water Formation

Georgia is located in the central part of the Caucasus segment of the Alpine-Himalayan orogenic belt. This segment was formed in the Phanerozoic as a result of the accretion of Paleo- and Neotethys complexes on the southern edge of the Scythian platform of the Eurasian continent. The territory of Georgia covers all major structural units of the Caucasian collisional orogen: the Greater and the Lesser Caucasus fold systems and the intermountain depressions lying between them. This area includes geological structures and heterogeneous rock types ranging in age from the early Paleozoic to the Quaternary (Gamkrelidze et al. 2019) (Fig. 1).

Alpine collision produced a particularly complex mosaic geology in Georgia. Rock assemblages derived from former Tethyan domains converged, underwent multiple deformation phases, and were dissected by deep-seated and near-surface fault and fracture systems. Post-collisional magmatic activity began during the Paleogene and was especially intense in southern Georgia. During the Eocene, the Adjara Trialeti volcanogenic belt developed along the southern margin of the Transcaucasian microplate within a back arc to arc setting. From the late Miocene, basaltic to andesitic volcanism intensified in this belt, forming the Samtskhe Javakheti volcanic plateau. Quaternary volcanism also occurred in the central Greater Caucasus, associated with the Keli volcanic area and the Kazbegi volcanic center. During the Quaternary, active deformation and relief growth contributed to development of major intermontane basins, including the Rioni, Kartli, Mtkvari, and Alazani basins (Fig. 2).

This tectono-magmatic framework is fundamental to the abundance and diversity of mineral and thermal waters in Georgia. Deep fault systems provide pathways for recharge and circulation, inherited heat from magmatic centers supports warming at depth, and high precipitation supplies the recharge needed to sustain large groundwater flows. As a result, spring occurrence, water temperature, and chemical type are strongly controlled by both lithology and deep fault architecture (Figs. 1 and 2). Mineral and thermal waters are therefore unevenly distributed across the country, and many spring groups coincide with regional faults and structural nodes, underscoring the decisive role of tectonic controls.

Lithology provides an overlay on the underlying structure. Along the southern slope of the Greater Caucasus, mineral springs are common while thermal waters are scarce. This may reflect the dominance of older basement and early to middle Mesozoic rocks and the absence of young magmatic heat sources at shallow levels. In the Kazbegi region, despite Quaternary volcanism, surface thermal waters are also limited, whereas mineral waters are abundant. Abkhazia is notable for numerous mineral springs aligned along the main Caucasus fault system, including well known waters such as Sakeni and Avadhara.

A different pattern is observed in the Adjara Trialeti segment of the Lesser Caucasus, which is dominated by

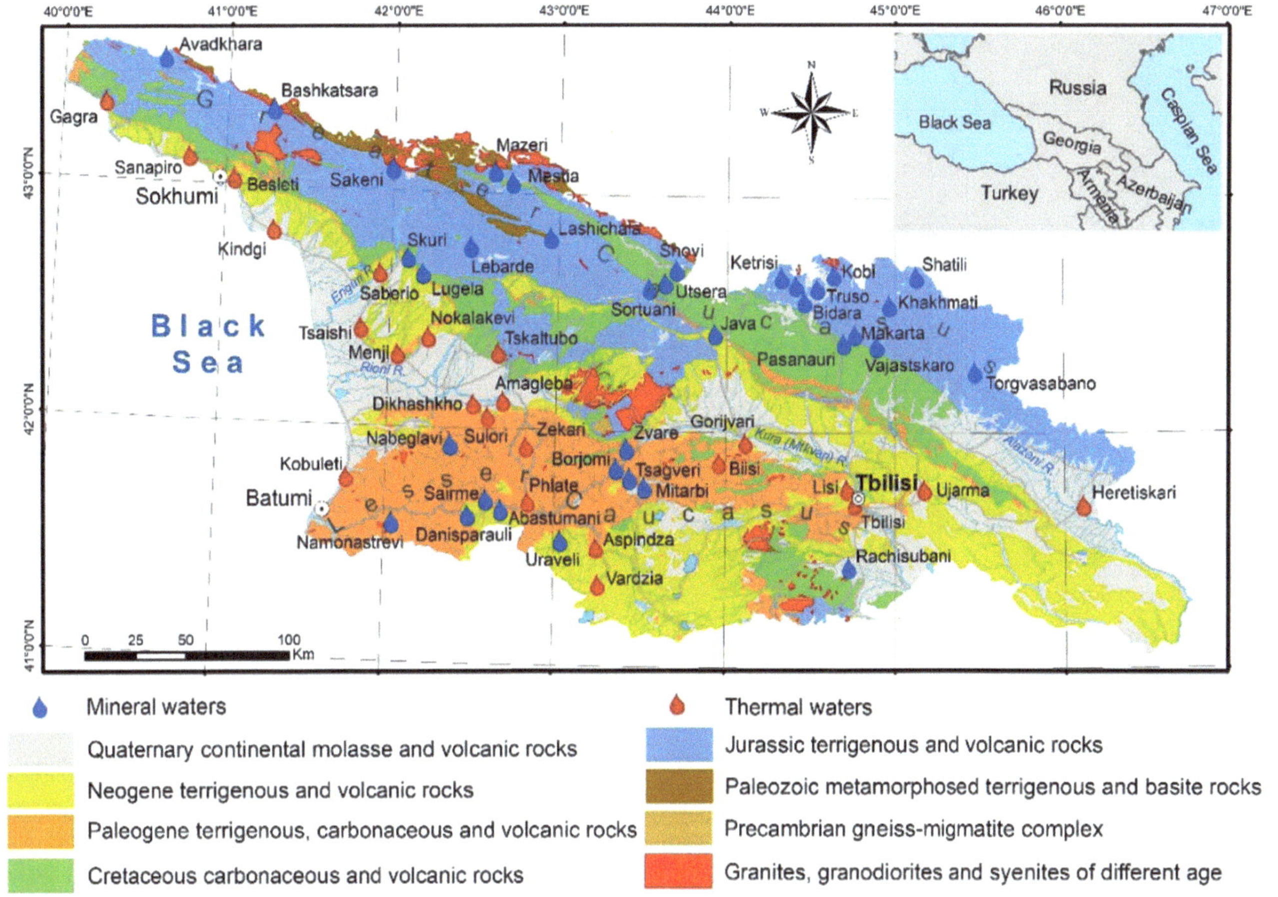

Fig. 1 Distribution of mineral and thermal waters on a simplified geological map of Georgia. The geological framework of the map is modified after Gudjabidze et al. (2003)

younger Eocene volcanogenic and plutonic associations. Here both mineral and thermal waters are widespread (Fig. 1), including internationally known mineral waters such as Borjomi, Nabeglavi, and Sairme, as well as the thermal waters associated with Tbilisi. The Adjara Trialeti rift related structure developed along the southern margin of the Transcaucasian microplate and involved interaction between mantle melts and crustal components. It is plausible that some chemical characteristics of the waters, including potassium rich signatures where observed, are influenced by water-rock interaction within these lithologies and their alteration products (Okrostsvaridze et al. 2021). More generally, mineral and thermal water composition is strongly influenced by the parent rocks through which waters circulate (e.g. Porowski 2019).

Based on chemical composition, prior studies distinguish several broad mineral water provinces in Georgia, including carbon dioxide rich waters along parts of the Greater Caucasus, methane nitrogen and hydrogen sulfide waters associated with intermontane depressions, and mixed nitrogen carbon dioxide and methane waters linked to the Lesser Caucasus domain (Buachidze 1987; Tsertsvadze 2017). This spatial separation broadly mirrors Georgia's geological structure and supports the interpretation that lithology and tectonic setting jointly control water type.

Source and circulation mechanisms are closely tied to deep faults and their associated fracture networks. These structures facilitate infiltration of meteoric waters to depth, heating and mineralization through prolonged interaction with bedrock, and return flow to the surface. In geoheritage terms, this tectonically guided circulation system represents a living connection between active geology and everyday human experience, expressed through springs that can be mapped, visited, interpreted, and sustainably managed as nationally significant geosites.

3 Mineral Waters

Mineral waters are commonly defined as waters whose dissolved chemical components provide therapeutic or health-related properties and whose circulation includes interaction with subsurface rocks. Many mineral waters remain within the crust as part of groundwater systems,

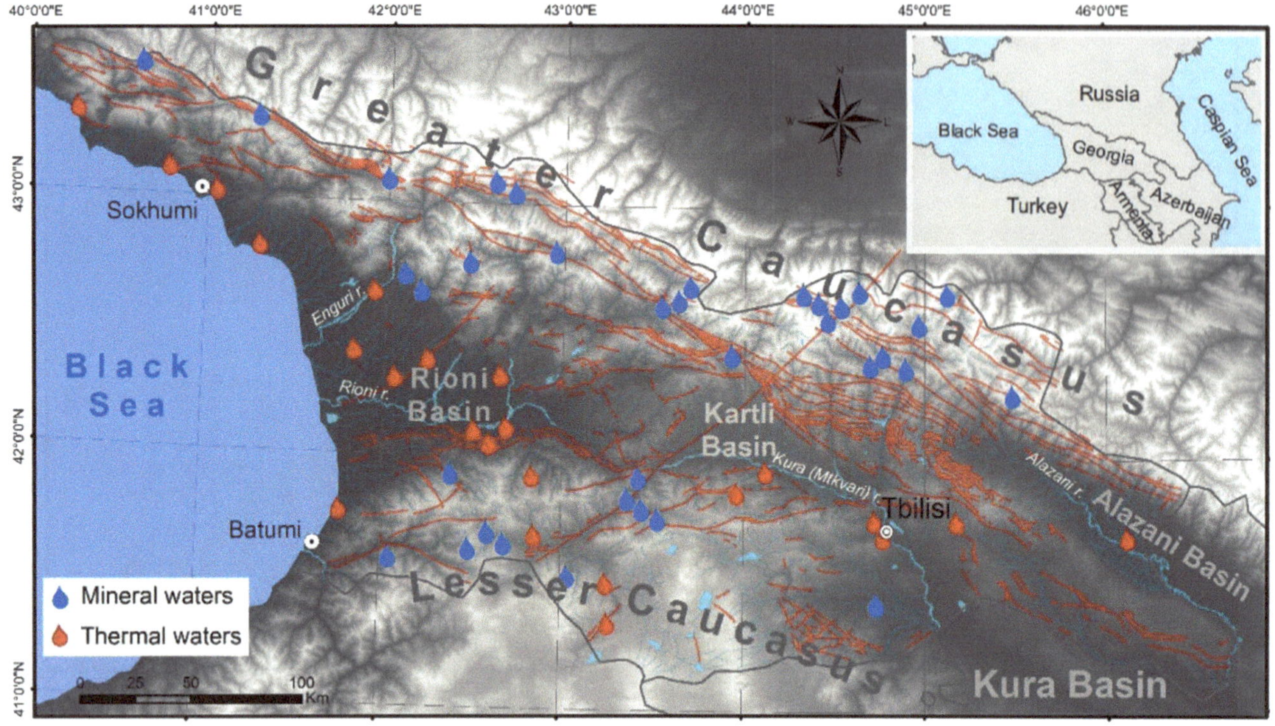

Fig. 2 Spatial distribution of mineral and thermal waters on a simplified tectonic map of Georgia, illustrating their close association with major fault systems, tectonic lineaments, and structural basins. Tectonic framework modified after Gudjabidze et al. (2003)

while others discharge naturally at springs, including along deeply incised valleys and fault controlled corridors. Mineral waters are used directly at spring sites and resort facilities for balneological treatment and wellness, and many are also bottled for commercial distribution. Globally, mineral waters are bottled and sold in many countries and have become a major economic sector linked to place identity and consumer culture (LaMoreaux and Tanner 2001).

Georgia is particularly rich in mineral waters. Published inventories report up to approximately 1500 natural mineral water springs, together with additional engineered or well captured sources. Well known examples include Borjomi, Nabeglavi, Sairme, Zvare, Lugela, Ughara, Kobi, and Avadhara, among many others. The range of mineral water types is broad, from cold, low-mineralized carbonated bicarbonate alkaline waters to highly mineralized chloride calcium waters that may be warm to hot in certain settings (e.g. Tsulukidze 2004; Gvakharia 2006; Tsertsvadze 2017; Zautashvili and Mkheidze 2022). This diversity reflects differences in recharge pathways, residence time, water-rock interaction, and structural control, consistent with the geological framework described above.

Several mineral water resorts operate in Georgia, including Borjomi, Nabeglavi, Sairme, and Utsera. In parallel, commercial bottling is strongly developed for Borjomi, Nabeglavi, and Sairme, and these brands are widely marketed in Georgia, across the region, and internationally. Notably, the principal springs of these three major deposits are associated with the Adjara Trialeti domain of the Lesser Caucasus, where Eocene volcanogenic and plutonic associations and associated sedimentary units provide favorable lithologic and structural conditions for mineral water circulation.

Borjomi. Borjomi is the oldest and most widely known mineral water resort in Georgia. It is situated at approximately 900 m elevation in the Borjomula River valley within a forested landscape of broadleaf and coniferous vegetation. Multiple springs and production wells are present, and water temperatures are typically warm, approximately 35 to 40 C. The waters are commonly described as bicarbonate sodium type, and their composition includes a broad suite of dissolved constituents that underpin the long-standing therapeutic reputation of the site. Bottled Borjomi is exported widely and has become a major component of Georgia's mineral water identity. The precise formation mechanisms and circulation model for Borjomi waters remain an active topic of discussion in the literature (e.g. Tsulukidze 2004; Tsertsvadze 2017; Gamkrelidze et al. 2021), and detailed debate is beyond the scope of this chapter.

In geoheritage terms, Borjomi illustrates how mineral waters can become a focal point where geology, landscape,

Fig. 3 A view of the current Borjomi Mineral Waters Park. In the background is the pavilion of Spring #1. *Photo* © Georgian travel

Fig. 4 The eastern facade of the Romanovs' summer residence in the Likani district of Borjomi resort. *Photo* © Georgian travel

and human history are tightly intertwined. Archaeological remains indicate that spring waters were used for therapeutic purposes in the area since at least the early historical period, including evidence of water conveyance and bathing infrastructure. From the mid nineteenth century onward, Borjomi gained prominence as an imperial era therapeutic and recreational resort, which catalyzed development of park and spa infrastructure (Fig. 3). By 1850 a mineral water park had been established, and by the mid-1850s the first bottling facilities were operating. In the late nineteenth century, a Romanov summer residence was built in the valley in a manner intended to harmonize with the surrounding landscape, and early energy development followed as part of the resort transformation (Fig. 4). This layered history demonstrates how mineral water sites can become long lived cultural landscapes anchored in a specific geological setting.

Today, Borjomi remains an active resort area with modern hotels, wellness facilities, and visitor infrastructure, while bottling and export continue at large scale. The combination of an unusual hydrogeologic system, a recognizable landscape setting, and a deep cultural narra-

Fig. 5 A general view of the central part of the Sairme resort. Here, on the right bank of this river, are the natural mineral waters sources. *Photo* © Georgian travel

tive makes Borjomi a clear example of living geoheritage in Georgia. It is therefore an important reference site for national geotourism routes and for the broader Key Geoheritage Area and UNESCO Global Geopark frameworks discussed elsewhere in this book.

Sairme. Sairme is located in western Georgia on the northern slope of the Adjara Trialeti ridge, in the pine and deciduous forest valley of the Tsablaristsqali River at approximately 950 m elevation. The springs were discovered by hunters in the late nineteenth century, who followed deer trails to the water. The name Sairme is commonly interpreted as meaning a place where deer gather, reflecting the long-observed link between wildlife movement, springs, and local knowledge.

The mineral springs of Sairme are carbon dioxide bearing hydrocarbonate sodium calcium waters and, like Borjomi, occur in an attractive landscape and favorable climate. Scientifically, Sairme is notable because within a very small area, tens of meters, several springs discharge waters with differing chemical composition and reported therapeutic properties. In addition, thermal waters occur in the broader area, with a thermal spring site located a few kilometers north of the main mineral spring cluster, supporting spa and sanatorium use.

Together these waters create a resort landscape in which different springs, and combinations of waters, are used for different therapeutic and restorative purposes. Several sanatoriums and hotels operate year-round, and bottling of Sairme water for commercial distribution began in 1945 and continues today (e.g., Zautashvili and Mkheidze 2022). In geoheritage terms, Sairme is a strong example of living hydrogeologic processes expressed directly in the landscape, linked to local history and modern wellness culture. The site is therefore a natural candidate node for georoute development and for the Key Geoheritage Area and UNESCO Global Geopark frameworks for the country (Fig. 5).

Nabeglavi. The Nabeglavi mineral water springs are located in the western part of the Adjara Trialeti ridge, in the Guria region, at approximately 470 m elevation in the Gubazouli River valley. Like Borjomi and Sairme, Nabeglavi is set in a favorable landscape and climate with dense forest cover and proximity to the Black Sea. The natural mineral waters are commonly described as Borjomi type and correspond chemically to carbonated hydrocarbonate sodium waters (e.g. Tsertsvadze 2017).

The springs were identified in the early twentieth century through local observation, similar to the Sairme discovery. Livestock repeatedly visited the forest springs to drink, drawing attention to the water among local residents. The perceived healing properties of the waters, the warm climate, and the scenic setting created strong conditions for development of a balneological and climatic resort and for later expansion into commercial bottling.

Today, Nabeglavi supports resort use and modern bottling infrastructure, and its products are widely consumed in Georgia and exported internationally (Fig. 6). Together, the mineral waters of Borjomi, Sairme, and Nabeglavi represent a distinctive Georgian hydrogeologic heritage associated with the Adjara Trialeti domain and expressed as both landscape and culture. As such, they are important reference sites for national geoheritage programming, candidate geopark narratives, and geotourism routes.

Fig. 6 A general view of the Nabeglavi mineral water bottling plant and its surrounding environment. *Photo* © Georgian travel guide

4 Thermal Waters

Georgia is not as rich in natural thermal springs as it is in mineral waters. Nevertheless, thermal waters have played and continue to play an important role in Georgian life and culture over centuries of use (e.g., Batonishvili 1745; Tsulukidze 2004). The founding traditions of the Georgian capital, Tbilisi, and its long-established sulfur bath culture provide a well-known example, discussed further below (Buntebarth et al. 2006). This chapter also highlights the radon bearing thermal waters of Tskaltubo, which are genetically distinct and are discussed here primarily from a geoheritage perspective, with detailed scientific treatment available in Okrostsvaridze et al. (2025).

Currently, published inventories indicate that Georgia contains a limited number of natural thermal spring localities and a much larger number of wells from which thermal waters are extracted. Reported well temperatures vary widely, ranging from approximately 40 C to 100 C (Zautashvili and Mkheidze 2022). Thermal waters are used locally as a heat source and also represent a potential renewable energy resource, although development remains limited relative to the estimated potential (Sangin et al. 2018).

Thermal waters emerge naturally at the surface primarily within the Adjara Trialeti domain of the Lesser Caucasus, where Eocene age volcanogenic and tectonic structures provide circulation pathways. Examples include Dikhasho, Abastumani, Tbilisi, and Ujarma. Drilled thermal waters occur in additional settings, including Kobuleti, Aspindza, Vardzia, Bisi, Lisi, and other locations. Notable among these is the Bisi thermal mineral water in the Ateni valley (Shida Kartli), with stable temperatures reported near 38 to 39 C and elevated fluoride concentrations. The Lisi wells near Tbilisi are also noteworthy, as they supply several city districts with sulfur bearing thermal water. The highest temperature well waters are reported from the Colchis plain, where deep wells yield waters in the approximate range of 60 C to over 100 C (Zautashvili and Mkheidze 2022).

On the Greater Caucasus, surface thermal waters are generally absent, including in areas influenced by young volcanism. However, mineral spring systems can still produce striking hydrogeologic and geochemical expressions. In the Truso valley of the Kazbegi district, for example, numerous high discharge hydrocarbonate sulfate calcium mineral springs occur, and extensive altered rocks create the impression of an active geothermal landscape, even though spring temperatures are modest and may reflect rapid cooling during ascent (Fig. 7). Calcium rich waters actively precipitate travertine over a broad area, and local carbon dioxide degassing is intense in places. A small feature known locally as Abano Lake is characterized by vigorous gas release that can create a boiling appearance and small splashes at the surface. Local accounts report that under calm conditions gas can pond in low areas and cause wildlife mortality. These features highlight Truso valley as a compelling geohazard and geoheritage locality where fluid flow, mineral precipitation, and gas emissions can be observed directly.

From a geoheritage perspective, special attention is also warranted for the sulfurous thermal waters of Dikhasho, located near the town of Vani in the Imereti region of western Georgia. Geologically, this area lies within a

Fig. 7 Truso Velley. In the foreground is a mineral water stream and travertines. *Photo* © Georgian travel guide

Fig. 8 Hydrogen sulfide thermo-mineral water–Dikhashkho, Imereti region. *Photo* © Kutaisi Travel

complex tectonic transition between the Adjara Trialeti domain and the Transcaucasian microplate. The waters have a stable temperature reported near 38 °C year round and form several visually attractive pools (Fig. 8). Dikhasho has long been associated with therapeutic use and, today, attracts tourists and visitors because the geological process and its human use can be experienced directly in the landscape.

Overall, Georgia's thermal waters form a smaller but emblematic component of the country's living geoheritage. They illustrate how fault-controlled circulation, variable crustal heat, and rock water interaction produce localized sites where geology becomes visible and culturally meaningful. These sites are also natural candidates for inclusion in Georgia's emerging geoheritage programme, including Key Geoheritage Areas and candidate geopark and georoute narratives discussed elsewhere in this volume.

4.1 Sulfur Hot Springs and Baths of Tbilisi

According to legend, King Vakhtang Gorgasali founded Tbilisi in the second half of the fifth century because of warm sulfur springs at this site. The name Tbilisi derives from the Georgian word tbili, meaning warm. These springs may be less visually dramatic than some of the major landscapes described elsewhere in this volume, but they have played a prominent role in the life of the city and in Georgian history for more than 1500 years. The bathhouses built on these springs became a distinctive cultural center of the old city. The Baths District also illustrates the long-standing tradition of coexistence in Tbilisi, where a church, a synagogue, and a mosque occur within a small area. Today, the Baths District, together with the Dabakhana river canyon, remains one of the city's best-known landmarks and a focal point of tourism (Figs. 9 and 10).

Geologically, Tbilisi lies within the eastern segment of the Adjara Trialeti Eocene fold-thrust belt, the northernmost orogenic domain of the Lesser Caucasus. The belt formed during the Eocene and extends approximately 350 km from the Tbilisi area to the Black Sea. Adjara Trialeti is composed mainly of trachytic and trachyandesitic volcanic sedimentary rocks, intruded by syenite, monzonite, and quartz monzonite plutons (e.g. Okrostsvaridze et al. 2021). The warm sulfur springs discharge along a deep tectonic fault that cuts these rocks and forms a small canyon on the right bank of the Mtkvari River. The Dabakhana (Leghvtakhevi) River flows through this canyon and, in the final reach before joining the Mtkvari, warm sulfur springs emerge on both banks over a distance of approximately 200 to 300 m.

Chemically, the spring waters are described as bicarbonate sulfate sodium and bicarbonate sulfate calcium types, with typical temperatures in the range of 30 to 40 C (Tsulukidze 2004; Buntebarth et al. 2006; Tsertsvadze 2017). Discharge is sufficiently large that the Dabakhana valley is commonly pervaded by a sulfur odor. Although much of the area is now occupied by bathhouses and associated buildings that reduce the visible area of natural discharge, the springs remain hydrologically active.

From a geoheritage perspective, the Tbilisi sulfur springs and bath system provide a clear example of how a geologic process, expressed at the surface, can shape settlement patterns and create a distinctive cultural, social, and economic landscape. As an accessible urban geosite, the Baths District also offers strong potential as a central node within future Key Geoheritage Area and UNESCO Global Geopark and georoute narratives, linking the Lesser

Fig. 9 General view of the Tbilisi Sulfur Baths (domed brick buildings). In the background is the old Tbilisi district, and above is the historical Narikala fortress. *Photo* © Georgian travel guide

Fig. 10 General view of the Dabakhana River valley. Warm sulfur springs flow into this valley on both banks. Above right, the entrance to the sulfur baths of the so-called "Colorful Bath" and the balcony houses of Old Tbilisi can be seen. *Photo* © Tbilisigovege

Caucasus geodynamic setting directly to everyday human experience.

4.2 The Radon-Containing Thermal Springs of Tskaltubo

Within Georgia's wider thermal water network, the Tskaltubo spa resort in western Georgia stands out as the only known locality where warm, low-mineralized springs containing radon discharge within a low relief karst depression near sea level. Written accounts of the healing properties of the waters date back at least to the early medieval period, and the resort became the main balneological center during the Soviet era, when grand sanatoriums and spa facilities were constructed for visitors from across the Union. Although many of these buildings are now partly abandoned or in ruins, they remain an important part of the cultural landscape and contribute to the site's well-defined sense of place.

Tskaltubo is located in the Imereti region roughly 9 km northwest of Kutaisi and about 90 km inland from the Black Sea, between the Rioni and Tskhenistskali river valleys. The springs occur at elevations approximately 100 to 150 m above sea level, within a mild humid subtropical climate. The resort is often described as an abandoned paradise because of the contrast between the monumental spa architecture and the current partial disuse of the Soviet era complexes, even as some facilities continue to operate and various rehabilitation concepts have been proposed (Fig. 11).

Fig. 11 Ruins of one of the Soviet-era sanatoriums in the subtropical forest of Tskaltubo. *Photo* © Kutaisi Travel

Fig. 12 One of the modern radon water baths at the Tskaltubo Spa resort. *Photo* © Kutaisi Travel

Brief Geology of the Tskaltubo Area. Geologically, Tskaltubo lies near the transition between the southern slope of the Greater Caucasus and the Colchis lowland, in an area of gently folded Mesozoic rocks cut by faults and dykes. The local stratigraphy includes Jurassic volcanogenic units, a thick Lower Cretaceous limestone aquifer, and younger volcanic and intrusive rocks. Quaternary sediments fill parts of the depression and the Tskaltuboskali river valley. Historical engineering works altered the natural spring discharge area by lining and infilling parts of the former riverbed to improve capture and management of spa waters (Fig. 12). As a result, much of the original natural discharge is now masked, and direct observation of springs in their former natural state is limited. Today, radon bearing warm waters are experienced primarily through the managed spa system rather than as open spring outcrops.

Radon bearing warm waters. The Tskaltubo waters are characterized by modest mineralization and warm temperatures suitable for balneological use. Radon is the feature that distinguishes Tskaltubo from other Georgian thermal sites, and it is closely tied to the subsurface geology and circulation pathways. In brief, the most widely discussed conceptual framework involves deep circulation through fractured and karstified limestone aquifers combined with structural focusing along faults and dyke related barriers that channel upflow into the depression. A key question has long been the source of radon enrichment and the mechanism of warming, including the unusual observation that waters may warm toward the surface rather than follow a simple deep geothermal model.

The detailed analytical evidence, measurements, and the proposed revised conceptual model for radon sourcing and heating are presented in Okrostsvaridze et al. (2025). In the context of this geoheritage volume, we emphasize the interpretive outcome rather than reproducing the full technical argument. The main conclusion is that Tskaltubo's radon bearing warm waters are best understood as a product of multiple interacting geological factors, including a deep source domain with elevated natural radioactivity, efficient vertical transport pathways through faults and karst systems, mixing with groundwater circulating in the thick limestone aquifer, and structural focusing that delivers these waters into the low relief depression (Fig. 13).

Geoheritage, KGA and UGGp significance. The Tskaltubo radon bearing healing springs are a geoheritage site where deep geological processes, mineral waters, and a long spa tradition intersect. As a landscape, Tskaltubo integrates three levels of heritage in one place: the subsurface geologic system that produces a rare water type, the surface expression of that system through an internationally known spa resource, and the architectural and social legacy

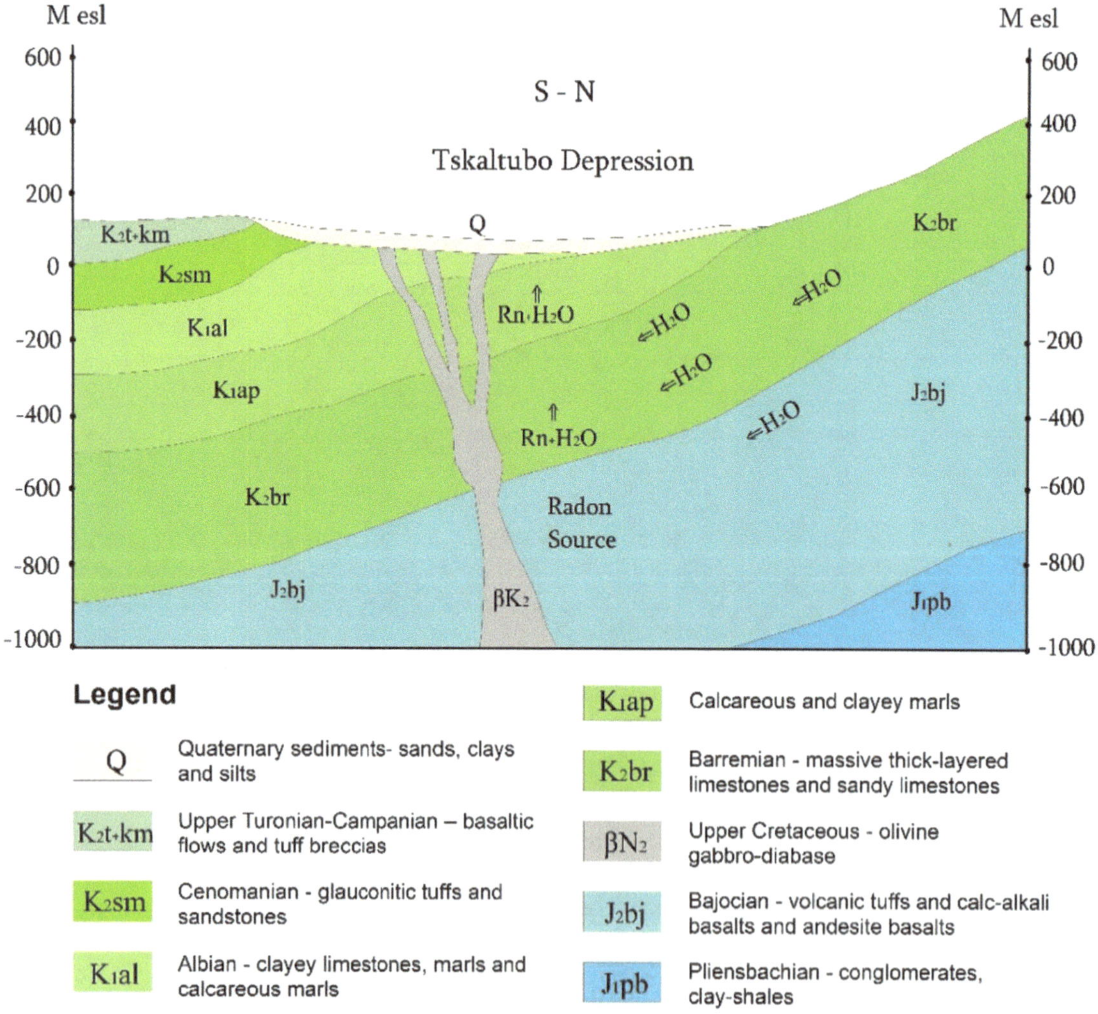

Fig. 13 Conceptual geological model of the formation of radon-containing healing springs, of the Tskaltubo depression (Okrostsvaridze et al. 2025)

of a twentieth century resort city set within a subtropical park.

Within Georgia's emerging Key Geoheritage Area inventory, Tskaltubo is a strong candidate for classification as an active hydrogeological KGA representing radon bearing warm mineral waters in western Georgia. It is the only documented Georgian locality where this combination of attributes occurs in a low relief near sea level setting. The site also sits naturally within a broader Imereti geoheritage corridor that includes caves and karst systems, river gorges, and low mountain relief landscapes, allowing integrated interpretation of water, rock, and landscape evolution.

From the UNESCO Global Geopark perspective, Tskaltubo is unlikely to form a standalone geopark, but it has clear potential as a primary site within a wider Imereti geopark proposal or thematic georoute. Its scientific value lies in the clearly interpretable linkage between geology and water, its educational value in the accessibility of the resort and park setting, and its cultural value in the long tradition of healing waters and the iconic sanatorium architecture. A future Imereti geopark or geotourism network could integrate Tskaltubo as the principal node in a healing waters and karst landscapes trail, linking spa culture with underlying geological processes in a way that is distinctive to Georgia.

Tskaltubo demonstrates how a rare hydrogeologic phenomenon can become a nationally significant cultural landscape. The site's radon bearing warm waters reflect deep crustal controls and focused groundwater circulation, while the resort and its architectural legacy record the ways societies build meaning, identity, and economy around geological processes. For Georgia's national geoheritage programme, Tskaltubo is therefore both a main hydrogeological geosite and a practical candidate KGA, with strong potential to anchor geopark and georoute storytelling in the Imereti region.

5 Conclusion

Mineral and thermal waters in Georgia are not only a hydrogeologic phenomenon. They are also a surface expression of deeper geological evolution within an active collision orogen. Their distribution and characteristics reflect the combined influence of collisional tectonics, deep fault systems that provide circulation pathways, post-collisional magmatism and inherited heat, and the development of intermontane and lowland basins that organize groundwater flow. This combination of geological prerequisites has produced an unusual diversity of mineral and thermal waters that can be understood as a living component of Georgia's geoheritage.

In contemporary Georgia, mineral waters such as Borjomi, Nabeglavi, and Sairme represent more than a health-related resource and economic product. They are also part of cultural identity and place meaning. In Georgian folklore and historical memory, healing waters are expressed through the motif of the water of immortality. While folkloric in form, this motif reflects the real social and therapeutic roles that springs have played in daily life, settlement patterns, and national narratives over long time periods.

From a geoheritage perspective, Georgia's mineral and thermal waters should be viewed as dynamic natural systems with combined scientific, educational, cultural, and social value. Many sites also function as accessible geosites where subsurface processes can be interpreted directly through springs, baths, and resort landscapes. In current conditions, special attention should be given to protection and sustainable management, including maintaining water quality and discharge integrity, managing visitor pressure at popular sites, and ensuring that development and extraction do not compromise the long-term functioning of these hydrogeologic systems. These steps are essential if mineral and thermal waters are to remain a durable pillar of Georgia's national geoheritage program and its future Key Geoheritage Area and geopark and georoute initiatives.

References

Batonishvili V (1745) Description of the kingdom of Georgia. Publishing house Sabchota Sakartvelo, Tbilisi, 1973, pp 1–162 (in Georgian)

Buachidze SG (1987) Hydrogeology of Georgia. Metsniereba, Tbilisi, pp 1–187 (in Russian)

Buntebarth G, Chelidze T, Melikadze G (2006) The hydrothermal system of Tbilisi, Georgia. In: urban groundwater management and sustainability. NATO Science Series, vol 74, pp 417–426

Gamkrelidze I, Okrostsvaridze A, Maisadze F, Basheleisvili L, Boichenko G, Skhirtladze I (2019) Main features of geological structure and geotourism potential of Georgia, the Caucasus. Modern Environ Sci Eng 5(5):422–442. https://doi.org/10.15341/mese(2333-2581)/05.05.2019/010

Gamkrelidze I, Okrostsvaridze A, Koiava K, Maisadze F (2021) Geotourism potential of Georgia, the Caucasus. Springer, pp 1–81

Gvakharia V (2006) Investigation of the mineral waters of Georgia, vol 2. Georgian Engineering News, pp 263–269

Gudjabidze GE, Gamkrelidze IP, Abesadze GN et al (2003) Geological map of Georgia 1:500,000 scale. Georgian State Department of Geology and National Oil Company, Tbilisi, Georgia

Javakhishvili D (1939) On the hydrogeological work carried out on the territory of Tskaltubo in 1937–1939. Report of the Geological Department of Georgia, Tbilisi, pp 1–142 (in Russian)

LaMoreaux P, Tanner J (2001) Springs and bottled water of the world: ancient history, source, occurrence, quality and use. Springer, Berlin, Heidelberg, New York, pp 1–279. http://books.google.com/books?id=sjEoBmfU

Okrostsvaridze A, Chang YH, Chung SL, Rabayrol F, Boichenco G, Gogoladze S (2021) Paleozoic xenoliths in Eocene plutons: evidence for the destruction of pre-Jurassic crystalline basement beneath the Adjara-Trialeti belt, lesser Caucasus. Geol Carpath 72(4):299–315

Okrostsvaridze A, Tormey D, Gabrielashvili D (2025) New insight into formation of Tskaltubo radon-containing healing springs, Georgia, the Caucasus. J Geol Earth Sci 11(1):11–22. https://doi.org/10.18178/ijges.11.1.17-22

Porowski A (2019) Mineral and thermal waters. In: Environmental geology. Springer Publisher, pp 149–163. Chapter 3

Tsertsvadze N (2017) Mineral waters of Georgia. Nekeri Publishing House, Tbilisi, pp 1–242. (in Russian)

Tsulukidze IP (2004) Groundwater of Georgia–the white wealth of the 21st century. Min J 4:20–23. (In Georgian)

Sangin S, Buntebarth G, Weller A, Melikadze G (2018) Temperature gradient measurements in hydrothermal areas of Georgia. Int J Terr Heat Flow Appl Geotherm 1(1):14–17. https://doi.org/10.31214/ijthfa.v1i1.7

Zautashvili B, Mkheidze B (2022) Hydrogeology and mineral waters. In: Bolashvili N, Neidze V (eds) The physical geography of Georgia. Geography of the physical environment. Springer, Cham. https://doi.org/10.1007/978-3-030-90753-2_3

Geoheritage Factors in the Development of Viticulture in Georgia: The Homeland of Wine

Avtandil Okrostsvaridze, Daniel Tormey, Giorgi Boichenko, and Rabi Gabrielashvili

Abstract

Georgia is increasingly recognized by modern scholars as the probable homeland of wine, a claim supported by extensive archaeological findings, ancient written records, and the presence of over 500 endemic grape varieties. Recent research highlights that, beyond favorable climatic conditions, the development of viticulture is also shaped by geological factors such as soil and subsoil composition, tectonic structure, and regional geodynamic regimes. Together, these elements create the "terroir effect," a complex system that significantly influences the characteristics of wine. This study explores the role of geology in the development of viticulture and traditional winemaking in Georgia. It demonstrates that, in addition to its advantageous climate, Georgia's geological diversity—arising from its location within the collage-structured Caucasus orogen, which includes the Greater and Lesser Caucasus and the Transcaucasian massif—provides essential conditions for high-quality wine production. The resulting diversity of lithologies and weathering products has given rise to a wide variety of soils, which in turn support Georgia's rich viticultural heritage. The analysis reveals a spatial correlation between tectonic junctions and the distribution of unique endemic grape varieties that yield exceptional wines such as Khvanchkara and Kindzmarauli. These zones are often enriched with macro-and microelements due to underlying fault systems, enhancing soil fertility. Additionally, the paper emphasizes the importance of local clay deposits, which are critical for producing qvevris—the traditional clay vessels essential to Georgia's ancient winemaking technique. The existence of high-quality clay sources has played a crucial role in sustaining this cultural and technological heritage.

Keywords

Georgia · Geoheritage · Influences · Viticulture · Winemaking · Qvevri

1 Introduction

There is broad scholarly agreement that wine has played a pivotal role in the development of human civilization and continues to serve as a central element in ritual and celebratory practices worldwide. Its cultural significance is reflected in sacred texts, including the Bible, where it is mentioned alongside water and described as a "gift from God." Debate over the geographic origins of wine has persisted for centuries. A growing body of research supports the view that Georgia is the likely birthplace of viticulture (e.g. Johnson 1989; McGovern et al. 2017). As a medicine, social lubricant, mind-altering substance, and highly valued commodity, wine became the focus of religious cults, medicines, cuisines, economies, and society in the ancient Near East. It was among humans' first agricultural practices. Wine culture subsequently spread around the globe. Viniculture illustrates human ingenuity in developing horticultural and winemaking techniques, such as domestication, propagation, selection of desirable traits, wine presses, suitable containers and closures (McGovern et al. 2017). Georgia is identified as the earliest wine making area by archaeological evidence, ancient textual references in Persian, Greek, and Georgian sources, and the extraordinary diversity of indigenous grape varieties found here.

The earliest biomolecular archaeological and archaeobotanical evidence for grape wine and viniculture from

A. Okrostsvaridze (✉) · G. Boichenko · R. Gabrielashvili
Institute of Earth Sciences, Ilia State University, Tbilisi, Georgia
e-mail: avtandil.okrostsvaridze@iliauni.edu.ge

D. Tormey
Catalyst Environmental Solutions, Santa Monica, USA

IUCN World Commission on Protected Areas, Gland, Switzerland

A. Okrostsvaridze and D. Tormey (eds.), *Geoheritage of Georgia*, Geoheritage, Geoparks and Geotourism,
https://doi.org/10.1007/978-3-032-27639-1_25

Georgia, ca. 6000–5800 BC during the early Neolithic Period, comes from archaeological, archaeobotanical, climatic, and chemical methods to newly excavated pottery fabrics and amphorae materials from two sites in Georgia in the South Caucasus (McGovern et al. 2017). More recent genetic analysis validates these findings and includes grape domestication in Western Asia. These data also identify challenging climate conditions in the Pleistocene ice ages as causing fragmentation of the habitat for wild grape ecotypes leading to a wide variety of grape types and geological associations, particularly in Georgia. Domestication occurred after the conclusion of the last ice age in the Holocene, at the beginning of human agricultural practice, to yield grape cultivation leading to wine production.

Georgia's qvevri winemaking tradition is found nowhere else in the world. Rather than age wine in oaken casks, Georgian wine is aged in clay pots: the region's unique qvevri winemaking tradition is found nowhere else in the world. These large-capacity jars, some of the earliest pottery made in the Near East, likely served as combination fermentation, aging, and serving vessels. They are the most numerous pottery type at many sites comprising the so-called "Shulaveri-Shomutepe Culture" of the Neolithic period, which extends into western Azerbaijan and northern Armenia. The discovery of early sixth millennium BC grape wine in this region is crucial to the later history of wine in Europe and the rest of the world (McGovern et al. 2017).

In addition, some linguists propose that the word "wine" in various languages around the world originates from the Georgian language "ghwino", which serves as further evidence supporting Georgia's role in this ongoing debate (e.g. Gamkrelidze and Ivanov 1995).

Moreover, the Georgian language contains numerous specific words for wine—such as Machari (newly fermented wine), Zedashe (sacramental wine), Shumi (wine fermented without pomace), and Shamani (wine from a second pressing of the pomace)—pointing to an ancient, deeply rooted winemaking tradition. Hugh Johnson, who was the first to suggest that Georgia may be the homeland of wine, emphasizes that great wines have their roots in the customs, traditions, and history of their home ground (Johnson 1989). He also poses a rhetorical question that highlights the cultural depth of winemaking: "Why have some places, and some people, made great wine, and what in the history of the process explains the difference between the great and the merely good?"

In response to this question, this paper puts forth that for Georgian wine, the answer lies both in the natural environment in which the vineyard is cultivated, both climatic and geologic: its terroir. And it lies in the the vessel in which the wine is fermented likewise influences its quality: in Georgia this is the unique use of the qvevri container.

Considering Georgia's terroir, the territory of Georgia is located between the Black and Caspian Seas, lying between 40° and 45° northern latitudes—an ideal climatic belt for viticulture. The country is protected from cold northern air masses by the Greater Caucasus range, while the Lesser Caucasus shields it from the hot air masses of the south. In the central part of Georgia lies the Surami Ridge, oriented meridionally, with elevations ranging around 2000 meters. To the west of this ridge, the land descends toward the Black Sea, dropping by approximately 2000 meters; to the east, it slopes toward the Caspian Sea, with a drop of about 1700 meters up to the Azerbaijani border.

As a result, Georgia features two distinct climatic zones: the west, influenced by the humid air masses of the Black Sea, and the east, affected by the dry air masses from the Caspian Sea. Due to these conditions, the most favorable zones for viticulture are found at elevations between 500 and 800 meters above sea level in western Georgia, and between 400 and 700 meters in the eastern part of the country.

And Georgia's geology is tremendously complex leading to a wide variety of rock and soil types for a relatively small area. The bedrock geology and the resulting soils beneath the vineyard are equally important to the character of the produced wine. This includes the geodynamic setting of Georgia, the complexity of which yields a wide variety of chemical and mineral composition, and structure in which a given region is located. In this chapter, we examine the geological factors that, to a significant extent, determine the optimal conditions for the advancement of viticulture. All of the above factors are combined in the French language into one word, "terroir," (or land), which, together with the microclimatic zone, includes the relief of the area, botanical environment, the bedrock geology and resulting soils. The role of geology is much more complex than the bedrock structure, which we will discuss in detail below.

Moreover, as mentioned above, winemaking in Georgia has historically been linked to the use of qvevri, so it is important to analyze the quality and quantity of clay deposits in this region. Without high-quality clay, from which qvevris are made, the Georgian winemaking tradition would not have developed as it has. Qvevri's shape, porosity, and mineral content influence fermentation and maturation, linking the geological to the viticultural.

There is no doubt that favorable climatic conditions combined with advantageous geological factors contributed to the emergence of indigenous wild grape varieties in the territory of Georgia. These were later so thoroughly cultivated by the local population that they were even endowed with religious and sacred significance. Subsequently, at the beginning of the fourth century AD, the adoption of Christianity as the official religion in Georgia further stimulated the development of winemaking in the country, much as in in Europe. Accordingly, Christianity and the highly developed culture of winemaking represent Georgia's closest connection to European civilization. The connection between Georgia's landscapes and its wines is both physical and deeply cultural, rooted in traditions passed down for millennia.

The objective of this paper is to summarize the ancient history, traditions, or primacy of Georgian wine production, and then to evaluate the associated geological factors—including qvevri clay—that have collectively defined its uniqueness. The study shows that, in addition to favorable climatic conditions for winemaking, Georgia also possesses suitable geological conditions for producing high-quality wine, as well as sufficient reserves of clay of appropriate quality for the production of high-quality qvevris. Consequently, Georgian wine currently holds an honorable place among the world's finest wines.

2 A Brief History of Georgian Winemaking

Despite the development of viticulture and winemaking, there are neither written nor oral records indicating when vine cultivation began in Georgia. In this country, information about the ancient history of viticulture is encoded only in archaeological artifacts and various branches of art, including architectural monuments in the form of vine and grape carvings. This most likely indicates that viticulture and winemaking in Georgia have a long history and were, in fact, an inseparable part of life. This is confirmed by the great diversity and number of cultivated vine varieties in Georgia. It may seem incredible, but it is an established fact that the total number of cultivated vine varieties across all regions of Georgia exceeds 50 (Annual Report 2025).

The archaeological sites provide valuable information about the ancient history of winemaking in Georgia. In southern Georgia, during the archaeological excavations of the "Khramis Gora" site of the Shulaveri Neolithic culture of the sixth millennium BC—conducted in the 1970s (Kiguradze 1986)—a large quantity of fragmented clay vessels were discovered, among them, with high probability, wine storage containers adorned with ornamental depictions of grapes (Fig. 1). Undoubtedly, this unique archaeological artifact is significant evidence of the ancient history of winemaking in Georgia.

Substantial insights into the earliest history of Georgian winemaking derive from the Gadachrili Gori archaeological site of the aforementioned culture. Situated some 50 km south of Tbilisi, its excavations have been under way since 2015. Here, numerous clay-vessel fragments, chemical biomarkers of wine and vine pollen have been recovered. Their concurrence strongly suggests that the site's inhabitants were already producing wine around 8000 years ago. Organic acids characteristic of wine, extracted from the walls of these ceramic vessels, were radiocarbon-dated at the Weizmann Institute (Israel) using the C^{14} method. The resulting dates span 6000–5800 BC, roughly 600 years earlier than the previously oldest identified wine traces in the Zagros Mountains (e.g. McGovern et al. 2017). This constitutes compelling evidence that Georgia is very likely one of the world's earliest centres of wine origin.

Fig. 1 Wine vessel with a relief image of grapes. Archaeological site of the "Khramis Didi Gora", sixth millennium BC. Georgian National Museum. Photo © Mirian Kiladze

Later, in the early Middle Ages, records by ancient Greek, Persian, and Arab travelers began to mention the well-developed state of viticulture and winemaking in Georgia. According to these accounts, vineyards in Georgia were initially trained on tall trees, a practice that was gradually replaced by vineyards cultivated on lower trees and short trellises.

After Christianity was declared the state religion of Georgia in the early fourth century, viticulture and winemaking developed even more extensively. However, this field experienced several major periods of crisis in Georgia. The first is associated with nearly 300 years of Mongol domination in the 12th–15th centuries; the second—with the invasions by Iran in the seventeenth century; and the third—with the occupation of Georgia by Russia in the nineteenth century. It is noteworthy that all three conquerors, in their attempts to destroy the Georgian state, not only carried out numerous raids and brought immense suffering,

Fig. 2 Rkatsiteli vineyards in Kakheti, on the Alazani Valley plain. Photo © 2025AgroNews.ge

but also massively uprooted vineyards. This reality serves as yet another piece of evidence of the great economic importance that the cultivation of the wine held for Georgian statehood, and the importance for heritage designation in keeping the tradition alive through these external challenges.Despite numerous periods of crisis, winemaking today is one of the main branches of agriculture in Georgia and is developing intensively.

According to Georgia's National Wine Agency (Annual Report 2025), 25 wines of local origin currently have international registration. Of these, 19 are produced in eastern Georgia, in the Kakheti region.

In the Alazani and Iori Valleys of this region, vineyards of the best local grape varieties in Georgia are cultivated. Among them, notable varieties include Rkatsiteli, Saperavi (Fig. 2), Kakhuri Mtsvane, Kisi, Khikhvi, and others.

According to data from the aforementioned agency, in 2022 Georgia exported wine worth 253 million USD to 69 countries, ranking 19th in the world among wine-exporting countries. In 2024, 320,000 tons of grapes were processed for wine production in this small country, which will likely lead to a significant increase in exports in the future (Annual Report 2025). Against the background of growing global competition in the wine market, the production of qvevri wine in Georgia is perhaps the most promising direction—especially since the country possesses large reserves of high-quality qvevri clay.

In Georgia, grapes are not used solely for making wine or alcoholic beverages—the culture of grape usage is much broader and more diverse. For example, in this country, grape juice is fermented for a long time and transformed into a thick liquid called badagi. Badagi can be stored for extended periods and, when diluted with water, can be consumed as a beverage. However, its primary use is in the preparation of churchkhela, an ancient Georgian confection. Churchkhela (Fig. 3) is made from badagi, wheat flour, and walnuts, cooked at high temperatures. It is considered a delicacy, yet it is also highly caloric and can be stored for more than a year (Fig. 3). The tradition of making churchkhela is nearly as old as Georgia's recorded history and thus serves as additional evidence of the ancient origins of viticulture in the country.

The tradition of making *churchkhela* is nearly as old as the history of Georgia itself and thus serves as additional evidence of the country's ancient winemaking heritage. Notably, *churchkhela* played an important role in preserving the cultural identity of this small nation among great empires, as it was one of the primary food supplies for Georgian soldiers during military campaigns.

3 The Georgian Qvevri

The Georgian word *"qvevri"* refers to a clay vessel fired at high temperatures, designed exclusively for fermenting and storing wine, and traditionally buried underground. It has a characteristic egg-like shape, unique to this vessel type, with a minimum capacity of 20–30 liters and a maximum that can reach 4000–5000 liters (Fig. 4). There is no definitive historical record of when qvevri production began; however, the vessel's technologically refined form serves as indirect evidence of a long evolutionary development. While many ancient civilizations used fired clay vessels for various purposes, the practice of burying such vessels underground specifically for winemaking and wine storage is only used in Georgia—and remains in use to this day. This distinctive feature defines the qvevri's uniqueness and, along with the traditional Georgian method of fermenting wine on the grape pomace (*chacha*) inside the qvevri, has significantly

Fig. 3 Georgian churchkhela and one of its main ingredients—walnuts. Photo © 2025AgroNews.ge

Fig. 4 Classical Georgian qvevris of various sizes. Photo © Georgian Wine Association

Fig. 5 A modern Georgian *Marani*. Photo © 2025AgroNews.ge

Fig. 6 An ancient Georgian satsnakheli—a traditional vessel used for pressing grapes. Photo © Georgian wine association

strengthened the widely supported view that Georgia is the cradle of wine.

In Georgia, *qvevris* are traditionally buried in the ground within specially designated, meticulously maintained wine-making structures called *marani* (Fig. 5). These subterranean vessels, which may contain grape juice together with grape pomace (*chacha*) or with only a small amount of it, provide ideal technological conditions for natural wine fermentation. These conditions include a nearly constant underground temperature (12–14 °C), which is essential for both fermentation and aging. The egg-like shape of the *qvevri* promotes optimal dynamics for fermentation and maturation of the grape juice, while the porous, kiln-fired clay of the vessel ensures aeration during these active chemical processes. It is worth noting that *qvevris* are traditionally handmade—a laborious and demanding craft that requires exceptional knowledge and experience. One might say that their makers are true masters, producing a unique cultural artifact that deserves deep respect.

Due to its unique qualities, the Georgian *qvevri* was inscribed in 2013 by UNESCO on the Representative List of the Intangible Cultural Heritage of Humanity, recognizing it as a cultural monument in need of safeguarding. Currently, Georgian *qvevri* wine has achieved wide international recognition and popularity. As a result, *marani*-style cellars equipped with Georgian *qvevris* have begun to appear in numerous countries, where traditional *qvevri* wine-making practices are being adopted.

The *qvevri* phenomenon enables the production of tannic yet "soft" wines through a unique method known as the Kakhetian technique—a method with no true analogue anywhere else in the world. Kakheti, located in eastern Georgia, is the country's most prominent wine-making region. The essence of the method can be summarized as follows: in a *satsnakheli* (a traditional winepress used before transferring the juice into the *qvevri,* see Fig. 6), the pressed grape juice is poured into the *qvevri,* along with the full amount of remaining grape pomace *(chacha).*

Once placed in the open *qvevri*, this mixture quickly begins alcoholic fermentation. After the fermentation process is complete, the grape pomace (*chacha*) settles to the bottom of the vessel, at which point the *qvevri* is sealed. In early spring, the wine is drawn from the *qvevri* and aged for about one year in various containers, including oak barrels. The result of this process is a unique, tannic *qvevri* wine made according to the Kakhetian method, distinguished by its exceptional taste and aroma. These qualities are most likely the result of ionic exchange between the wine and the clay walls of the *qvevri.* However, *qvevri* wine would not exist without Georgia's remarkable geological heritage—its natural deposits of *qvevri*-grade clay.

4 Qvevri Clay Deposits in Georgia

Clay is a soft, water-rich secondary aluminosilicate rock that forms either through the weathering of rocks or via hydrothermal alteration. Chemically, clay is primarily composed of the following oxides: SiO_2, Al_2O_3, Fe_2O_3, TiO_2, CaO, MgO, K_2O, and Na_2O. These oxides form various clay minerals, of which there are dozens; however, the principal clay minerals are kaolinite ($Al_2Si_2O_5(OH)_4$) and montmorillonite $(Na,Ca)_{0.33}(Al,Mg)_2(Si_4O_{10})(OH)_2{\cdot}nH_2O$ (e.g. Galán, Ferrell 2013). Clay mineral particles generally do not exceed 0.01 mm in size. However, as modern Raman spectroscopy has shown, most particles are significantly smaller, typically ranging between 2 and 4 micrometers (Kloprogge 2017).

Clay typically contains terrigenous or relict minerals such as quartz, feldspars, and micas, whose combined content should generally not exceed 35–40% of the rock. In addition to these, clays often include gypsum-group minerals with the chemical formula $CaSO_4{\cdot}2H_2O$, which exhibit high water absorption capacity (55–65%). Gypsum minerals are among the most undesirable components in qvevri-grade clay, as they do not vitrify during firing and undergo significant volumetric expansion when exposed to moisture—factors that can compromise the structural integrity of the vessel. Another undesirable component is organic carbon, which produces so-called "black spots" during firing and weakens the mechanical strength of the ceramic. Soluble salts also pose a threat, as their dissolution can lead to cracking or structural degradation of the qvevri. In some cases, qvevri clay may also contain trace amounts of radioactive elements, which are geochemically unfavorable. Overall, the geochemical quality of clay is directly related to the concentration of these harmful components—the lower their abundance, the higher the suitability of the clay for ceramic production.

Clay dissolves and softens easily in water, making it possible to shape it into virtually any form. Most importantly, when fired at high temperatures, it undergoes a transformation and becomes solid and durable. Technologically, it is after firing that clay becomes impermeable to water, environmentally inert, and resistant to external influences. Due to these unique properties, clay can be used to produce qvevris as well as a wide range of other ceramic products, including roof tiles and bricks.

Georgia hosts numerous significant and diverse clay deposits (Okrostsvaridze 2014), however, as noted earlier, qvevri clay must possess a specific set of distinctive properties. According to historical sources, many qvevri clay deposits were once actively exploited throughout the country, but a few were particularly renowned—such as the Anthoki deposit in Kakheti, the Utslevi deposit in Kartli, and the Shrosha deposit in Imereti (e.g. Batonishvili 1745). Unfortunately, only a few deposits are currently in use for qvevri clay extraction in Georgia. Among the most well-known are Vardisubani, Tkemlovani, and Satsablia.

The Vardisubani qvevri clay deposit is located in eastern Georgia in the Kakheti region, on the eastern slope of the Tsivi-Gombori Ridge, within the Telavi Municipality. The deposit represents the weathering product of Upper Cretaceous–Paleogene terrigenous-sedimentary rocks that make up this ridge. It is noteworthy that in Kakheti, qvevri clay deposits are exclusively associated with the Tsiv-Gombori Ridge. The Vardisubani clay is predominantly fine-grained and composed mainly of clay minerals. Feldspar content reaches up to 9%, while quartz content is approximately 35%. The concentrations of major oxides in this deposit fall within the following percentage ranges: SiO_2 from 63.5% to 67.3%, Al_2O_3 from 15.2% to 15.92%, TiO_2 from 0.78% to 0.81%, Fe_2O_3 from 6.99% to 7.29%, CaO from 1.51% to 1.9%, MgO from 2.02% to 2.52%, K_2O from 2.52% to 2.62%, and Na_2O from 1.05% to 1.1% (Uebel and Petzold 2017).

The Tkemlovani qvevri clay deposit is located in western Georgia, in the Chiatura Municipality of the Upper Imereti region. It represents the weathering product of Upper Cretaceous–Paleogene terrigenous-sedimentary rocks. The deposit is mainly composed of medium-grained minerals, with quartz and feldspar content ranging between 38% and 49%. Among the feldspars, potassium feldspar is dominant. The proportion of clay minerals ranges from 46% to 52%, with kaolinite prevailing within the clay fraction. The chemical composition of the Tkemlovani deposit differs from that of Vardisubani. The major oxide contents in Tkemlovani clays fall within the following percentage ranges: SiO_2 from 56.67% to 62.98%, Al_2O_3 from 16.74% to 18.19%, TiO_2 from 0.78% to 0.97%, Fe_2O_3 from 5.67% to 7.70%, CaO from 0.43% to 1.32%, MgO from 1.72% to 2.85%, K_2O from 1.54% to 2.852%, and Na_2O from 0.22% to 1.07% (Uebel and Petzold 2017).

The Satsable qvevri clay deposit is located in western Georgia, in the Zestafoni Municipality of the Imereti region. The deposit formed through the weathering of Paleozoic microcline granites of the Dzirula Massif (Okrostsvaridze and Tormey 2013) and is composed of fine-grained minerals. Unlike the Vardisubani and Tkemlovani clay deposits, the Satsable deposit shows an increased quartz content, reaching up to 44%. The concentration of clay minerals is also elevated, ranging from 51% to 58%. Among the clay minerals, fine-grained illite predominates, in contrast to the kaolinite-rich clays of the other two deposits. The major oxide contents in the Satsable clay fall within the following percentage ranges: SiO_2 from 55.87% to 62.60%, Al_2O_3 from 16.84% to 18.87%, TiO_2 from 1.02% to 11.24%, Fe_2O_3 from 8.45% to 12.05%, CaO from 0.22% to 0.74%, MgO from 0.74% to 1.50%, K_2O from 0.96% to 1.01%, and Na_2O from 0.06% to 0.37% (Uebel and Petzold 2017).

Although all three discussed deposits contain qvevri-grade clays, they differ geochemically from one another. The Vardisubani deposit exhibits the highest concentration of SiO_2, ranging from 63.5% to 67.3%, but also the lowest concentration of Al_2O_3, which varies between 15.2% and 15.92%. Iron oxide (Fe_2O_3) is present in the clays of all three deposits, though its concentration is highest in the Satsable deposit, where it exceeds 12%. It is worth noting that the reddish color of fired qvevris is primarily due to the presence of iron oxide in the clay.

Qvevris also contain various trace elements in differing concentrations, which can influence the sensory characteristics of the wine. The fact that the chemical composition of the qvevri has a notable effect on the taste of the wine is further evidenced by the requirement that qvevri clay must be kneaded exclusively with clean water. Otherwise, despite the high-temperature firing process, the vessel may continuously impart an unpleasant taste to the wine.

5 Soils of Georgia

Alongside climatic conditions, soils are among the most crucial factors in the development of viticulture and winemaking. Soils are the products of rock weathering and erosion and therefore represent a direct component of geological heritage. Within the small territory of Georgia (69,700 km^2), rocks of extremely diverse ages (ranging from the Early Paleozoic to the present) and chemical compositions (from ultrabasic to acidic, including carbonate rocks and others) are concentrated (e.g. Okrostsvaridze et al. 2022). As a result, the country's soils are remarkably diverse, which directly affects agricultural yields, including grape cultivation. Notably, the territory of Georgia began its final uplift from the shallow seas fringing the ancient Tethys Ocean during the Late Neogene (e.g. Wang et al. 2015), making its soils geologically very young. For instance, the soils of the Alazani Valley—where the majority of Georgia's vineyards are located—are less than one million years old.

The study of soils in Georgia has a long history. The first written description appears in the work of the renowned Georgian scholar Vakhushti Batonishvili (1745), titled "Description of the Kingdom of Georgia." Scientific investigation of Georgian soils began in the late nineteenth century, primarily driven by the need to select suitable soils for vineyards. Over time, both individual researchers and scientific institutions have studied Georgian soils; however, these studies have largely been conducted without considering the "bedrock" formations that underlie them—despite the fact that such rocks have a direct impact on soil types and on the processes of their subsequent evolution.

Currently, a vast amount of information is available on the soils of Georgia, both in Georgian and in the international scientific literature (e.g. Urushadze et al. 2016; Jorbenadze et al. 2017; Macharashvili 2019). Georgia is characterized by a great variety of soil types on its small territory, which includes many soils of the world. This can be explained by the enormous variety of soil forming factors within short distances. Soils of Georgia are distinguished by diversity related to mixed character of soil formation. Geological, geomorphologic and climatic conditions change there on comparatively short distances what conditions rich spectrum of soil formation together with diverse vegetation, animal world and surface age (Jorbenadze et al. 2017).

The territory of Georgia is predominantly characterized by yellow soils, yellow-podzolic soils, gley (swamp) soils, brown soils, gray-brown soils, saline soils, and chernozem soils. Among these, chernozems are particularly widespread and are notable for their high fertility. They are primarily used for cultivating field crops and perennial plants, including vineyards (AgroNews.ge 2025).

6 Geological Factors Supporting the Development of Viticulture in Georgia

Under suitable climatic conditions, vineyards can be cultivated almost anywhere on our planet. However, the quality of wine produced varies greatly by region—ranging from low to medium, good, and in some cases, truly unique wines distinguished by exceptional flavor and aroma. Tthis variability is determined by several key factors: the quality of wine grapes, the winemaking techniques employed, and the historical expertise of winemakers. The quality of grapes, in turn, is shaped by a combination of microclimatic zones, surrounding vegetation, topography, soil structure, composition and color, and geological factors. The combined influence of these elements on wine quality is referred to as the terroir effect, a concept that encompasses the distinctiveness of every wine-producing region or zone (e.g. Johnson 1989; Wilson 1998; MacNeil 2001; Frankel 2014; Suthren et al. 2006; Huggett 2006; Swinchatt et al. 2018; Barnes 2018; Menert 2020; Charles 2022).

As recent studies suggest, the geological factors that define terroir are far more complex than soil structure or composition alone. The geodynamic and tectonic conditions of the substrate on which a vineyard is established can have a significant impact on grape quality (e.g. Nadiradze 2013; Robinson 2006; Swinchatt et al. 2018).

There is no doubt that a close geochemical and mineralogical relationship exists between bedrock and soil. The formation of the substrate is governed by complex geological processes and is influenced by both exogenous factors (such as weathering and transport) and endogenous processes. Specifically, as a result of degassing from the Earth's deeper zones in tectonically active areas, the substrate—and consequently the soil—becomes enriched with various macro-and microelements. Wines produced from grapes grown on such soils often exhibit distinctive flavor profiles and aromatic qualities.

In brief, the Earth's crust is not homogeneous; it varies in thickness—from approximately 5 to 80 kilometers—and is composed of diverse rock types. Beneath it lies the geodynamically and geochemically active upper mantle, extending to depths of around 700 kilometers. Magmatic and hydrothermal processes within this mantle contribute to the thinning, destruction, or, conversely, the generation of new crustal material. As a result, tectonic fractures and a dense network of faults of varying scales form within the crust. The high-temperature mantle is in a state of continuous degassing. Along these regional fractures and fault zones, active chemical elements migrate upward from the Earth's interior, gradually penetrating the substrate and eventually reaching the soil. This vertical migration of elements is especially intense in rifted zones, where the crust is extended and thinned. Thus, in geodynamically active regions, both the substrate and the overlying soil become enriched with a broad spectrum of chemical elements. This geochemical enrichment directly affects the characteristics of agricultural produce cultivated in such environments, including grapes, and ultimately imparts distinctive qualities to the wines produced from them.

Georgia is located within the young Alpine orogenic system, and its deep geological structure is markedly heterogeneous and its crust is intersected by numerous regional and local tectonic faults. Which is why it is characterized by a very complex geomorphological structure (Gamkrelidze et al. 2019). Despite that, almost everywhere in Georgia, historically vineyards were grown.

Although the finest wine vineyards are located in the eastern part of the country, primarily in the Kakheti region, within the Alazani and Iori river valleys. Among the local grape varieties cultivated here, the most widespread are Rkatsiteli, Saperavi, Kakhuri Mtsvane, Kisi, Khikhvi, and Budeshuri Tetri, along with several others. Among international varieties, Cabernet Sauvignon is the most commonly grown. In addition to traditional Kakhetian-style wines, many high-quality European-style wines are also produced from these grape varieties. For instance, the well-known Georgian white wine Tsinandali is produced by blending Rkatsiteli and Kakhuri Mtsvane, while the red wine Mukuzani is made from a combination of Saperavi and Cabernet grapes.

In Kakheti, vineyards are primarily cultivated at elevations ranging from 300 to 700 meters above sea level, on alluvial humus-carbonate and black soils of the Alazani and Iori river valleys. As previously mentioned, this region is renowned for its high-quality wines; however, it also features specific wines with Protected Designation of Origin (PDO) that are considered truly unique and distinguished by exceptional sensory characteristics. One such example is Kindzmarauli, a naturally semi-sweet red wine, produced exclusively from vineyards located in a specific part of the Alazani Valley, to the northwest of the village of Kindzmarauli.

The Alazani Valley is situated between the Greater Caucasus range and the Tsiv-Gombori Ridge and represents a deep depressional structure filled with Quaternary sediments. It lies at elevations ranging from 550 to 200 meters above sea level, sloping and widening toward the southeast over a distance of approximately 160 kilometers, reaching a maximum width of 37 kilometers at its southeastern end (Fig. 7). The valley is traversed along its entire length by the Alazani River, which is fed by numerous tributaries originating from both the Greater Caucasus and the Tsiv-Gombori Ridge. These tributaries actively supply the valley with large volumes of Quaternary sediments.

The Alazani Valley is characterized by a moderately humid subtropical climate, as it is well protected from cold air masses by natural barriers to the west and north. Two main soil types are distinguished within the valley. On the left bank of the Alazani River, soils are primarily derived from erosional material originating from Lower Jurassic clay shales of the Greater Caucasus. In contrast, the right bank is mainly influenced by erosional inputs from Oligocene–Neogene argillites, sandstones, and marly limestones of the Tsiv-Gombori Ridge. The left bank is dominated by non-carbonate meadow soils, while the right bank features alluvial carbonate meadow soils.

Geological data on the Alazani Valley are limited and often contradictory, despite its significance as a major tectonic unit within the Caucasus orogen. In the 1960s, the valley was interpreted as an active subsidence zone (geosyncline), characterized by intense sediment accumulation (Maruashvili 1964). According to this work, the formation of the Alazani Valley depression began in the Pliocene, within the foreland of the Greater Caucasus, and continues to this day—an assertion supported by various stratigraphic, petrological, and geomorphological evidence. To the southeast, in Azerbaijan, the Alazani depression continues as the Agrichay depression, and together they constitute an Alazani–Agrichay geotectonic unit (Maruashvili 1964).

A factual basis for this assumption is provided by the work of Philippe et al. (1989), according to which Quaternary deposits in the axial part of the Alazani depression reach a thickness of up to 800 meters, underlain by thinned Meso–Paleogene sediments (C–D cross-section) (Fig. 8). These sedimentary layers rest upon a thinned marginal basin crust, which is most likely oceanic or transitional in nature (Philippe et al. 1989). This idea is further supported by the results of geophysical studies conducted by Mumladze et al. (2015), which suggest that subduction of an oceanic-type slab is currently ongoing beneath the Greater Caucasus, and that oceanic-type crust is also present beneath the Alazani depression. According to the most recent geotectonic research, the Alazani depression is interpreted as a piggyback basin (Alania et al. 2016).

If we synthesize the results of the geological studies discussed above, it is highly probable that the present-day Alazani Valley depression represents a rifting structure undergoing extension, thinning, and subsidence due to the northeastward subduction of oceanic-type crust. This process leads to intensive sediment accumulation within

Fig. 7 A fragment of the Alazani Valley as seen from the Gombori Ridge. In the background, the main range of the Greater Caucasus is visible. Photo © Lasha Gabelia

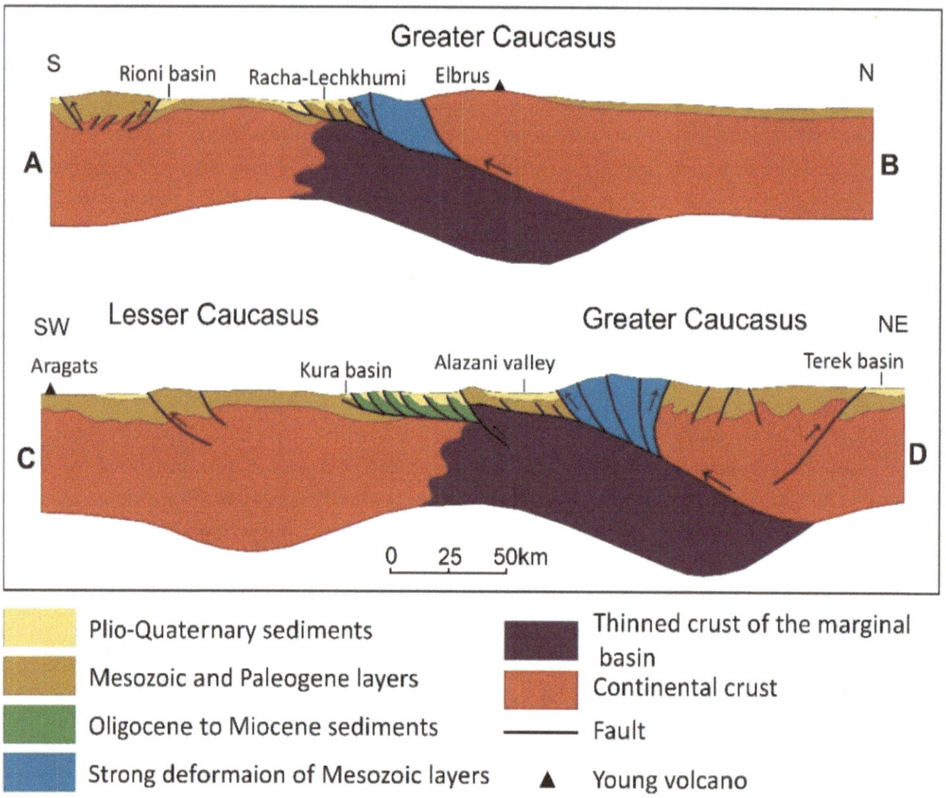

Fig. 8 Deep cross-sections of the Caucasus orogen: A–B through the western segment, and C–D through the eastern segment (Modified after Philip et al. 1989)

the structure. At the same time, the thinned and fractured crust beneath the Quaternary deposits of the Alazani Valley facilitates the migration of deeply sourced, degassed active chemical elements such as CO_2, H_2O, H_2, H_2S, and various microelements. As a result, these elements enrich the groundwater and soil. Naturally, the influx of these elements is more intense in areas where tectonic faulting is more active, especially within tectonic junctions. Additionally, it is known that thermal flows from the mantle are also released along deep tectonic faults.

The best study area for exploring this idea is the geological structure of the vineyard zone producing the unique Kindzmarauli Saperavi wine in the Alazani Valley (Fig. 9). The vineyards cover approximately 600 hectares and are planted on the right bank of the Durujis River, at elevations ranging from 250 to 500 meters above sea level. Although

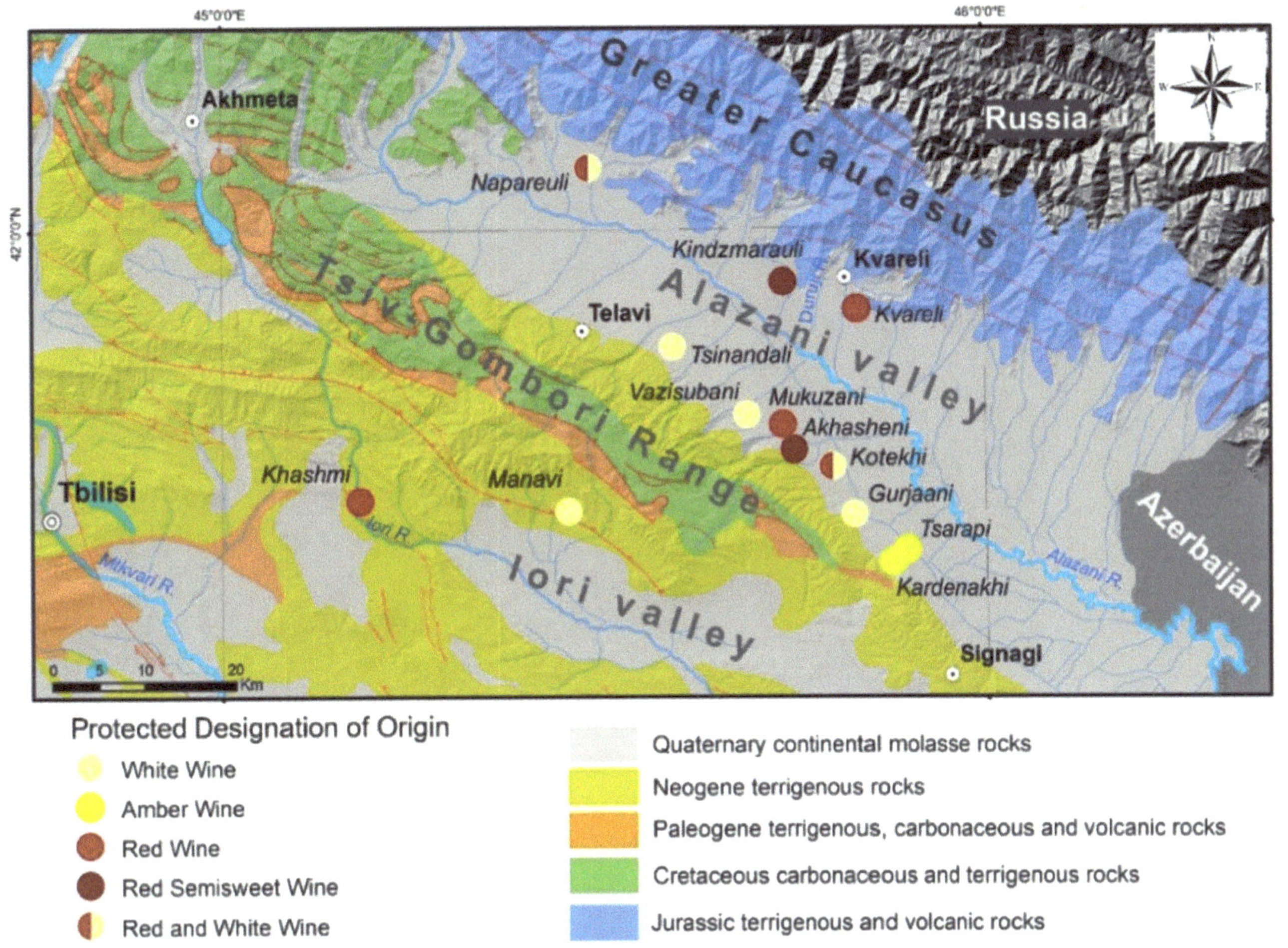

Fig. 9 Geological map of eastern Georgia, Kakheti region showing the Protected Designation of Origin (PDO) zones for wine. The geological framework is modified after Gudjabidze et al. (2003)

there are many Saperavi vineyards in the Alazani Valley, no other wine made from this grape variety exhibits the same distinctive aroma and taste characteristics as those from this specific zone.

Although the Kindzmarauli vineyards are planted on Quaternary deposits, regional analysis indicates the presence of a strong tectonic junction in this area, where a sublatitudinal fault zone intersects with the Transcaucasian submeridional fault. The Transcaucasian fault forms the course of the Duruji River, within whose zone the rocks undergo intense fracturing, disintegration and weathering. For this reason, the local population has named the river carrying this material "Black Duruji" (the source of the Duruji River). Notably, due to these conditions, the southern slope of the Greater Caucasus along the Duruji River is prone to powerful and hazardous landslides, posing a constant threat to the town of Kvareli.

In contrast to the Alazani Valley, the vineyards of the well-known origin-designated wines, Khashmi Saperavi and Manavi Mtsvane, located in the Iori Valley, are situated on Neogene formations, where the tectonic structure is clearly exposed at the surface (see Fig. 9). These vineyard zones are intersected by a sublatitudinal regional fault, which in turn is crosscut by the submeridional fault in the Khashmi area, and by the Tokhliauri submeridional fault in the Manavi area. As a result, the vineyards of Manavi Mtsvane and Khashmi Saperavi are located in tectonic junction zones, which likely contributes to the uniqueness of the wines produced there.

In Georgia, another prominent area known for wines with unique Protected Designation of Origin (PDO) characteristics is the Racha-Lechkhumi region. Located in western Georgia on the southern slope of the Greater Caucasus, this region lies within a deep depression bordered by Cretaceous and Jurassic formations and is intersected by both submeridional and sublatitudinal fault systems (Fig. 10). As illustrated in the A–B profile of Fig. 8, this depression, similar to the Alazani Basin, rests above a series of thinned Meso-Paleogene sediments. Beneath these deposits lies the

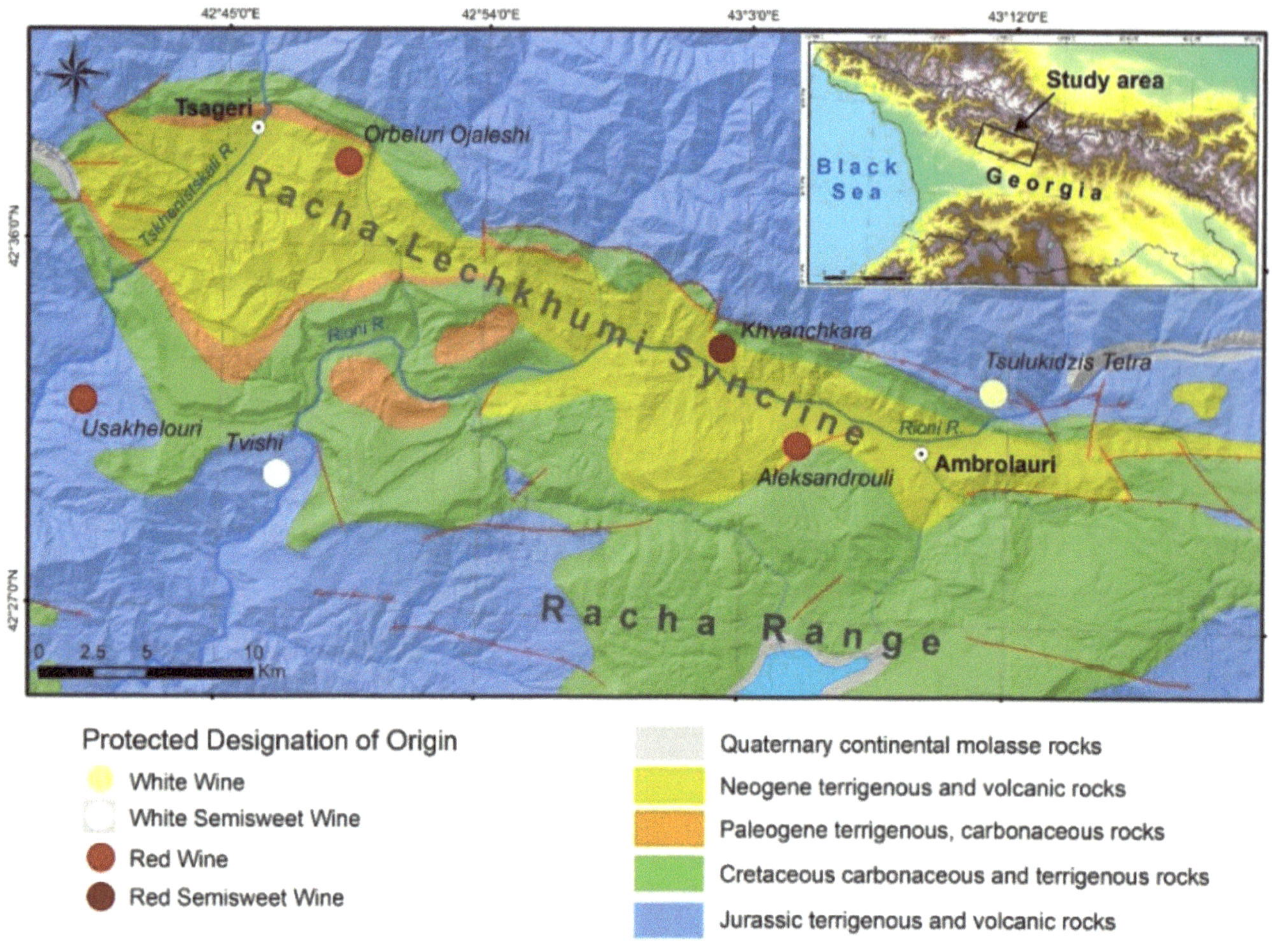

Fig. 10 Geological map of Racha-Lechkhumi region showing the protected designation of origin (PDO) areas for wines. (PDO—Protected Designation of Origin). The geological framework is modified after Gudjabidze et al. (2003)

subsided basin crust, which is likely oceanic or transitional in origin (Philipe et al. 1989).

The Racha-Lechkhumi viticultural zone is situated at elevations ranging from 450 to 750 meters above sea level and extends over approximately 40 kilometers. It is renowned for producing both red and white wines with distinctive taste profiles. Among these, notable examples include the naturally semi-sweet red wine Khvanchkara, as well as the red wines Aleksandrouli, Ojaleshi, and Usakhelouri, and the white wines Tvishi and Tsulukidze's Tetra. The most prominent among them is Khvanchkara, produced from Aleksandrouli and Mujuretuli grape varieties cultivated in the Khvanchkara community. Khvanchkara is a deep ruby-colored wine characterized by its velvety texture, refined structure, pleasant sweetness, and unique natural aroma.

According to M. Nadiradze (2013), a system of tectonic faults with latitudinal orientation passes through the specific viticultural zone of Racha-Lechkhumi. This system intersects with meridional faults, resulting in significant rock dislocation and subsequent enrichment with various chemical elements. In his view, these geological features are inherited by the soils developed on the surface of these rocks and, consequently, influence the characteristics of the wines produced from grapes grown in these vineyards.

7 Discussion

The discussion focuses on two unique and specific wine-producing regions in Georgia, Kakheti and Racha-Lechkhumi. Many regions possesses suitable climatic, geographical, and geological conditions for the development of viticulture and winemaking; however, Kakheti and Racha-Lechkhumi stand out among them. As illustrated in Map 8, the Earth's crust beneath both zones is thinned, and the underlying rocks are intensely dislocated. This observation supports our hypothesis that tectonic dislocation alone is not sufficient to account for the existence of unique viticultural zones; rather, the crust beneath such areas must also be thinned to allow for the intense upward migration of deep-sourced chemical elements in hydrothermal waters.

Moreover, the uniqueness of wine also depends on the presence of fault systems with high permeability, typically found at tectonic junctions. The best examples of such conditions are the Kindzmarauli, Khashmi Saperavi, Manavi Mtsvane, Khvanchkara, and Tvishi wines, whose vineyards are situated within tectonic junction zones. It is noteworthy that wines produced from vineyards with similar climatic conditions, soils, exposure, grape varieties, and other comparable factors may still be of high quality, yet they do not attain the same quality as those produced in these geological zones.

Scientific study of the relationship between viticulture and geology is still in its early stages. Naturally, many aspects remain open to debate and require further clarification and refinement. At this stage of research, we believe that climatic conditions play a leading role in shaping grape varieties and the development of viticulture, while geological factors primarily determines the sensory qualities of wine. Georgia's diverse terroirs arise from a complex interplay of geological, climatic, and cultural factors that have shaped its winemaking heritage.

8 Conclusion

The research confirms that Georgia—the cradle of wine—possesses all the essential preconditions, including climatic, geographical, and geoheritage factors, that enabled the emergence of viticulture and the subsequent development of winemaking technologies.

With its geological conditions—defined by a complex interplay of soil chemistry, structure, and color; substrate composition and configuration; the tectonic setting of underlying rocks; and the region's overall geodynamic framework—Georgia represents an outstanding environment for the development of viticulture and winemaking, including the production of unique wines. Furthermore, the presence of suitable geoheritage clay deposits has enabled the local production of clay vessels, including qvevris, which laid the foundation for the country's distinctive qvevri winemaking tradition.

Heritage has kept Georgian winemaking alive through outside challenges. More recently, the Georgian *qvevri* was inscribed in 2013 by UNESCO on the Representative List of the Intangible Cultural Heritage of Humanity, recognizing it as a cultural monument in need of safeguarding. The source clay is as aspect of Georgia's geoheritage. Georgia's geological complexity is also distinctive and supports Georgia's ancient winemaking traditions, traditions that are part of the Georgian soul. As such, this geoheritage, which sets the stage for the winemaking tradition, also deserves recognition and safeguarding. Overall, the study demonstrates that Georgia possesses not only favorable climatic conditions but also a complex set of geoheritage factors that have played a fundamental role in the development of viticulture and the creation of wines distinguished by their unique regional character.

References

AgrNews.ge (2025). https://agronews.ge/2025-tslis-sartsqhavi-sezoni-gaikhsna/

Alania MV, Chabukiani OA, Chagelishvili R, Enukidze O, Gogrichishvili OK, Razmadze A, Tsereteli N (2016) Growth structures, piggy-back basins and growth strata of the Georgian part of the Kura foreland fold-thrust belt: implications for late Alpine kinematic evolution. Geol Soc Lond Spec Publ 428(1). https://doi.org/10.1144/SP428.5

Annual Report (2025) Georgian National wine agency. https://wine.gov.ge

Barnes A (2018) Napa geology: how Napa Valley was formed and its soils. https://aroundtheworldin80harvests.com/2018/04/01/napa-geology-napa-soils/

Batonishvili V (1745) Description of the Kingdom of Georgia. Sabchota Sakartvelo Publishing, Tbilisi, 1973, pp 1–162 (in Georgia).

Charles N (2022) Good land for wine: how geology can influence the quality of wine. Fr Geol Surv. https://www.brgm.fr/en/article

Frankel C (2014) Land and wine: the French Terroir. Univ Chic Press. https://doi.org/10.1080/09571264.2014.968769

Galán E, Ferrell RE (2013) Genesis of clay minerals. Dev Clay Sci 5:83–126. https://doi.org/10.1016/B978-0-08-098258-8.00003-1

Gamkrelidze TV, Ivanov VV (1995) Indo-European and the Indo-Europeans: a reconstruction and historical analysis of a proto-language and a proto-culture. Vols I–II. Mouton de Gruyter, Berlin/New York. https://doi.org/10.1515/9783110815030

Gamkrelidze I, Okrostsvaridze A, Maisadze F, et al. (2019) Main features of geological structure and geotourism potential of Georgia, the Caucasus. Modern Environ Sci Eng 5(5):422–442. https://doi.org/10.15341/mese(2333-2581)/05.05.2019/010

Gudjabidze G, Gamkrelidze I, Abesadze G, et al. (2003) Geological map of Georgia. Tbilisi, Georgia, scale 1:500,000

Huggett RJ (2006) Geology and wine: a review. Proc Geol Assoc 117(2):239–247. https://doi.org/10.1016/S0016-7878(06)80012-X

Johnson H (1989) Vintage: the story of wine. Simon & Schuster, New York, pp 1–480

Jorbenadze TL, Urushadze FT, Urushadze TT, Kunchulia OI (2017) Physical properties of the soils of Georgia. Ann Agrar Sci 15(2):224–234

Kiguradze T (1986) Neolithische siedlungen von kvemo kartli, Georgien. Mater Zur Allg Und Vgl Archäologie 29:159–166

Kloprogge JT (2017) Raman spectroscopy of clay minerals. Dev Clay Sci 8:150–199. https://doi.org/10.1016/B978-0-08-100355-8.00006-

Macharashvili L (ed) (2019) The soils of Georgia. Springer, Cham, pp 1–179. https://link.springer.com/book/10.1007/978-3-030-11182-1

MacNeil K (2001) The wine bible. Workman Publishing, New York, pp 1–190

Maruashvili L (1964) Physical geography of Georgia. Tsodna Publishing, Tbilisi, pp 1–345 (in Georgian)

McGovern P, Jalabadze M, Batiuk S at al. (2017) Early Neolithic wine of Georgia in the South Caucasus. Anthropology 114(48):E10309–EE1031. https://doi.org/10.1073/pnas.1714728114

Menert DL (2020) Wine and terroir. in: geochemical. Perspectives 9(1):99–119

Mumladze T, Forte MA, Eric S, Cowgill SE, Trexler CC, Nathan A, Niemi AN, Yıkılmaz BM, Louise H, Kellogg HL (2015) Subducted, detached, and torn slabs beneath the greater Caucasus. GeoResJ 5:36–46. https://doi.org/10.1016/j.grj.2014.09.004

Nadiradze V (2013) Georgian unique wines and the connection of tectonics. Georg Min J 2(31):40–41. (in Georgian)

Okrostsvaridze A (2014) General dynamics and current situation of the mining industry of Georgia. CENN Report, Tbilisi, pp 1–65

Okrostsvaridze A, Akimidze K, Bluashvili D, Gogoladze S, Boichenco G, Gabrielashvili R (2022) Thorium—the energy source of the future and its ore occurrences in Georgia. Universal Publishing, Tbilisi, pp 1–224 (in Georgian)

Okrostsvaridze A, Tormey D (2013) Phanerozoic continental crust evolution of the inner Caucasian microplate: the Dzirula uplift. Episodes 36(1):31–39

Philip H, Cisternus A, Gvishiani A, Gorshkov A (1989) The Caucasus: an actual example of the initial stages of a continental collision. Tectonophysics 161:1–21

Robinson J (ed) (2006) The Oxford companion to wine, 3rd edn. Oxford University Press, Oxford, pp 1–695

Suthren RJ, Fowler MB, Guion PD (2006) Terroir, wine and geoscience fieldwork education in Languedoc-Roussillon, southern France. In: Macqueen RW, Meinert LD (eds) Fine wine and Terroir: the geoscience perspective. Geoscience Canada Reprint Series vol 9. pp 233–246

Swinchatt JD, Howell DG, MacDonald SI (2018) The scale-dependence of wine and terroir: examples from coastal California and the Napa Valley. Elements 14(3):179–184. https://doi.org/10.2138/gselements.14.3.179

Uebel M, Petzold I (2017) Elements and physical-chemical characteristics of pitcher clay and the process of making pitcher. Petit, Tbilisi, pp 1–99

Urushadze FT, Blum W, Kvrivishvili T (2016) Classification of soils on sediments, sedimentary and andesitic rocks in Georgia by the WRB system. Ann Agrar Sci 14(4):351–355

Wang SJ, Lee YH, Chung SL, Okrostsvaridze A (2015) Cenozoic mountain building process in the greater Caucasus, from subduction to collision. AGU Fall Meeting Abstracts #T21A–45

Wilson JE (1998) Terroir: the role of geology, climate and culture in the making of French wines. Q J Eng Geol Hydrogeol 33(4):350–350. https://doi.org/10.1144/qjegh.33.4.350

Epilogue: Georgia's Future Includes Celebration and Protection of its Geoheritage

Daniel Tormey and Avtandil Okrostsvaridze

Abstract

This concluding chapter synthesises the main messages of *Geoheritage of Georgia* and outlines how the country can move from recognising individual geosites to building a coherent national geoheritage programme. It argues that Georgia's geoheritage cannot be understood as a simple list of caves, volcanoes, deserts, fossil sites and mining districts, but as an interconnected system structured by four major belts: the Greater Caucasus, the Inner Caucasian (Transcaucasian) Microplate, the Lesser Caucasus volcanic–metallogenic domain and the Black Sea–Colchis margin. Within this framework, Key Geoheritage Areas (KGAs) provide the scientific backbone, grouping sites into coherent areas where geology underpins biodiversity, geotourism and cultural identity. On top of this KGA base sit UNESCO Global Geoparks, World Heritage Sites, protected areas and national georoutes and geocultural routes, which together can form a network of designations and territories across all regions of Georgia. Finally, the chapter situates Georgia within the global geoheritage community, emphasising the roles of the UNESCO Global Geoparks programme and the emerging IUCN KGA initiative, and explains how international engagement can both benefit from and support Georgia's efforts. The chapter closes with an invitation to policy-makers, scientists, managers, communities and the wider public to see geoheritage as a strategic asset and a tangible expression of the "Georgian soul", and suggests that if future visitors better understand how geology, life and culture fit together in this small, remarkable country, the work captured in these pages will have achieved its purpose.

Keywords

Georgia · Geoheritage · UNESCO Global Geoparks · Geoheritage programme

1 Introduction

Georgia's story has always been a story of land. From the high, glaciated peaks of the Greater Caucasus to the volcanic uplands of the Lesser Caucasus, from karst plateaus and deep river gorges to temperate rainforests and semi-desert steppe, the country's landscapes are inseparable from its history and identity. This volume has shown that Georgia's geoheritage is not a niche interest reserved for specialists; it is a strategic asset that underpins science, conservation, culture, and sustainable development.

This concluding chapter looks ahead. It argues that Georgia's future must include the celebration and protection of its geoheritage through a coherent national programme, and that the work assembled in this volume provides both the intellectual groundwork and a practical roadmap for that journey.

2 From a List of Sites to a National System

One of the clearest messages emerging from this volume is that Georgia's geoheritage cannot be understood as a simple list of sites. The country already has hundreds of documented geosites and natural monuments, as well as world-renowned areas such as Vardzia, Sataplia, Kazbegi, Svaneti, Tusheti, Racha, Borjomi–Kharagauli, Mtirala, Javakheti, the Colchis Lowlands and Vashlovani. Read individually, these

D. Tormey (✉)
Catalyst Environmental Solutions, Santa Monica, USA
e-mail: dtormey@ce.solutions

IUCN World Commission on Protected Areas, Gland, Switzerland

A. Okrostsvaridze
Institute of Earth Sciences, Ilia State University, Tbilisi, Georgia

A. Okrostsvaridze and D. Tormey (eds.), *Geoheritage of Georgia*, Geoheritage, Geoparks and Geotourism,
https://doi.org/10.1007/978-3-032-27639-1_26

chapters offer vivid portraits of caves, volcanoes, fossil beds, deserts, petrified forests and mining landscapes. Read together, they reveal a national system of geoheritage.

Chapters 2 and 3 interpret that system through the language of Key Geoheritage Areas (KGAs). Rather than treating every outcrop or locality as isolated, the KGA approach groups sites into coherent areas where Georgia's geology is of clear international significance and underpins biodiversity, geotourism and cultural identity. In this framework, the Imereti karst and cave belt, the Greater Caucasus high-peak zone, the Lesser Caucasus volcanic highlands, the Artvin–Bolnisi metallogenic belt, the Colchis coastal lowlands, and the Vashlovani badlands and desert landscapes become recognisable building blocks of a national geoheritage map.

On top of this KGA backbone sit international and national instruments: UNESCO Global Geoparks as territorial designations linking geoheritage with local development; World Heritage inscriptions where Outstanding Universal Value is demonstrated; national parks and other protected areas; and a growing network of georoutes and geocultural routes that connect geosites with communities, crafts, and hospitality. Chapters 2 and 3 show that when seen through this lens, Georgia is a country with a portfolio of candidate territories that, over time, can form a network of geoparks and related designations across all regions.

The task now is to move from understanding this system on paper to embedding it in policy, institutions and practice.

3 Towards a National Geoheritage Programme

The chapters in this book collectively outline the elements of a national geoheritage programme for Georgia. Such a programme would not replace existing efforts in nature conservation, cultural heritage or tourism, but would give them a stronger geological foundation and a clearer, shared language.

Several components are central:

- **A national KGA inventory and map** Building on the work presented here, Georgia can formalise a KGA map that identifies a finite set of Key Geoheritage Areas across the country. This map should sit alongside existing Key Biodiversity Areas and protected area networks, making geodiversity visible in the same way that biodiversity already is.
- **Legal and policy recognition of geoheritage** Many chapters point to the need for clear legal definitions of geosites and geoheritage areas, improved integration into spatial planning, and explicit recognition of geological values in environmental impact assessment, land-use zoning and disaster-risk reduction strategies.
- **A Georgian Geopark and Geoheritage Council** Chap. 3 proposes a small national body, hosted under the Georgian National Commission for UNESCO, that brings together ministries, municipalities from candidate territories, academic and heritage institutions, and geotourism entrepreneurs. Such a council could steer a national roadmap, maintain a shared readiness matrix, coordinate UNESCO Global Geopark nominations and revalidations, and ensure that geopark development reinforces, rather than duplicates, existing World Heritage and protected areas.
- **Integration with education and research** Georgia's geosites are natural classrooms. The fossil forests of Goderdzi, the dinosaur tracks of Sataplia, the Gryphaea mass burial at Kldeisi, the natural freezers of the Caucasus foothills, and the Argonauts' placer-gold landscapes offer ready-made case studies for schools and universities. Systematically embedding geoheritage into curricula, field training, and outreach will strengthen both scientific literacy and national pride.
- **Monitoring, risk management and climate resilience** Many geoheritage sites are also geohazard and climate-sensitive sites: glacial valleys threatened by warming, coastal and deltaic systems exposed to sea-level rise and erosion, unstable slopes, caves and freezers sensitive to increased visitation. A national programme should link geoheritage monitoring with hazard assessment and climate adaptation, so that protecting the record of past Earth change also informs responses to present and future change.
- **Partnerships with communities and local economies** Throughout the volume, a recurring theme is the "mining–geoheritage paradox": mining districts and industrial landscapes are both sources of impact and repositories of scientific, technological and social history. Similarly, sacred springs, wine cellars, natural freezers and cave monasteries are places where geology and daily life are intertwined. A national programme must therefore be built with communities, not merely around them, finding ways for geoheritage to support livelihoods while respecting local values and knowledge.

Together, these components transform geoheritage from a collection of interesting places into a coordinated national endeavour.

4 Priorities across Georgia's Landscapes

The thematic and regional chapters in this book show that geoheritage is not evenly distributed, but certain landscape clusters emerge as priorities for early action.

- **High peaks and active processes of the Greater Caucasus** Glacial landforms, high-relief valleys and active geohazards combine with tower villages, transhumance routes and polyphonic song traditions, making this belt a priority for integrating geoheritage into mountain protected area management, geohazard communication and carefully designed high-altitude geotourism.
- **Volcanic highlands of the Lesser Caucasus** The Vardzia–Javakheti volcanic plateau, Neogene basins and Goderdzi fossil forests together tell stories of explosive volcanism, lahar burial, long-lived lakes and forest ecosystems, and are natural focal points for a volcanic geopark and education on hazards, water resources and climate change.
- **Karst and caves of Imereti and other regions** Sataplia, Prometheus, Tskaltubo and numerous lesser-known caves and karst systems host fossils, speleothems, underground rivers and mineral springs, and are major tourism draws that require careful visitor management, integration with spa and healing-water traditions, and protection as fragile subterranean habitats.
- **Mining landscapes and metallogenic belts** The Artvin–Bolnisi zone, Chiatura, the Svaneti ore districts and other mining areas vividly illustrate the mining–geoheritage paradox and offer opportunities to tell stories of ore formation, mining technology, social history and environmental restoration, and to demonstrate best practices in coexistence between extraction and conservation.
- **Coastal and lowland systems** The Colchis Lowlands, Black Sea coast and associated wetlands are places where Quaternary sea-level change, sedimentation and human land use intersect, important not only for geoheritage but also for biodiversity and climate mitigation, and best treated as integrated geo–bio–cultural landscapes.
- **Semi-deserts, badlands and steppe** Vashlovani and similar landscapes demonstrate erosion, soil formation, fossiliferous sequences and the interface between arid land use and fragile geomorphology, ideal for illustrating how geological processes continue to shape land and livelihoods today.
- **Sacred waters, natural freezers and wine landscapes** Mineral and thermal springs, natural ice caves and freezers, and vineyard terroirs are recurring motifs that link geology directly with health, belief and gastronomy, and belong equally in discussions of geoheritage, sacred hydrology and cultural landscapes.

These clusters overlap and interact, but together they provide a spatial blueprint for concentrating early investments in geosite conservation, interpretation and geopark development.

5 Georgia in the Global Geoheritage Community

Georgia does not embark on this path alone. The country is engaging with several international frameworks that give its geoheritage work both visibility and support. Through active participation in UNESCO and IUCN processes, and through contributions of Georgian scientists and practitioners to global best-practice guidance, Georgia can bring the lessons of its collision-zone landscapes, mining paradox, sacred waters and geocultural traditions into international debates. This volume is part of that contribution: it documents, analyses and interprets Georgia's geoheritage in a way that is accessible to decision-makers, researchers and practitioners elsewhere. In turn, international engagement brings benefits back to Georgia: peer review of geopark and World Heritage proposals, access to training and networking, increased visibility for geotourism destinations, and alignment with global agendas on climate, risk reduction and sustainable development.

6 The Role of this Volume: Groundwork and Roadmap

This book has aimed to do three things:

First, it **documents** Georgia's geoheritage in breadth and depth. From regional overviews of the Caucasus collision zone to detailed chapters on specific geosites, fossils, caves, volcanoes, deserts, mining districts and water-related geoheritage, it assembles dispersed knowledge into a single, accessible reference.

Second, it **frames** that geoheritage within a coherent national and international context. The introduction and Chap. 2 and 3 provide the conceptual tools: the four pillars of collision-zone geology, geotourism, biodiversity and the Georgian soul; the KGA–UGGp–World Heritage triad; the evaluation framework and readiness assessment; and the idea of a phased network of geoparks connected by georoutes and embedded in a wider conservation and development strategy.

Third, it **points forward**. The candidate territory analysis and the thematic chapters together suggest specific priorities for action: territories where a first wave of geopark nominations is realistic; regions where foundational work on inventories, partnerships and infrastructure is needed; and cross-cutting themes, such as mining landscapes, sacred hydrology, natural freezers and wine, that can anchor distinctive Georgian narratives.

In that sense, this volume is both **groundwork and roadmap**. It does not attempt to prescribe a single path, but it does show that multiple, credible paths exist, and that they are connected.

7 An Invitation

Shota Rustaveli wrote that "stone yields at last to water—to rivers spilling down from the high mountains." In a geoheritage context, this line can be read as both a geological and a social metaphor. Over time, persistence changes landscapes; over time, consistent care and attention can change how a country relates to its land.

The work presented in this book is a beginning, not an end. It is an invitation:

- to **policy-makers**, to see geoheritage as a strategic asset worth integrating into national planning, not as a marginal concern;
- to **scientists**, to continue refining inventories, KGA delineations, and thematic studies, and to share their insights beyond academic journals;
- to **protected area and cultural heritage managers**, to align geodiversity with biodiversity and cultural values in management plans and visitor experiences;
- to **municipalities and communities**, to consider how their local landscapes fit into the national geoheritage picture, and how geoparks and georoutes can support their futures;

Georgia's future will be shaped by many forces: economic, political, climatic. This book argues that whatever those futures may be, they should include the celebration and protection of Georgia's geoheritage: as scientific evidence of Earth's history, as the physical stage for extraordinary biodiversity, as the foundation of geotourism and local livelihoods, and as a tangible expression of the Georgian soul.